T0400938

POLYMER SYNTHESIS

MATERIALS SCIENCE AND TECHNOLOGIES

Additional books in this series can be found on Nova's website
under the Series tab.

Additional E-books in this series can be found on Nova's website
under the E-books tab.

POLYMER SCIENCE AND TECHNOLOGY

Additional books in this series can be found on Nova's website
under the Series tab.

Additional E-books in this series can be found on Nova's website
under the E-books tab.

MATERIALS SCIENCE AND TECHNOLOGIES

POLYMER SYNTHESIS

E. KOWSARI
EDITOR

Nova Science Publishers, Inc.
New York

NOTICE TO THE READER

Library of Congress Cataloging-in-Publication Data

Polymer synthesis / editor, E. Kowsari.
p. cm.
Includes index.
ISBN 978-1-61324-672-6 (hardcover)
1. Polymer engineering. 2. Polymerization. 3. Polymers--Industrial applications.
I. Kowsari, E. (Elaheh)
TP1087.P654 2011
668.9--dc23

2011015642

Published by Nova Science Publishers, Inc. ✛ *New York*

CONTENTS

PREFACE

This book presents an overview of research on advanced synthesis polymers over the past decade. This special issue, contributed by various authors, focuses on recent advances of the field, which handle the cutting-edge aspects of the advanced technology. The contributions in these twelve chapters summarize some major efforts in this area.

Chapter 1 - Recent examples for synthesis of new polyolefins containing polar functionalities by coordination insertion copolymerization using group 4 metal complex (Ti, Zr) catalysts have been reviewed. In particular, examples by adopting the approaches by (a) direct copolymerization of olefin with polar monomer using coordination insertion methods, and (b) controlled incorporation of reactive functionalities (and the subsequent introduction of polar funtionalities under mild conditions) have been described. Our recent efforts for precise synthesis of polyolefins containing polar functionalities by (i) efficient incorporation of reactive functionality by copolymerization of ethylene with nonconjugated diene using nonbridged half-titanocene catalysts and (ii) subsequent chemical modifications under mild conditions have been introduced.

Chapter 2 - The employment of tetradentate $[M(phen)(CN)_4]^{2-}$ (M = Ru, Os) building blocks with lanthanide cations (Ln = Gd^{3+}, Pr^{3+}, Nd^{3+}, Er^{3+}, Yb^{3+}) and ancillary polypyridine ligands (phen = 1,10′-phenanthroline, terpy = 2,2′:6′,2″-terpyridine, bpm = 2,2′-bipyrimidine) in designing novel luminescent multi-dimensional coordination polymers is systematized and discussed. The geometries of the resulting cyanide-bridged coordination polymers include linear, ladder and tube-like chains, chains of squares, as well as two-dimensional structures. Particular emphasis is given to the relationships between structural features of the M−CN−Ln coordination networks and Ln(III) fragments to generate luminescence.

Chapter 3 - Ionic liquids have attracted extensive research interest in recent years as environmental benign solvents due to their favorable properties like non-inflammability, negligible vapour pressure, reusability and high thermal stability. Poly(ionic liquid)s, the polymers made from ionic liquid monomers, have received much research interests for their potential applications such as gas separation materials, catalytic membranes, polymer electrolytes, and ionic conductive materials. Poly(ionic liquid)s contain anion cation pairs and therefore have a relatively high density of strong dipoles, which makes them promising candidates for different applications. Their physical properties can be tuned to specific design criteria by adjusting their chemical structure including the cation, anion, or their combination. Recent developments of the synthesis and applications of poly(ionic liquid)s containing either imidazolium or non imidazolium such as ammonium cations have been highlighted in this review chapter.

Chapter 4 - Aromatic polyimides constitute one of the most important classes of high-performance polymers exhibited a number of outstanding properties, such as high thermal and mechanical properties, high optical transparencies along with chemical and solvent resistance. These excellent combinations of properties make them suitable for a wide range of applications, from engineering plastics in aerospace industries to membranes for fuel-cell applications and gas or solvent separation. Incorporation of fluorine in high performance polymers like poly(ether imide)s is a subject of great research as it brings about dramatic improvements in several properties of the polymers. Polymers containing fluorine in the form of pendent trifluoromethyl ($-CF_3$) groups showed increased solubility, lower dielectric constant and water uptake, higher glass transition temperature, higher thermal and thermo oxidative stability, better optical transparency, higher gas-permeability and flame resistance in comparison to their non-fluorinated analogues. Extensive researches have been directed towards synthesis of $-CF_3$ substituted diamine or dianhydride monomers and their polymerization followed by property evaluations of the resulting polymers to understand the structure-property correlation in these polymers. The present article provides a comprehensive review, particularly on $-CF_3$-substituted aromatic polyimides that have been developed in the last decade as low dielectric constant polymers and as membrane based applications like gas separation, pervaporation and fuel cell membranes. A major effort towards development of novel polyimides and their gas transport properties that have been devoted by our group since several years thoroughly covered in this article.

Chapter 5 - With the rapid development of synthetic chemistry, many new molecular structures can be designed to investigate the role of polymer topology on the physical and chemical properties of macromolecules in traditional, block, hyperbranch, and, in particular, dendritic copolymers. Hyperbranched polymers are highly branched macromolecules with three-dimensional dentritic architectures. Due to their unique physical and chemical properties and potential applications in various fields from drug-delivery to coatings, interest in hyperbranched polymers is rapidly growing, as confirmed by the increasing number of publications. This chapter reviews the synthesis, modifications, and applications of hyperbranched and hyper grafted polymers, focusing on the recently developed novel synthetic strategies for hyperbranched polymers.

Chapter 6 - In recent years the synthesis of conducting polymers with controlled morphology is one of the big deals in the polymer science and technology. By the emergence of nanotechnology, researchers become more interested in studying the unique properties of nanoscale materials. There are different ways to produce polymeric nanofibers, as example the polymerization of aromatic monomers into media having large organic acids. These acids form micelles upon which the monomer is polymerized and doped. Fiber diameters are observed to be as low as 30-60 nm and are highly influenced by reagent ratios. Uniform nanofibers (from 30 to 200 nm) can also be obtained when the polymerization is done at an aqueous-organic interface. It is hypothesized that migration of the product into the aqueous phase can suppress uncontrolled polymer growth by isolating the fibers from the excess of reagents. Thus, the template-free methods, such as interfacial, seeding and micellar can be employed as different "bottom-up" approaches to obtain pure polymeric nanofibers. The possibility to prepare nanostructured conducting polymers by self-assembly with reduced post-synthesis processing warrants further study and application of these materials, especially in the field of electronic nanomaterials. The notable applications include in tissue engineering, biosensors, filtration, wound dressings, drug delivery, and enzyme immobilization. In this chapter this amazing new area of polymeric nanofibers will be reviewed concerning the state-or-art results of synthesis, spectroscopic characterization and applications. The discussion will be centered in the previous and new results obtained by our group, using mainly resonance Raman and X-ray absorption techniques applied to the study of polyaniline nanofibers. Special attention will be given in the role of the synthetic pathways in the control of the electrical, thermal and mechanical properties and also the morphological aspects of the polymeric material. The main goal of this work is to contribute in the rationalization of some important results obtained in the open area of polymeric nanofibers.

Chapter 7 - Although electronically conducting polymers have emerged as one of the most highlighted research fields in macromolecular science and engineering, the first preparation of conducting polymer was actually published in the 19th century. [1] In this pioneer work 'aniline black' (now we know it is polyaniline) was synthesized by the oxidation of aniline, but its electronic properties were not recognized. Natta's group first synthesized polyacetylene in 1958, and they found that its conductivity fell in the range of a semiconductor (10^{-11} to 10^{-3} S/cm). [2] Then an inorganic polymer polysulfur nitride (SN)x was produced with a high conductivity of the order of 10^3 S/cm. [3] It was not until Shirakawa, MacDiarmid and Heeger revealed metallic conductivity of p-doped crystalline polyacetylene films that the development of various organic conducting polymers became a booming research area. [4, 5] The three scientists were awarded the Noble Prize in Chemistry in 2000 for their great contribution in this field. [6-8] The conductivity of most conjugated polymers results from the presence of alternating single and double bonds along the macromolecular chain, which can delocalize the π-bonded electrons along the molecular backbone. [9] Except for polyacetylene, [10] current intensively studied conducting polymers include polyaniline, [11] polypyrrole, [12] poly(phenylenevinylene), [13] polythiophene, [14] polyfluorene, [15] polycarbazole, [16] polyphenylene, [17] poly(aryleneethynylene), [18] and their derivatives.

Polythiophene (PT) is one of the most important intrinsic conducting polymers. The unsubstituted PT is a rigid conjugated π system dominated by α-α' linkages of thiophene rings, which has excellent charge transport properties and environmental stabilities. [19] It has shown great potential in the field of organic electronics [20] especially in the production of photovoltaic modules. [13] Many conducting polymer films including PT have been prepared for the applications of photovoltaic devices, [21] chemical sensors [22] and field effect transistors. [23] Those films were usually casted from corresponding polymer solution. However, unsubstituted PT is essentially infusible and insoluble thus limiting its practical applications to a large extent.

The state-of-the-art solution for this insolubility problem is to incorporate kinds of substituents onto the PT backbone, such as alkyl, [24] alkoxyl, [25] perfluoroalkyl, [26] amine, [27] carboxyl, [28] and zwitterionic groups, [29] making PT soluble in organic solvents, water or supercritical fluids. Although this method was initially pursued to solve the processing problem, it has been widely recognized as an approach for tuning the electronic and optical properties of conducting polymers. The recent advances in molecular design for substituted PT and its copolymers aiming for various application especially photovoltaic device productions have been reported in several excellent reviews. [30-32] Furthermore, the incorporation of ionized groups on the backbone can make PT a self-doped conducting polymer, and a monograph on this topic is available. [9] Despite all these merits, the current synthetic procedures for these PT derivatives are usually complex and expensive, which generally need stringent conditions and several individual steps. Therefore they are not easily accessible for industrial scaling up. Also some toxic solvents are usually involved which are not environmentally favorable. In terms of organic photovoltaic applications, the material stability under continuous illumination has to be taken into account. [33] The soluble side chains may be passive in terms of light harvesting and charge transport, [34] and they may also make the materials soft and allow for the intrusion of moisture or other small molecules, [35, 36]. Recently, Bjerring et al prepared unsubstituted PT films by solution processing thermocleaving method. [37]

Another choice to stress the processing problem is PT dispersible micro- or nano-structures, including nanoparticles, microspheres, nanowires, and so on. Not only can these micro- or nano-structures retain the intrinsic properties of PT, but also they can introduce other bonus advantages, e.g. huge specific surface area especially favored by gas sensor application. The template synthesis has long been the primary choice for achieving this objective. [38] It is well known for its accurate morphology control, but removing the template after the synthesis is necessary. On the other hand, heterogeneous polymerization methods, including emulsion polymerization, miniemulsion polymerization, microemulsion polymerization, dispersion polymerization etc., have served decades for the preparation of conventional polymer nanoparticles and microspheres. These methods are robust, feasible, and suitable for both one-pot and continuous reaction process of industrial scale production. Considering the extensive application of conducting polymers in the future, the scaling up capability of synthetic methods seems necessary. Dispersible polyaniline and polypyrrole nanoparticles and nanofibers have been synthesized by oxidative emulsion polymerization with the doping of organic acid. [39-42] In comparison, the reports on how to prepare PT micro- or nanostructures by heterogeneous polymerization are rare.

In most cases, the oxidative polymerization of thiophene takes place in organic solvents such as chloroform and acetonitrile, [43] with the monomer soluble in continuous phase and the polymer precipitating out during the polymerization. If there is no effective stabilizing mechanism for the dispersed phase, this scenario is similar to precipitation polymerization. Trans et al pointed out that with this method one can only obtain coagulated structure with irregular morphology. [44] Instead of direct polymerization from thiophene monomer, they initiated the polymerization from thiophene oligomer and 1-D PT nanowires were formed. Nonetheless, how to achieve well-dispersed PT micro- or nanoparticles in one step by templateless oxidative polymerization starting from inexpensive thiophene monomer, which seems more favored by mass production, still remains a challenge.

The first part of this chapter will focus on the preparation of PT nanoparticles by Cu (II) catalyzed oxidative emulsion polymerization in aqueous medium, where the effect of different metal salt oxidant on the rate of reaction, the particle morphology and the properties of PT samples will be discussed. The second part will describe how to overcome the present difficulty in the preparation of well defined PT microspheres in organic solvents by oxidative dispersion polymerization. The diameter control by surfactant concentration and solvent power tuning will also be discussed.

Chapter 8 - In materials science, sonochemistry is mostly used for the fabrication of nanomaterials, but it has also been used for the polymerization of monomers. Recently, great efforts are being made to improve chemical and electronic properties of conductive polymers, especially via the organization of polymeric conductive chains as fibers, hollow tubes, and spheres. Conductives such as polyaniline with organized morphology can be used for advanced applications in electronic devices and molecular sensors as a result of its high surface area and low cost. In order to achieve these objectives, a few authors have been using ionic liquids (ILs), organic salts that are liquid at low temperature (typically lower than 100 °C), as synthetic media to prepare nanostructured conductive polymer, especially IL derived from the imidazolium cation. In this chapter, an IL-assisted sonochemical method is reported for the synthesis of conductive polymer and nanocomposites showing controlled conductivity. This method avoids the use of conventional oxidants and metal complexes. The effect of the ultrasonic irritation time and frequency on the morphology, conductivity, and yield are discussed

Chapter 9 - Room-temperature mechanochemical route to produce different materials is widely regarded as a simple and efficient synthetic method. The deployment of this synthetic strategy to produce conducting polymers was first realized only in 1960. The extension of this method to prepare some polyaniline type conducting polymers began in late 1990. However, utilization of this method for preparing nanostructured conducting polymers was discovered very recently. In this chapter, we intend to discuss some nuances behind the mechanochemical synthesis of poly(2-amino diphenylamine). The influence of oxidant namely ammonium persulphate and ferric chloride, inorganic doping acids on the properties of poly (2-amino diphenylamine) will be briefly explained on a comparative basis. A comprehensive account on the physicochemical properties of as prepared polymers and newer research findings from our research results would also be vividly presented. The promising areas of application of the mechanochemically prepared polymers are to be suggested.

Chapter 10 - In this work, we discuss the relationship between the structure of biodegradable polymers and their ability to recover their shape. Polymers were synthesized using ε-caprolactone and L(DL)-lactide with different architectures. Experiments and simulations were run to discover the correlations among the crystallinity of polymers, polycaprolactone and polylactide PCL, PLA segment lengths, the molecular weight of blocks and shape recovery.

The elastic properties of biodegradable thermoplastic polymers are based on physical crosslinking, which appears to be caused by the separation of amorphous and crystalline phases. Therefore, macromolecules of biodegradable copolymers should contain both amorphous and crystalline phases. Elasticity can be achieved using block or multiblock structures of the polymer.

Initially, we present the results of cyclic behavior in shape recovery for random copolymers and copolymers with multiblock structures, which can be described by the hypothetical structures (PLA-co-PCL)-b-(PCL-co-PLA)-b-(PLA-co-PCL) and (PLA-co-PCL)-b-(PCL-co-PLA), respectively. Analyzing the constructed model, we found that shape recovery was positively correlated with PLA-co-PCL molar mass (hard block) but inversely correlated with PLA crystallinity and block length. However, the molar mass, crystallinity and block length of PLA were directly connected. Therefore, we attempted to increase PLA molar mass but decrease crystallinity. A small amount of ε-caprolactone as the comonomer was added during the PLA synthesis for this purpose. In this case, the block structure was lost, and the formation of complex multiblocks occurred. The best results were achieved for diblock copolymers with the following theoretical structure: (PLA-co-PCL)-b-(PCL-co-PLA) with 30% ε-caprolactone and 30% L-lactide in the PCL block. Each block has molar mass of 20 kDa. This multiblock polymer had 89% shape recovery.

Several attempts were made using transesterification reactions, which allow access to high levels of irregular structures in macromolecules that should help improve shape recovery. To increase the influence of transesterification reactions, a hard PLA block was synthesized, followed by PLA-co-PLA random block synthesis. A polymer with the targeted structure PCL-PLA-PCL 20-40-20 kDa, which has 35% ε-caprolactone in the middle block and 25% L-lactide in the PCL block, was revealed to have the best shape recovery at 90%. It is worth noting that this polymer retained this property at up to 300% elongation. One additional positive feature of this sample was the high speed of shape recovery estimated by the size of the cyclic test loop. Furthermore, the received recovery characteristics were similar or even surpassed the properties of industrial biodegradable elastomers.

Next, we tried to synthesize the triblock structure PLA-b-(PCL-co-PLA)-b-PLA, in which the central soft block is represented by an amorphous copolymer with a random structure and side hard blocks of macromolecules synthesized from PLA homopolymers. To accomplish this, a series of polymers were synthesized where the molar mass of the middle block was included as the variable in the design of the experiment. The molar mass of the PLA hard block and the ε-caprolactone:L-lactide molar ratio was also varied. The best shape recovery was recorded for a polymer with the structure PLA-PCL-PLA 5-100-5 kDa with 50% L-lactide in the middle block, reaching 92% after 100% sample stretching. We further discovered that the molar mass of the middle block significantly influenced shape recovery. However, the closest relationship was revealed between shape recovery and PLA segment length.

Reducing the influence of the middle block crystallinity on shape recovery by replacing L-lactide with DL-lactide in triblock copolymers was also investigated. In this case, three parameters of the model were varied: the molar mass of the middle block, molar ratio (ε-caprolactone:DL-lactide) in the middle block and molar mass of the hard block, which was synthesized from PLA homopolymers. Two resulting polymers revealed good shape recovery: polymers with structures PLA-b-(PCL-co-PDLA)-b-PLA 5-40-5 kDa (50 %LA in PCL) and 5-20-5 kDa (with 75 % LA in PCL) attained 93 and 92% shape recovery after 100% sample stretching. Moreover, the polymer with the optimized structure PLA-b-(PCL-co-PDLA)-b-PLA 10-40-10 kDa (50 %LA in PCL) reached 96% shape recovery.

Chapter 11 - Recent growing use of plastics has led to the design and development of new degradable thermoplastic materials that are more "friendly" to the environment. Among the various biodegradable plastics available, there is a growing interest in polyhydroxyalka-noates (PHAs) or green plastics, due to their properties which resemble those of conventional petrochemical based polymers. PHAs are polyesters of various hydroxyalkanoate monomers accumulating as energy/carbon storage materials by granular inclusions in the cytoplasm of various bacterial cells under unfavorable growth condition along with the presence of excess carbon source. Their production from renewable resources and their complete biodegrade-ibility give PHAs promising advantages from an environment point of view. All PHAs are completely degradable to carbon dioxide and water. PHAs have versatile applications in packaging films, disposable items etc. PHAs have been recognized as a good tool for biodegradable polymers. The high production cost of PHA due to the substrate cost, constricts their industrial applications. Several processes for production of PHA from inexpensive carbon sources and waste products have been investigated to utilize bountiful organic compounds present in the waste. If waste products are used as substrate for the production of PHA, dual advantage of reducing waste disposal cost and production of value-added products could be actualized. Although, PHAs have been recognized as a good candidate for biodegradable polymers, their high production cost limits their industrial application. For the economical production of PHAs, various bacterial strains, either wild-type or recombinant strains along with new fermentation strategies have been developed through metabolic engineering for the production of PHAs with high concentration and productivity.

The most significant factor for the increased cost of production in PHA is mainly through the substrate as a carbon and energy source. Several processes for PHAs production from cheap carbon sources and waste products have also been investigated in order to utilize abundant organic compounds in waste. If waste products are used as a substrate for the production of PHA, dual advantage of reducing waste disposal cost and production of value-added products can be realized. This review paper will be focused on the production of PHA by microorganisms from cheap and inexpensive carbon sources, biosynthetic pathways, as well as factors affecting the production and its composition will also be discussed.

Chapter 12 - Cyanobacteria are prokaryotic oxygenic photoautotrophs with different morphologies, ranging from unicellular to colonial and filamentous forms, found in almost every conceivable habitat on earth. These organisms with a short generation time need some simple inorganic nutrients such as phosphate, nitrate (not in case of N_2-fixers), magnesium, sodium, potassium and calcium as macro-, and Fe, Mn, Zn, Mo, Co, B and Cu as micronutrients for their growth and multiplication. Further, these organisms can successfully be cultivated in wastewaters due to their ability to use inorganic nitrogen and phosphorus for their growth, and wastewaters such as effluents from farm-yards, fish-farms, sewage treatment plants, etc. are rich sources of N and P. Therefore, wastewater treatment involving cyanobacteria is quite attractive because of their ability to transform waste into useful biomass using sunlight as the energy source. Thus, these tiny photoautotrophs can be considered as alternative hosts for low-cost production of polyhydroxyalkanoates (PHAs), the most appropriate materials for alternative plastics. The photoautotrophic production of PHAs in cyanobacteria may encompass the high cost incurred due to feeding large amount of organic carbon and continuous oxygen supply during bacterial fermentation.

An overview is presented here on the progresses and possibilities of producing PHAs from cyanobacteria. The majority of efforts on photoautotrophic production of PHAs using CO_2 as the sole carbon source by cyanobacteria are available, but the contents, in general, are very low and amount less than 10% of dry cell weight (dcw) with sole exception to *Synechococcus* sp. MA19, where a much higher poly-β-hydroxybutyrate (PHB) content, 55% (dcw) was reported. Some cyanobacteria can accumulate PHB when grown mixotrophically with acetate, glucose or other exogenous carbon sources; accumulation more than 40% (dcw) was recorded for *Nostoc muscorum* and *Synechocystis* PCC 6803. Recent study of our laboratory showed that a N_2-fixing cyanobacterium, *Aulosira fertilissima* can accumulate PHB up to 77% (dcw) under P-deficiency with acetate. *Nostoc muscorum* was also found to accumulate poly(3-hydroxybutyrate-co-3-hydroxyvalerate) P(3HB-co-3HV) co-polymer ~40% (dcw) under specific growth condition. Uncouplers like CCCP and DCCD stimulated PHB accumulation, whereas supplementation of DCMU to the photoautotrophically-grown cultures suppressed PHB accumulation.

Analysis of material properties of these polymers by mechanical tests, surface analysis and differential scanning calorimetry (DSC) exhibited comparable material properties with the commercial polymers. Addition of exogenous carbon although found essential for stimulation of PHAs synthesis in cyanobacteria, the magnitude was more than one order lower (~ 10 time less) as compared to that of heterotrophic bacteria. Thus, low-cost production of PHAs polymers from cyanobacteria seems to be challenging, but a feasible approach.

PART 1.
SPECIAL DEBATES ON POLYMER SYNTHESIS

In: Polymer Synthesis
Editor: E. Kowsari

ISBN 978-1-61324-672-6
© 2012 Nova Science Publishers, Inc.

Chapter 1

PRECISE SYNTHESIS OF POLYOLEFINS CONTAINING POLAR FUNCTIONALITIES BY OLEFIN COORDINATION COPOLYMERIZATIONS USING GROUP 4 METAL CATALYSTS

Kotohiro Nomura* and Soliman Mehawed Abdellatif
Department of Chemistry, Graduate School of Science and Engineering, Tokyo Metropolitan University, 1-1 Minami Osawa, Hachioji, Tokyo 192-0397, Japan

ABSTRACT

Recent examples for synthesis of new polyolefins containing polar functionalities by coordination insertion copolymerization using group 4 metal complex (Ti, Zr) catalysts have been reviewed. In particular, examples by adopting the approaches by (a) direct copolymerization of olefin with polar monomer using coordination insertion methods, and (b) controlled incorporation of reactive functionalities (and the subsequent introduction of polar funtionalities under mild conditions) have been described. Our recent efforts for precise synthesis of polyolefins containing polar functionalities by (i) efficient incorporation of reactive functionality by copolymerization of ethylene with nonconjugated diene using nonbridged half-titanocene catalysts and (ii) subsequent chemical modifications under mild conditions have been introduced.

1. INTRODUCTION: BACKGROUND

Polyolefin is one of the most important commercial synthetic plastics in our daily life, and the market capacity still increases even in the conventional polymers such as polyethylene [HDPE (High Density Polyethylene), LLDPE (Linear Low Density Poly-ethylene)], polypropylene (Isotactic Polypropylene, PP). Recently, considerable attention has been paid to produce new polyolefins with specified functions called "fine polyolefins"

---

* Corresponding Author: tel.: +81-42-677-2547, fax: +81-42-677-2547, E-mail: ktnomura@tmu.ac.jp.

exemplified as COC (cyclic olefin copolymer, optical materials) and others. We recognize that design of the efficient transition metal complex catalysts that precisely control olefin polymerization should play an essential key role for the success,[1-10] as demonstrated by the recent progress in newly designed catalysts called half-metallocenes, [2d,3,9,10] non-metallocenes. [4] Precise control in the copolymerization is an important method that usually allows the alteration of the (physical, mechanical, and electronic) properties by varying the ratio of individual components, and the catalysts exhibiting notable activities with better comonomer incorporations in the (ethylene) copolymerizations are thus very important. This is because that new polymers generally can be prepared by incorporation of new comonomers (ex. sterically encumbered olefins etc.) in the copolymerization.

1) Direct copolymerization approaches

(a) Radical copolymerization (conventional)

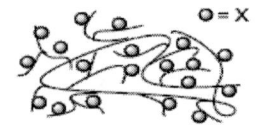

○ = X

(i) ultra high ethylene pressure, high temperature
(ii) highly branched by radical migration
(iii) difficult to control composition, molecular weight

(b) Transition metal catalyzed coordination insertion copolymerization

(1) direct copolymerization by late transition metal catalysts (Pd etc.)

(2) Copolymerization with protected comonomer by early transition metal catalysts

M = Zr etc.
Protection should be required to avoid the deactivation

R = protecting group

Presence of interaction between functional group and metal center

(c) Living radical copolymerization (ATRP, Nitroxide, RAFT etc.)

Atom Transfer Radical Polymerization (ATRP)

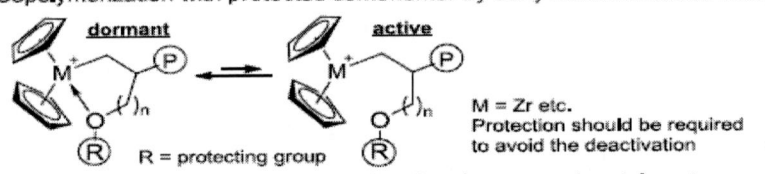

Presence of an equilibrium between dormant and active species for conducting low radical concentration (to avoid termination by radical coupling)

Scheme 1. Approaches for synthesis of polyolefins containing polar functionalities by "direct" copolymerization. [13b]

2) Post functionalization approaches

(a) Free radical functionalization (grafting) of polyolefins (conventional)

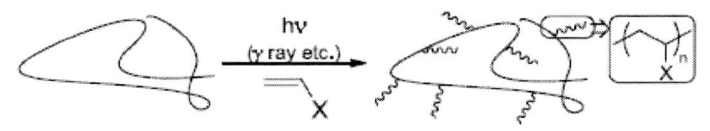

(i) Required severe, harsh conditions (under irradiation)
(ii) Difficult to control composition, molecular weight etc.

(b) Transition metal catalyzed direct functionalization (C-H bond activation)

(1) Regiospecific functionalization and subsequent oxidation

(2) Alkane transfer dehydrogenation

R = ^{t}Bu, ^{i}Pr
n = 2 or 4

(c) Controlled incorporation of reactive functionalities
(by transition metal catalyzed coordination polymerization)

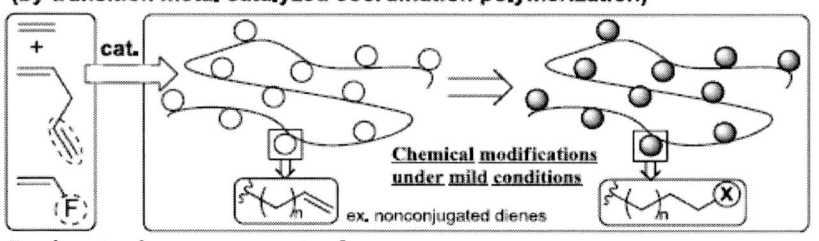

Chemical modifications under mild conditions

ex. nonconjugated dienes

F: polymerizable group with functionality

Scheme 2. Approaches for synthesis of polyolefins containing polar functionalities by "post functionalization" approaches. [13b]

Precise, efficient synthesis of polyolefins containing polar functionalities attracts considerable attention [11-14] especially because of their promising amphiphilic nature. Two known strategies commonly employed for this purpose are, 1) *direct* copolymerization with functionalized (polar) monomers (Scheme 1) [11-13] and 2) *post* polymerization modification (Scheme 2). [11,13,14] However, the first approach would face difficulties, because the conventional radical copolymerization (ex. ethylene/vinyl acetate copolymerization) generally requires both ultra high pressure (ca. 1000 atm) and temperature affording *branched* random copolymers (Scheme 1a). Moreover, significant decrease in the catalytic actvitities and/or decomposition of the catalytically-active species owing to the catalyst poisoning and interaction of centered metal with functionalized monomers (generating adducts) would be present in transition metal catalyzed coordination insertion copolymerization (Scheme 1b), although recent progress would introduce promising possibilities. [12] It is difficult to prepare the copolymer with high olefin (ethylene, propylene) content by adopting the living radical copolymerization, because the equilibrium shift to the left (dormant) upon increasing olefin contents. [13] The conventional post modification approach (Scheme 2), still faces difficulties, because only limited chemistry like a free radical grafting reaction is available to activate the *completely saturated* aliphatic molecular structure (Scheme 2a). [14] In this book chapter, the authors briefly summarize recent update for syntheses of polyolefins containing polar functionalities by direct copolymerization using group 4 metal complex catalysts.

Recently, controlled incorporation of a reactive moiety that introduces polar functionalities through chemical modification (Scheme 2c), has been considered as the third approach. [13] In this chapter, progresses on this approach especially our recent successful demonstrations for efficient controlled introduction of polar functionalities into polyolefins by copolymerization with nonconjugated dienes have also been introduced. [15-17]

2. DIRECT COORDINATION INSERTION COPOLYMERIZATION

Many reports have been known for *direct* copolymerization of olefin with polar monomers, [12d,18-20] especially using late transition metal complex catalysts (especially nickel, palladium). [12,18,19] This is because that decrease in the activities due to the catalyst poisoning and interaction of centered metal with functionalized monomers are often observed in the coordination insertion copolymerization.

The results for copolymerization of ethylene with 10-undecen-1-ol using ordinary metallocenes are summarized in Table 1. [20h] The hydroxyl group in 10-undecen-1-ol was protected by pretreatment with methylaluminoxane (MAO) to prevent the deactivation of the catalytically active species. The catalytic activities significantly decreased upon increasing the concentration of 10-undecen-1-ol [(indenyl)$_2$ZrCl$_2$ - MAO catalyst]. [20h] The fact is apparently due to a strong interaction of zirconium with oxygen atom in the comonomer (10-undecen-1-ol); the molecular weight distribution became broad upon further addition (M_w/M_n = 5.0), which may suggest a formation of another catalytically-active species via decomposition of the active species with polar group. Although the comonomer content was improved by using bridged metallocenes such as [(CH$_2$)$_2$(indenyl)$_2$]ZrCl$_2$, [Me$_2$Si (indenyl)$_2$] ZrCl$_2$, notable improvements in the observed activities were not seen; the resultant copolymers possessed rather low molecular weights due to increased degree of β-hydrogen elimination after comonomer insertion.

As shown in Table 2, the comonomer incorporation in ethylene/allylalcohol copolymerization was affected by the protecting group employed (Scheme 3). Steric bulk of the protecting group would be required for the subsequent insertion (next monomer, ethylene) due to prevent strong interaction between oxygen and zirconium. [20u] The zirconium complex employed here exhibit better comonomer incorporation in ethylene/undecen-1-ol copolymerization, the resultant copolymers was treated with 2-bromoisobutyryl bromide to prepare the macronitiator. The subsequent atom transfer radical polymerization of n-butyl acrylate (BA) or methyl methacrylate (MMA) in the presence of Cu afforded amphiphilic graft copolymers in a relatively controlled manner. [20u] However, as exemplified in Table 2, it seems very difficult to prepare the copolymer with high comonomer content efficiently due to the low catalytic activity by interaction between oxygen and the zirconium.

Table 1. Copolymerization of ethylene with 10-undecen-1-ol by (indenyl)$_2$ZrCl$_2$, [(CH$_2$)$_2$(indenyl)$_2$]ZrCl$_2$, [Me$_2$Si(indenyl)$_2$]ZrCl$_2$ - MAO catalysts [20h]

catalyst	alcohol in feed $\times 10^2$ M	activity kg-polymer/mol-Zr·h	M_w $\times 10^{-4}$	M_w/M_n	content / wt%
(indenyl)$_2$ZrCl$_2$	0	15100	67.3	2.4	--
(indenyl)$_2$ZrCl$_2$	2.2	1400	38.6	2.9	
(indenyl)$_2$ZrCl$_2$	2.8	120	45.2	5.0	0.8
[(CH$_2$)$_2$(indenyl)$_2$]ZrCl$_2$	2.2	350	12.3	2.3	5.5
[Me$_2$Si(indenyl)$_2$]ZrCl$_2$	1.6	670	15.5	2.4	7.1

Reaction conditions: (Ind)$_2$ZrCl$_2$ = 1.6×10^{-5} mol dm^{-3}, [Al]/[Zr] = 4300, ethylene 2.0 bar, 25 °C.

Table 2. Copolymerization of ethylene with allyl alcohol in the presence of AlR$_3$ [20u]

AlR$_3$	Activity Kg-polymer/mol-Zr·h	M_w	M_w/M_n	content / mol%	number of OH end	in
AlEt$_3$	121	10700	2.38	0.33	1	n.d.
AlMe$_3$	40	31800	2.42	0.17	1	n.d.
Al^{i}Bu$_3$	91	60600	2.59	1.20	1	10
Al(n-C$_8$H$_{17}$)$_3$	84	32600	2.54	0.22	1	0.1

Reaction conditions: Zr 0.025 mmol, time 100-105 min, MAO 1.57 mmol, allyl alcohol 40 mmol, AlR$_3$ 48 mmol.

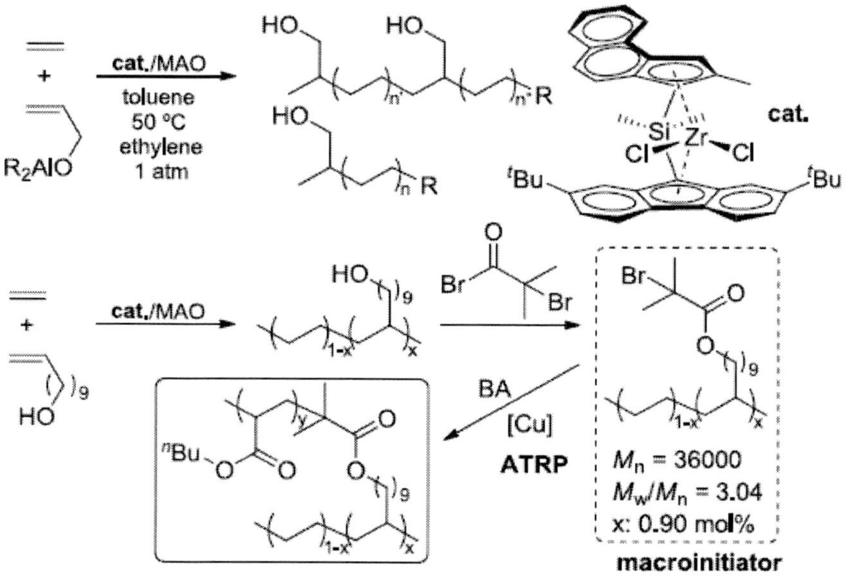

Scheme 3. Copolymerization of ethylene with allyl alcohol or undecen-1-ol in the presence of Zr complex – MAO – AlR$_3$ catalyst systems. [20u,y] Synthesis of amphiphilic graft copolymers by coupling of coordination polymerization and atom transfer radical polymerization (ATRP). [20y]

More recently, certain titanium complex containing bis(phenoxyimine) ligand, [2-{(4-^{t}BuC$_6$H$_4$)N=C}-6-Ph-C$_6$H$_3$]$_2$TiCl$_2$, incorporated 5-hexexe-1-yl-acetate in ethylene copolymerization in the presence of MAO, but the activity decreased upon increasing the commonomer content (15 kg-polymer/mol-Ti·h, 3.20 mol% vs 341 kg-polymer/mol-Ti·h, 0.90 mol%: ethylene 1 atm, 25 °C, 10 min). [20z] Therefore, more efficient catalyst for the desired copolymerization should be the promising target.

Efficient Introduction of Functional Group into Polyolefins by Direct Copolymerization Using Half-Titanocenes

We recently reported that copolymerization of ethylene with allyltrimethylsilane (ATMS) by half-titanocene containing aryloxo ligand **(1)** proceeded efficiently, affording high molecular weight copolymers with uniform molecular weight distributions (Scheme 4). [21] Note that the ATMS content was relatively close to 1-pentene content in poly(ethylene-*co*-1-pentene) conducted under the similar conditions. [10b] As shown in Table 3, the Cp-ketimide analogue, CpTiCl$_2$(N=C^{t}Bu$_2$) **(2)** showed the less efficient ATMS incorporation under the same conditions, but exhibited notable catalytic activity, affording high molecular weight copolymers. The activities by 1,2 increased upon increasing the ATMS concentration and the ethylene pressure, and the M_n values in the copolymers were independent upon the ATMS contents. The facts are different from those in the copolymerization by ordinary metallocenes, [22] because both the activities and the M_n values in the copolymers significantly decreased upon increasing the ATMS contents [22] due to that ATMS also plays a role as chain transfer reagent due to favored β-hydrogen elimination after bulky ATMS insertion.

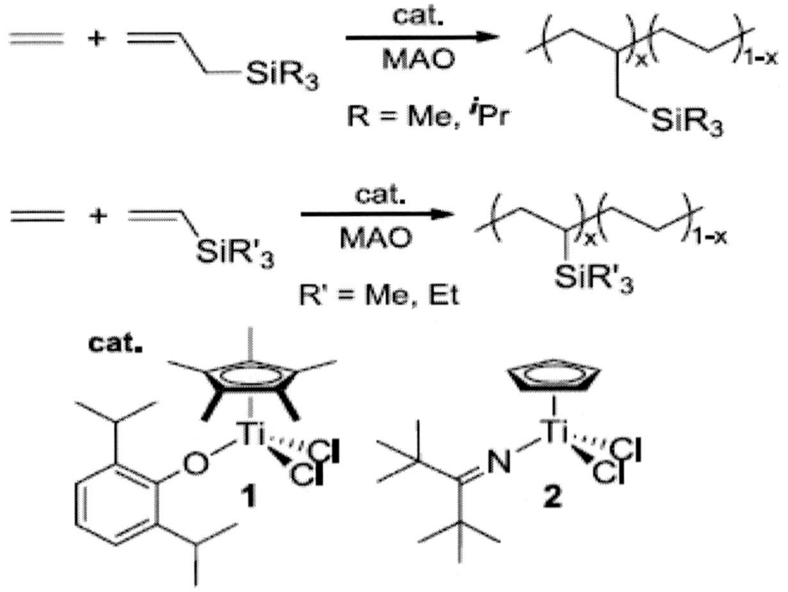

Scheme 4. Copolymerization of ethylene with allyltrialkylsilanes, vinyltrialkylsilanes. [21]

Table 3. Copolymerization of ethylene with $CH_2=CHCH_2SiR_3$ [R = Me (ATMS), ^iPr (ATIS)], and $CH_2=CHSiMe_3$ (VTMS) using $Cp*TiCl_2(O-2,6-^iPr_2C_6H_3)$ (1), $CpTiCl_2(N=C^tBu_2)$ (2), $[Me_2Si(C_5Me_4)(N^tBu)]TiCl_2$ (3), and Cp_2ZrCl_2 – MAO catalyst systems[a],[21]

Complex (μmol)	Comonomer (M)	activity[b]	$M_n{}^c \times 10^{-4}$	$M_w/M_n{}^c$	cont.[d]/ mol%
1 (0.50)	ATMS (0.52)	1840	2.37	2.43	29.5
1 (0.50)	ATMS (1.05)	3550	2.78	2.69	48.8
1 (0.50)	VTMS (1.15)	1870	30.5	1.9	5.1
2 (0.05)	ATMS (0.52)	45000	31.3	2.35	16.6
2 (0.05)	ATMS (1.05)	47500	28.3	2.55	30.3
2 (0. 50)	ATIS (0.69)	3480	13.0	1.97	17.5
2 (1.00)	VTMS (1.15)	3730	57.3	2.3	11.9
3 (0.25)	VTMS (1.15)	2280	36.7	2.5	10.4
Cp_2ZrCl_2 (10.0)	VTMS (1.15)	23	0.56	3.9	trace

[a] Cited from references 21, conditions: comonomer 5.0 or 10 mL; comonomer + toluene total 30.0 mL, dried MAO 3.0 mmol, ethylene 6 atm, 25 °C; 10 min. [b] Activity = kg-polymer/mol-Ti·h. [c] GPC data in o-dichlorobenzene vs polystyrene standards. [d] Comonomer content estimated by ^{13}C NMR spectra.

Vinyltrialkylsilanes should be considered as better comonomer in terms of the direct functional group introduction into polyethylene (or polypropylene) backbone as well of its use as cross-linking reagents to improve thermal properties. These sterically encumbered olefins are, however, very difficult to coordinate into the metal center in ordinary metallocenes, [23] and no reports for the copolymerization with vinyltrialkylsilane had been known until recently. Note that both the Cp*-aryloxo analogue (**1**) and the Cp-ketimide analogue (**2**) incorporated vinyltrimethylsilane (VTMS) into polyethylene (Scheme 4), and the resultant copolymers possessed high molecular weights with uniform molecular weight distributions. [21a] **Complex 5** exhibited both higher catalytic activities and better VTMS incorporations than **1**. efficient synthesis of high molecular weight copolymers with uniform compositions could be achieved by adopting **2**, although both the activity and the M_n values decreased upon increasing the VTMS contents. Later, we found that the copolymerization by the constrained geometry catalyst, $[Me_2Si(C_5Me_4)(N^tBu)]TiCl_2$ (**3**), also proceeded efficiently. [27k] The attempted copolymerization using Cp_2ZrCl_2 – MAO catalyst afforded linear polyethylene with low catalytic activity. [21a]

3. Synthesis of Polyolefins Containing Polar Functionalities by Controlled Introduction of Reactive Functionalities

Recently, an approach on controlled incorporation of a reactive moiety that introduces functionalities through chemical modification (Scheme 2c) [11,13] has been considered as an alternative route. A known classic model of this approach is the introduction of unsaturation as exemplified by using 5-ethylidene-2-norbornene, 1,4-hexadiene, and dicyclopentadiene (but showed inefficient incorporations in the copolymerization by ordinary catalysts). [11,24]

Introduction of unsaturated olefinic double bond into the side chain can be possible by adopting the copolymerizatioin of ethylene with 7-methyl-1,6-octadiene (MOD, scheme in Figure 2), [16,25] utilized by a difference in the reactivity of two olefinic double bonds (terminal vs trisubstituted). However, their incorporations by the ordinary catalysts (metallocenes etc.) were inefficient, affording (co)polymers with low molecular weights as well as rather broad molecular weight distributions, as exemplified in the previous section. [11]

Table 4. Copolymerization of ethylene with 1-octene (OC) or 7-methyl-1,6-octadiene (MOD) by Cp*TiCl$_2$(O-2,6-^{i}Pr$_2$C$_6$H$_3$) (1) - MAO catalyst [16]

cat.1 μmol	ethylene / atm	MOD or OC (conc./ M)	activity[a]	M_n^b ×10^{-4}	M_w/M_n^b	comonomer[c] / mol%
0.02	6	OC (1.06)	154000	13	2.1	27.9
0.02	4	OC (1.06)	79000	12	2.0	33.2
0.05	6	MOD (1.01)	97200	12	2.3	25.5
0.05	4	MOD (1.01)	46200	11	2.2	32.9

Reaction Conditions: 1-octene (OC) or 7-methyl-1,6-octadiene (MOD) + toluene total 30 mL, methylaluminoxane (MAO, prepared by removing toluene and AlMe$_3$ from ordinary MAO) 3.0 mmol, 25 °C, 6 min. [a] Activity in kg-polymer/mol-Ti·h. [b] GPC data in o-dichlorobenzene vs polystyrene standards. [c] OC or MOD in copolymer (mol%) estimated by ^{13}C NMR spectra.

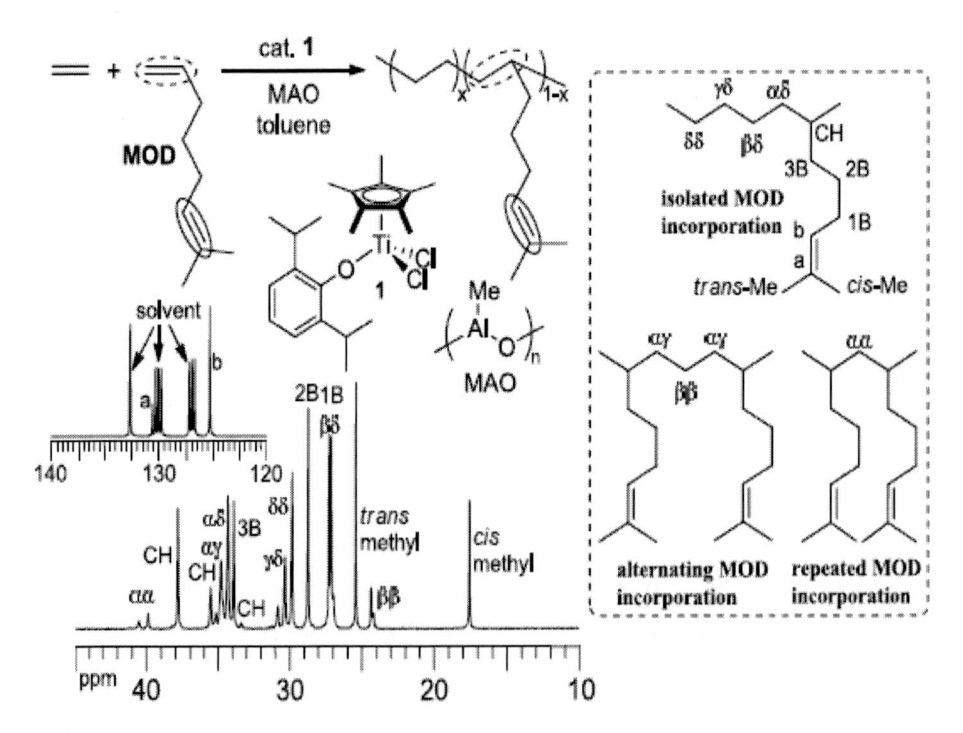

Figure 1. ^{13}C NMR spectrum for poly(ethylene-co-7-methyl-1,6-octadiene) prepared by Cp*TiCl$_2$(O-2,6-^{i}Pr$_2$C$_6$H$_3$) (1) - methylaluminoxane (MAO) catalyst system (in o-dichlorobenzene-d_4 at 110 °C, MOD 32.9 mol%). [16]

As described in the introductory, design of efficient transition metal catalysts for precise olefin polymerization plays an essential role for production of new polyolefins with specified functions. [1-10] We already demonstrated that *nonbridged* half-titanocenes containing anionic donor ligand (X) of type, Cp'TiCl$_2$(X) [Cp' = cyclopentadienyl group, X = aryloxo, ketimide etc.], display unique characteristics especially for copolymerization of ethylene with α-olefin, styrene, cyclic olefins, and with disubstituted α-olefins. [10,16,17,26,27] We reported that copolymerizations of ethylene with 7-methyl-1,6-octadiene (MOD) by Cp*TiCl$_2$(O-2,6-^{i}Pr$_2$C$_6$H$_3$) **(1)** - MAO catalyst system proceeded at remarkable rates with exclusive incorporation of monoolefins (without incorporating trisubstituted olefin), affording high molecular weight unsaturated poly(ethylene-*co*-MOD)s with high MOD contents (Table 4, Figure 1). [16] The MOD contents in the resultant copolymers were thus closely related to those in the ethylene/1-octene copolymerizations under the similar conditions.

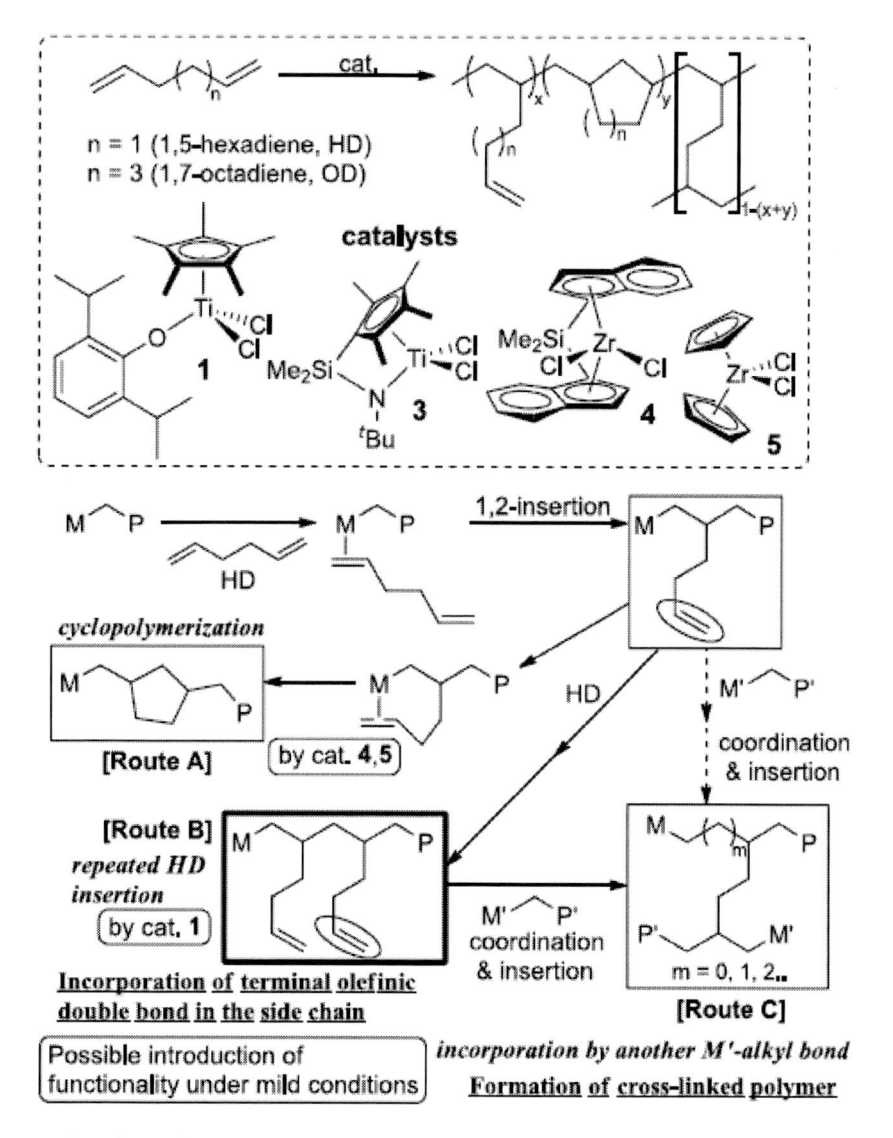

Scheme 5. Basic scheme for polymerization of nonconjugated diene (1,5-hexadiene, HD) using Ti, Zr complex catalysts. [28]

In order to introduce more reactive functionality (terminal olefin) into the side chain for the efficient functionalization process under mild conditions, we thus focused on nonconjugated diene like 1,5-hexadiene (HD) for synthesis of functionalized polyolefin by the favored repeated HD insertion, [28] although most of the reported examples using ordinary catalysts such as zirconocene (exemplified as 4 and 5, Scheme 5), [29,30] titanocene, [29c] half-zirconocene, [31a] and others [31b, 32] favored cyclopolymerization incorporating methylene-1,3-cyclopentane unit (Scheme 5 route A). The approach should introduce promising possibilities like, (i) incorporation of *terminal* olefinic double bond (route B) that would introduce polar functionality in a controlled manner by chemical modification under mild conditions and/or (ii) controlled incorporation of cross-linking (route C) that improves chemical-, heat-resistance. In order to achieve the above goal, selectivity of the repeated insertion (route B and C) against cyclization was improved by designing the complex catalyst. [28,32]

Polymerization of 1,5-hexadiene (HD) by Cp*TiCl$_2$(OAr) (**1**, Cp* = C$_5$Me$_5$, Ar =2,6-^{i}Pr$_2$C$_6$H$_3$), [Me$_2$Si(C$_5$Me$_4$)(N^{t}Bu)]TiCl$_2$ (**3**), [Me$_2$Si(indenyl)$_2$]ZrCl$_2$ (**4**), and Cp$_2$ZrCl$_2$ (**5**), were explored in the presence of MAO. Both **1 and 3** exhibited the notable catalytic activities, and the resultant poly(HD)s were insoluble. [28] Based on the DSC thermograms (sole glass transition temperature), ^{13}C CPMAS spectra, and the dynamic mechanical analysis (DMA), it turned out that 1 favored repeated **1,2**-insertion rather than cyclization affording polymers containing olefinic double bond (butenyl group) in the side chain with uniform distributions, whereas the polymerization by **4,5** favored cyclization under the same conditions. Linked half-titanocene (**3**) showed lower catalytic activities with favored repeated insertion, affording high molecular weight poly(HD)s which possessed *internal* olefinic double bonds by isomerization accompanied. [28b]

Note that exclusive repeated insertion of 1,7-octadiene (OD) could be achieved in OD polymerization using **1** - MAO catalyst, affording polymers containing terminal olefinic double bonds in the side chain (Table 5, 2). [15]

Table 5. 1,7-Octadiene (OD) polymerization by Cp*TiCl$_2$(O-2,6-^{i}Pr$_2$C$_6$H$_3$) (1) - MAO catalyst [15]

x >> y
x: 82->92%

cat. (μmol)	OD/ hexane / mL	OD / mmol/mL	time / min	TON[a]	M_n^b ×10^{-4}	M_w/M_n^b	double bond(%)[c]
0.3	15.0/-	6.35	20	4840	66.4	1.89	(98)[d]
0.3	15.0/-	6.35	20	4840	76.3	1.86	92
0.1	15.0/-	6.35	30	10900	81.1	1.63	88
0.1	10.0/5.0	4.23	30	9400	61.2	1.66	82
1.0	7.5/7.5	3.17	20	2060	37.8	2.00	85 (92)[d]

Reaction conditions: **complex 1** in toluene (1.0 mL), MAO (prepared by removing AlMe$_3$ and toluene from commercially available MAO) 2.0-5.0 mmol, 25 °C. [a] Molar amount of monomer consumed /mol-Ti. [b] GPC data in THF vs polystyrene standards. [c] Percentage of hexenyl group estimated by ^{13}C NMR spectra. [d] Percentage of hexenyl group estimated by ^{1}H NMR spectra.

The integration ratio of protons in the terminal olefins vs protons in the aliphatic region assumed that the polymerization occurred with an exclusive OD repeated insertion. Selectivity of the repeated insertion (percentage of the double bond, hexenyl group, based on the OD insertion) was affected by the OD concentration employed, but the selectivity was high even under low OD concentration; the fact should be a promising contrast to those observed by certain metallocenes which favored cyclization under similar conditions. [30] The observed higher selectivity in the OD polymerization than that in the HD polymerization would be considered as the difference of the proposed intermediate in which the other olefin coordinates to Ti after HD or OD insertion (formation of 5- or 7-membered ring after cyclization).

The copolymerizations of OD with 1-octene (OC) afforded high molecular weight copolymers with uniform molecular weight distributions, and the OD contents (degree of hexenyl group) estimated by ^{1}H NMR spectra could be varied by the OD/OC feed ratio. [15] The resultant copolymers were treated with 9-BBN and then NaOH/H$_2$O$_2$ aq. to give poly{OC-*co*-(7-octen-1-ol)} exclusively (98.9%) without decrease/increase in the M_n value. [15,33]

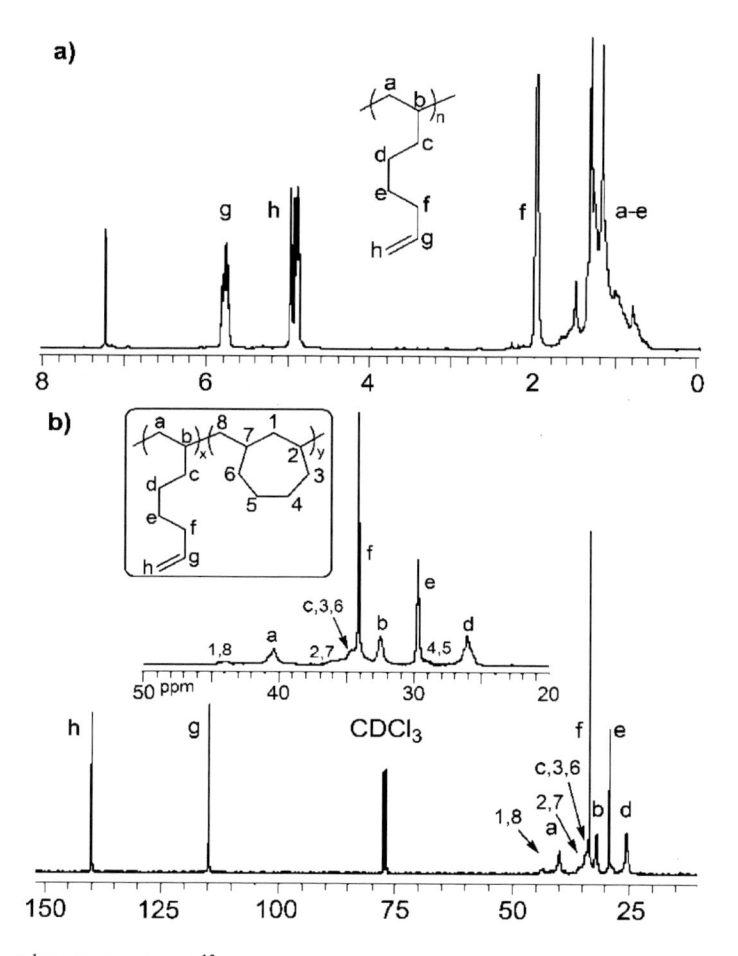

Figure 2. Selected ^{1}H NMR (a) and ^{13}C NMR (b) spectra for poly(OD)s (shown in Table 5, OD 15.0 mL under bulk conditions) prepared by **1** - MAO catalyst (in CDCl$_3$, 25 °C). [15]

The OH group in the copolymers was further treated with AlEt$_3$, and was then added ε-caprolactone (CL) to afford the graft copolymers, poly{OC-co-(7-octan-1-ol)-graft-poly(CL), via Al-alkoxide initiated ring-opening polymerization (ROP, Scheme 6). The M_n values increased at longer reaction hours consistently with unimodal molecular weight distributions, because the ROP occurred in a living manner. The resultant copolymers were identified by ^{1}H, ^{13}C (dept) NMR spectra, and DSC thermograms.

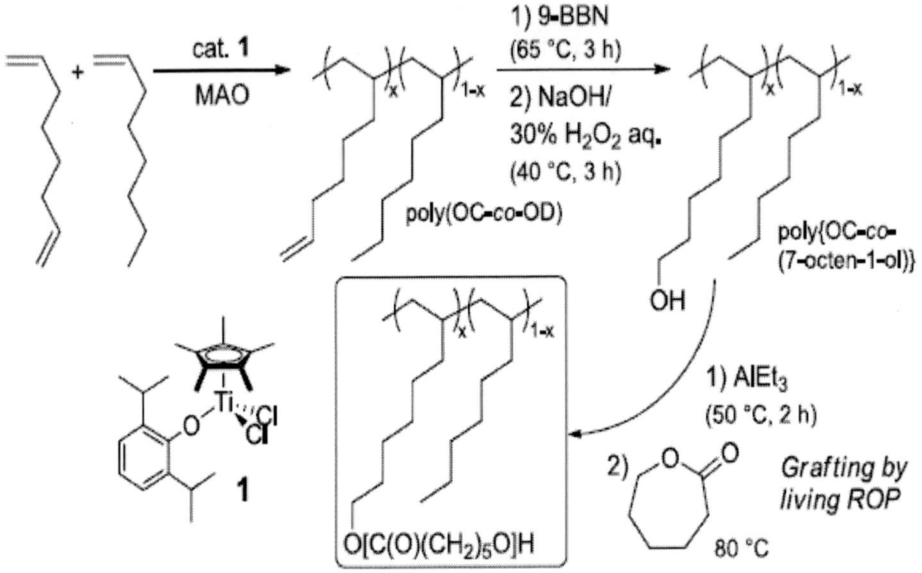

Scheme 6. Controlled introduction of polar functionality into polyolefins, poly(1-octene-co-1,7-octadiene). [15]

Table 6. Copolymerization of ethylene with 1-octene, 1,7-octadiene by Cp*TiCl$_2$(O-2,6-^{i}Pr$_2$C$_6$H$_3$) (1) - MAO catalyst [15,33]

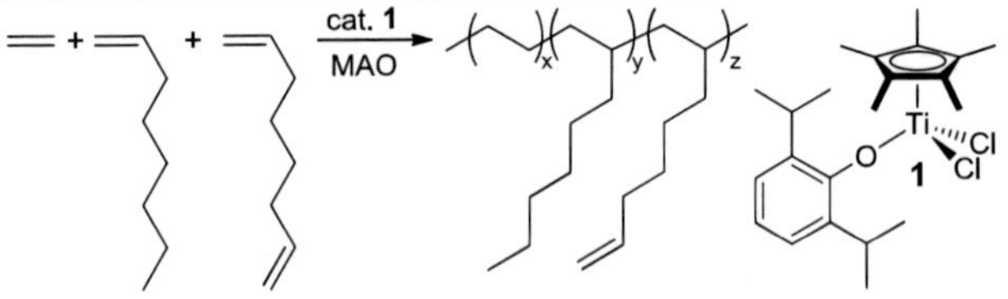

1 / μmol	ethylene / atm	1-octene / mL	1,7-octadiene / mL	activity kg-polymer/mol-Ti·h	M_n^a ×10^{-4}	M_w/M_n^a	OC/OD^b / mol%
0.03	6	10	5	61400	15.5	2.15	28.1/15.7
0.03	6	5	10	29400	13.0	1.98	12.6/25.0
0.03	6	9	1	55600	19.6	1.87	35.2/6.0
					22.0^c	1.93^c	

Reaction conditions: 1-octene + 1,7-octadiene + toluene total 30 mL, MAO 2.0 mmol, 10 min, at 25 °C. ^{a} GPC data in THF vs polystyrene standards. ^{b} Estimated by both ^{13}C NMR spectra (whole content) and ^{1}H NMR spectra (ratio of OC/OD). ^{c} After hydrogenation.

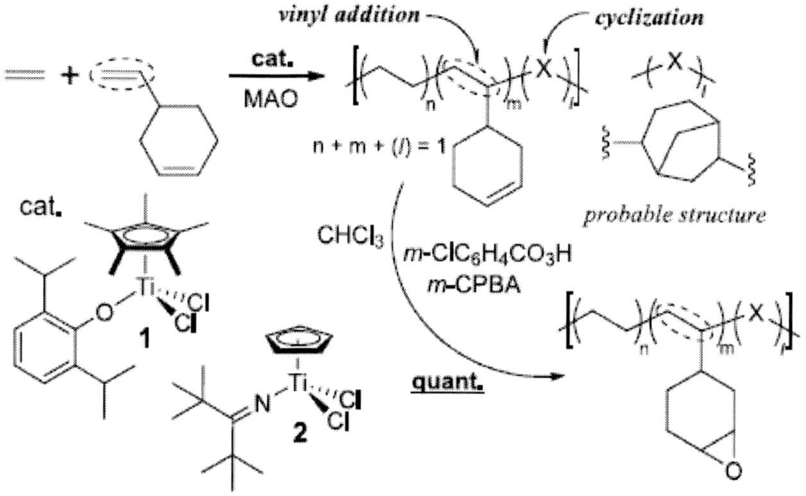

Scheme 7. Synthesis of cyclic olefin copolymers containing polar functionality. [17]

Importantly, copolymerization of ethylene with OC, OD were thus conducted, affording poly(ethylene-co-OC-co-OD)s containing olefinic double bond in the side chain (Table 6). [15,33] Note that remarkable activities (activity: 29400-102600 kg-polymer/mol-Ti·h) were also observed in the copolymerization, even in syntheses of the copolymers with high OC/OD contents (OC+OD total 37.6-43.8 mol%). [15,33] Also note that the resultant copolymers possessed high molecular weights with unimodal molecular weight distributions. The present approach should thus demonsrate a promising new possibility for preparation of functionalized polyolefins under mild conditions, and will be considered as the environmentally benign route for the desired polymers.

More recently, as shown in Scheme 7, we demonstrated that the copolymerizations of ethylene with vinylcyclohexene (VCHen), commercially produced by dimerization of butadiene, proceeded via vinyl addition affording high molecular weight copolymers containing cyclohexenyl side chains (with uniform compositions as well as with unimodal molecular weight distributions) accompanied with certain degree of side reaction (via intramolecular cyclization after VCHen insertion). [17] This approach could be possible, because both $Cp^*TiCl_2(O-2,6-^iPr_2C_6H_3)$ (1) and $CpTiCl_2(N=C^tBu_2)$ (2) exhibited efficient vinylcyclohexane (VCH) incorporation in the ethylene/VCH copolymerization, [27g] as well as exhibited negligible comonomer incorporation in ethylene/cyclohexene copolymerization. [27f] The Cp-ketimide analogue (2) showed the best catalyst performance in terms of both the activity and the selectivity (lowest degree of the subsequent intra-molecular cyclization). Quantitative epoxidation of the olefinic double bonds in the resultant copolymer could be achieved by using m-chloroperbenzoic acid under mild conditions; a facile, precise synthesis of functionalized polyolefin can be demonstrated by adopting this approach.

4. Summary and Future Outlook

As described above, precise synthesis of polyolefins containing polar functionality attracts considerable attention, and recent progress in living radical, metal-catalyzed coordination polymerization in addition to the direct terminal alkane C-H activation method offer new promising possibilities. Recently, approaches on controlled incorporation of a reactive moiety that introduces polar functionalities through chemical modification has been considered, and we demonstrated an introduction of terminal olefininc double bond by repeated insertion of 1,7-octadiene (OD) without cyclization/cross linking in polymerization of OD by using nonbridged half-titanocene catalysts. More recently, we demonstrated that our designed catalysts, *nonbridged* half-titanocesnes, are effective for direct copolymerization of ethylene with vinyltrialkylsilanes that cannot be achieved by ordinary metallocenes.[27k.l] Since several promising findings that should be very important from both academic and industrial viewpoints can be demonstrated, particular attention should be thus paid to explore the possibility for preparing fine polyolefins by adopting these precise polymerization techniques.

ACKNOWLEDGMENTS

Our research efforts were partly supported by Grant-in-Aid for Scientific Research (B) from the Japan Society for the Promotion of Science (JSPS, No.21350054). The author would like to express his heartfelt thanks to his former group members who contributed to the projects as the coauthors. S.M.A. thanks Egyptian Ministry of Higher Education for giving an opportunity to study under K.N. through Partnership and Ownership (ParOwn) Initiative program.

REFERENCES AND NOTES

[1]   Mason, A. F.; Coates, G. W. In *Macromolecular Engineering*, Matyjaszewski, K.; Gnanou, Y.; Leibler L., Eds.; Wiley-VCH: Weinheim, Germany, vol. 1, 2007; pp 217. (b) *Metal Catalysts in Olefin Polymerization, Topics in Organometallic Chemistry 26*, Guan, Z. Ed.; Springer Verlag: Berlin, 2009.

[2]   (a) Brintzinger, H. H.; Fischer, D.; Mülhaupt, R.; Rieger, B.; Waymouth, R. M. *Angew. Chem., Int. Ed. Engl.*, 1995, 34, 1143. (b) Kaminsky, W. *Macromol. Chem. Phys.* 1996, 197, 3907. (c) Kaminsky,W.; Arndt, M. *Adv. Polym. Sci.,* 1997, 127, 143. (d) Suhm, J.; Heinemann, J.; Wörner, C.; Müller, P.; Stricker, F.; Kressler, J.; Okuda, J.; Mülhaupt, R. *Macromol. Symp.* 1998, 129, 1.

[3]   (a) McKnight, A. L.; Waymouth, R. M. *Chem. Rev.* 1998, 98, 2587. (b) Braunschweig, H.; Breitling, F. M. *Coord. Chem. Rev.,* 2006, 250, 2691. (c) Cano, J.; Kunz, K. *J. Organomet. Chem.* 2007, 692, 4411.

[4]   (a) Britovsek, G. J. P.; Gibson, V. C.; Wass, D. F. *Angew. Chem., Int. Ed. Engl.* 1999, 38, 428. (b) Gibson, V. C.; Spitzmesser, S. K. *Chem. Rev.* 2003, 103, 283. (c) Bolton, P. D.; Mountford, P. *Adv. Synth. Catal.* 2005, 347, 355. (d) Nomura, K.; Zhang, S. *Chem. Rev.,* ASAP (web released on October 29, 2010, DOI: 10.1021/cr100207h).

[5]   *Frontiers in Metal-Catalyzed Polymerization* (special issue), Gladysz J. A.; Ed., *Chem. Rev.* 2000, 100(4). For example, (a) Ittel, S. D.; Johnson L. K.; Brookhart, M. *Chem. Rev.* 2000, 100, 1169. (b) Alt, H. G.; Köppl, A. *Chem. Rev.* 2000, 100, 1205. (*c*) Chen, E. Y. -X.; Marks, T. J. *Chem. Rev.* 2000, 100, 1391.

[6]   *Metallocene complexes as catalysts for olefin polymerization* (special issue), Alt, H. G.; Ed., *Coord. Chem. Rev.* 2006, 250(1-2), 1.

[7]   *Metal-catalyzed polymerization* (special issue), Milani, B.; Claver, C. Eds., *Dalton Trans.* 2009, (41), 8769.

[8]   (a) Coates, G. W.; Hustad, P. D.; Reinartz, S. *Angew. Chem. Int. Ed.* 2002, 41, 2236. (b) Domski, G. J.; Rose, J. M.; Coates, G. W.; Bolig, A. D.; Brookhart, M. *Prog. Polym. Sci.* 2007, 32, 30. (c) Sita, L. R. *Angew. Chem. Int. Ed.* 2009, *48*, 2464.

[9]   Stephan, D. W. *Organometallics* 2005, 24, 2548.

[10] (a) Nomura, K.; Liu, J.; Padmanabhan, S.; Kitiyanan, B. *J. Mol. Catal. A* 2007, 267, 1. (b) Nomura, K. *Dalton Trans.* 2009, 8811. (d) Nomura, K. *Dalton Trans.* accepted.

[11] Chung, T. C. In *Functionalization of Polyolefins*, Academic Press: San Diego 2002.

[12] (a) Boffa, L. S.; Novak, B. M. *Chem. Rev.* 2000, 100, 1479. (b) Chung, T. C. *Prog. Polym. Sci.* 2002, 27, 39. (c) Boaen, N. K.; Hillmyer, M. A. *Chem. Soc. Rev.* 2005, 34, 267. (d) Nakamura, A.; Ito, S.; Nozaki, K. *Chem. Rev.* 2009, 109, 5215.

[13] (a) Nomura, K.; Kitiyanan, B. *Cur. Org. Synth.* 2008, 5, 217. (b) Nomura, K. *J. Syn. Org. Chem., Jpn.* 2010, 68, 1150.

[14] (a) Doak, K.W. In *Encyclopedia of Polymer Science and Engineering*, Mark, H.F., Ed.; John Wiley and Sons: New York, 1986; Vol. 6, p386. (b) Moad, G. *Prog. Polym. Sci.* 1999, 24, 81.

[15] Nomura, K.; Liu, J.; Fujiki, M.; Takemoto, A. *J. Am. Chem. Soc.* 2007, 129, 14170.

[16] Itagaki, K.; Fujiki, M.; Nomura, K. *Macromolecules* 2007, 40, 6489.

[17] Itagaki, K.; Nomura, K. *Macromolecules* 2009, 42, 5097.

[18] Examples, (a) Klabunde, U.; Ittel, S. D. *J. Mol. Catal.* 1987, 41, 123. (b) Youkin, T. R.; Connor, E. F.; Henderson, J. I.; Friedrich, S. K.; Grubbs, R. H.; Bansleben, D. A. *Science* 2000, 287, 480. (c) Connor, E. F.; Younkin, T. R.; Henderson, J. I.; Hwang, S.; Grubbs, R. H.; Roberts, W. P.; Litzau, J. J. *J. Polym. Sci.: Part A: Polym. Chem.* 2002, 40, 2842.

[19] (a) Johnson, L. K.; Killian, C. M.; Brookhart, M. *J. Am. Chem. Soc.* 1995, 117, 6414. (b) Brookhart, M. S.; Johnson, L. K.; Killian, C. M.; Arthur, S. D.; Feldman, J.; McCord, E. F.; McLain, S. J.; Kreutzer, K. A.; Bennett, M. A.; Coughlin, E. B.; Ittel, S. D.; Parthasarathy, A.; Tempel, D. J. WO 9623010, 1996. (c) Mecking, S.; Johnson, L. K.; Wang, L.; Brookhart, M. *J. Am. Chem. Soc.* 1998, 120, 888. (d) Guan, Z.; Cotts, P. M.; McCord, E. F.; McLain, S. J. *Science* 1999, 283, 2059.

[20] Examples for transition metal catalyzed coordination copolymerization with monomers containing polar functionalities, (a) Vogl, O. *J. Macromol. Sci., Chem.* 1985, A22, 541. (b) Purgett, M. D.; Xie, S.; Bansleben, D. A.; Vogl, O. *J. Polym. Sci., Part A: Polym. Chem.* 1988, 26, 657. (c) Purgett, M. D.; Vogl, O. *J. Polym. Sci., Part A: Polym. Chem.* 1989, 27, 2051. (d) Ramakrishnan, S.; Berluche, E.; Chung, T. C. *Macromolecules* 1990, 23, 378. (e) Kesti, M. R.; Coates, G. W.; Waymouth, R. M. *J. Am. Chem. Soc.* 1992, 114, 9679. (f) Schulz, D. N.; Bock, J. *J. Macromol. Sci., Chem.* 1991, A28, 1235. (g) Aaltonen, P.; Lofgren, B. *Macromolecules* 1995, 28, 5353. (h) Aaltonen, P.; Fink, G.; Lofgren, B.; Seppala, J. *Macromolecules* 1996, 29, 5255. (i) Wilén, C. E.; Luttikhedde, H.; Hjertberg, T.; Näsman, J. H. *Macromolecules* 1996, 29, 8569. (j) Tsuchida, A.; Bollen, C.; Sernetz, F. G.; Frey, H.; Mülhaupt, R. *Macromolecules* 1997, 30, 2818. (k) Aaltonen, P.; Lofgren, B. *Eur. Polym. J.* 1997, 33, 1187. (l) Hakala, K.; Lofgren, B.; Helaja, T. *Eur. Polym. J.* 1998, 34, 1093. (m) Stehling, U. M.; Stein, K. M.;

Kesti, M. R.; Waymouth, R. M. *Macromolecules* 1998, 31, 2019. (n) Radhakrishnan, K.; Sivaram, S. *Macromol. Rapid Commun.* 1998, 19, 581. (o) Marques, M. M.; Correia, S. G.; Ascenso, J. R.; Ribeiro, A. F. G.; Gomes, P. T.; Dias, A. R.; Foster, P.; Rausch, M. D.; Chien, J. C. W. *J. Polym. Sci., Part A: Polym. Chem.* 1999, 37, 2457. (p) Stehling, U. M.; Stein, K. M.; Fischer, D.; Waymouth, R. M. *Macromolecules* 1999, 32, 14. (q) Goretzki, R.; Fink, G. *Macromol. Chem. Phys.* 1999, 200, 881. (r) Wendt, R. A.; Fink, G. *Macromol. Chem. Phys.* 2000, 201, 1365. (s) Hakala, K.; Helaja, T.; Lofgren, B. *J. Polym. Sci., Part A: Polym. Chem.* 2000, 38, 1966. (t) Hagihara, H.; Murata, M.; Uozumi, T. *Macromol. Rapid Commun.* 2001, 22, 353. (u) Imuta, J.; Kashiwa, N.; Toda, Y. *J. Am. Chem. Soc.* 2002, 124, 1176. (v) Hagihara, H.; Tsuchihara, K.; Takeuchi, K.; Murata, M.; Ozaki, H.; Shiono, T. *J. Polym. Sci., Part A: Polym. Chem.* 2004, 42, 52. (w) Hagihara, H.; Tsuchihara, K.; Sugiyama, J.; Takeuchi, K.; Shiono, T. *Macromolecules* 2004, 37, 5145. (x) Wendt, R. A.; Angermund, K.; Jensen, V.; Thiel, W.; Fink, G. *Macromol. Chem. Phys.* 2004, 205, 308. (y) Inoue, Y.; Matsugi, T.; Kashiwa, N.; Matyjaszewski, K. *Macromolecules* 2004, 37, 3651. (z) Terao, H.; Ishii, S.; Mitani, M.; Tanaka, H.; Fujita, T. *J. Am. Chem. Soc.* 2008, 130, 17636,

[21] (2) Liu, J.; Nomura, K. *Macromolecules* 2008, 41, 1070. (b) Nomura, K.; Kakinuki, K.; Fujiki, M.; Itagaki, K. *Macromolecules* 2008, 41, 8974.

[22] Byun, D.-J.; Shin, S.-M.; Han, C. J.; Kim, S. Y. *Polym. Bull.* 1999, 43, 333.

[23] Equilibrium constants for alkene coordination to $(MeC_5H_4)_2Zr^+(O^tBu)(ClCD_2Cl)$, Stoebenau III, E. J.; Jordan, R. J. *J. Am. Chem. Soc.* 2006, 128, 8162. No complexes were formed with vinyltrimethylsilane, *tert*-butylethylene.

[24] Chung, T. C.; Lu, H. L.; Li, C. L. *Macromolecules* 1994, 27, 7533.

[25] Previous example for copolymerization of MOD, (a) Hackmann, M.; Rieger, B. *Macromolecules* 2000, 33, 1524. (b) Song, F.; Pappalardo, D.; Johnson, A. F.; Rieger, B.; Bochmann, M. *J. Polym. Sci.: Part A: Polym. Chem.* 2002, 40, 1484. (c) Williamson, A.; Fink, G. *Macromol Chem. Phys.* 2003, 204, 1178.

[26] (a) Nomura, K. *Chin. J. Polym. Sci.* 2008, 26, 513. (b) Nomura, K. In, *Syndiotactic Polystyrene - Synthesis, Characterization, Processing, and Applications*, Schellenberg, J. Ed.; John Wiley and Sons, Inc.; Hoboken, New Jersey, USA, 2010; pp.60-91. (c) Nomura, K. *Catal. Surv. Asia* 2010, 14, 33.

[27] Selected examples for ethylene copolymerizations,[16,17] see: (a) Nomura, K.; Oya, K.; Komatsu, T.; Imanishi, Y. *Macromolecules* 2000, 33, 3187. (b) Nomura, K.; Komatsu, T.; Imanishi, Y. *Macromolecules* 2000, 33, 8122. (c) Nomura, K.; Okumura, H.; Komatsu, T.; Naga, N. *Macromolecules* 2002, 35, 5388. (d) Nomura, K.; Tsubota, M.; Fujiki, M. *Macromolecules* 2003, 36, 3797. (e) Nomura, K.; Itagaki, K.; Fujiki, M. *Macromolecules* 2005, 38, 2053. (f) Wang, W.; Fujiki, M.; Nomura, K. *J. Am. Chem. Soc.* 2005, 127, 4582. (g) Nomura, K.; Itagaki, K. *Macromolecules* 2005, 38, 8121. (h) Zhang, H.; Nomura, K. *J. Am. Chem. Soc.* 2005, 127, 9364. (i) Zhang, H.; Nomura, K. *Macromolecules* 2006, 39, 5266. (j) Nomura, K.; Wang, W.; Fujiki, M.; Liu, J. *Chem. Commun.* 2006, 2659. (m) Khan, F. Z.; Kakinuki, K.; Nomura, K. *Macromolecules* 2009, 42, 3767. (n) Kakinuki, K.; Fujiki, M.; Nomura, K. *Macromolecules* 2009, 42, 4585.

[28] (a) Nomura, K.; Hatanaka, Y.; Okumura, H.; Fujiki, M.; Hasegawa, K. *Macromolecules* 2004, 37, 1693. (b) Nomura, K.; Takemoto, A.; Hatanaka, Y.; Fujiki, M.; Okumura, H.; Hasegawa, K. *Macromolecules* 2006, 39, 4009.

[29] Reports concerning polymerization of 1,5-hexadiene, (a) Resconi, L.; Waymouth, R. M. *J. Am. Chem. Soc.* 1990, 112, 4953. (b) Coates, G. W.; Waymouth, R. M. *J. Am. Chem. Soc.* 1991, 113, 6270. (c) Coates, G. W.; Waymouth, R. M. *J. Am. Chem. Soc.* 1993, 115, 91. (d) Cavallo, L.; Guerra, G.; Corradini, P.; Resconi, L.; Waymouth, R. M. *Macromolecules* 1993, 26, 260. (e) de Ballessteris, O. R.; Venditto, V.; Auriemma, F.; Guerra, G.; Resconi, L.; Waymouth, R. M.; Mogtad, A. L. *Macromolecules* 1995, 28, 2383. (f) Naga, N.; Shiono, T.; Ikeda, T. *Macromolecules* 1999, 32, 1348. (g) Napoli, M.; Costabile, C.; Pragliona, S.; Longo, P. *Macromolecules* 2005, 38, 5493.

[30] Example for 1,7-octadiene polymerization, Naga, N.; Shiono, T.; Ikeda, T. *Macromol. Chem. Phys.* 1999, 200, 1466.

[31] (a) Jayaratne, K. C.; Keaton, R. J.; Henningsten, D. A.; Sita, L. R. *J. Am. Chem. Soc.* 2000, 122, 10490. (b) Schaverien, C. J. *Organometallics* 1994, 13, 69.

[32] (a) Synthesis of poly(HD) partially containing 1-vinyl-tetramethylene (VTM) unit, Hustad, P. D.; Coates, G. W. *J. Am. Chem. Soc.* 2002, 124, 11578. (b) Postpolymerization of VTM fragment in the poly(HD)s and copolymers with ethylene/propylene by cross metathesis, Mathers, R. T.; Coates, G. W. *Chem. Commun.* 2004, 422.

[33] These results were introduced in the Supporting Information in reference 15.

In: Polymer Synthesis
Editor: E. Kowsari

ISBN 978-1-61324-672-6
© 2012 Nova Science Publishers, Inc.

Chapter 2

Luminescent Coordination Polymers Based on [M(phen)(CN)₄]²⁻ (M = Ru, Os) Cyanometallate Building Blocks

Svetlana G. Baca[1] and Michael D. Ward[2]

[1]Institute of Chemistry, Academy of Sciences of Moldova, Chisinau, Moldova
[2]Department of Chemistry, University of Sheffield, Sheffield, UK

Abstract

The employment of tetradentate $[M(phen)(CN)_4]^{2-}$ (M = Ru, Os) building blocks with lanthanide cations (Ln = Gd^{3+}, Pr^{3+}, Nd^{3+}, Er^{3+}, Yb^{3+}) and ancillary polypyridine ligands (phen = 1,10'-phenanthroline, terpy = 2,2':6',2''-terpyridine, bpm = 2,2'-bipyrimidine) in designing novel luminescent multi-dimensional coordination polymers is systematized and discussed. The geometries of the resulting cyanide-bridged coordination polymers include linear, ladder and tube-like chains, chains of squares, as well as two-dimensional structures. Particular emphasis is given to the relationships between structural features of the M–CN–Ln coordination networks and Ln(III) fragments to generate luminescence.

Introduction

Functional coordination polymers, which represent infinite networks constructed from metal ions and specific organic ligands, represent strategic and revolutionary materials that find a wide range of applications in science and nanotechnology [1]. The design and investigation of these new inorganic-organic hybrid materials attract tremendous interest, investments and efforts around the world, and have recently been driven by the rapid growth of their practical use, especially in emerging fields such as information storage, green catalysis, energy conversion and storage, sensors, storage and transport of hydrogen [2–6]. Among potential applications, the study of luminescent materials is motivated by their great potential in the development of materials used in lighting and display devices, fiber-optic based telecommunications systems, medical diagnosis and imaging [7]. The most successful strategies for the design of coordination polymers are based on a building block approach [8]. Coordination polymers may be obtained by assembling of suitable pre-synthesized metal-containing building blocks and linking organic ligands. In such cases, varying the size, connectivity, charge and functionality of complex "building blocks" and/or ligands allows us to achieve control over both the structure and the targeted properties of the final product. For preparation of luminescent coordination polymers one of the most used strategies is to combine lanthanides with transition metal complexes which can absorb light, and then pass the resultant energy on to the Ln(III) ions to generate luminescence [9]. Recently, the tetracy-anometallate complexes $[M(phen)(CN)_4]^{2-}$ (M = Ru, Os; phen = 1,10′-phenanthroline), offering four externally-directed cyano groups potentially able to form cyano-bridged networks with lanthanide metals and with MLCT excited states that can act as energy donors to other species, has opened a new and very attractive field for creation of polynuclear photophysically-active assemblies [10–11]. This review focuses on the employment of $[M(phen)(CN)_4]^{2-}$ (M = Ru, Os) cyanometallate anions as building blocks in designing novel luminescent multi-dimensional coordination polymers. We describe our efforts towards the preparation of a series of coordination polymers based on these tectons with lanthanide cations (Gd^{3+}, Pr^{3+}, Nd^{3+}, Er^{3+} and Yb^{3+}) and phen, bpm or terpy diimine ancillary ligands (phen = 1,10′-phenanthroline, bpm = 2,2′-bipyrimidine, terpy = 2,2′:6′,2″-terpyridine) that gives a range of one- and two-dimensional polymers in which the d- and f-block components are linked by cyanide bridges (Table 1) [12–16].

STRUCTURAL DIVERSITY OF CYANIDE-BRIDGED COORDINATION POLYMERS

The cyanide-bridged coordination polymers based on tetracyanometallate building blocks have a variety of structural features ranging from simple one-dimensional chains to complex three-dimensional architectures. The reaction conditions, the nature of the ancillary polypyridyl ligands and the lanthanide cations used, the ratio of the different components used, and the assembly process all have an important effect on the structures in these systems. In general, cyanide-bridged coordination polymers can be prepared by two approaches. The first one is based on the interactions of $K_2[Ru(phen)(CN)_4]$ or $Na_2[Os(phen)(CN)_4]$ species with soluble lanthanide nitrates or chlorides in water/alcohol solution at room temperature. The second approach is similar to the first but also involves addition of the ancillary polypyridine-type ligand (phen, terpy or bpm) to coordinate to some of the vacant sites on the Ln(III) ion; these reactions are usually carried out in a mixed solvent system such as methanol-water, acetone-water or acetonitrile-water.

ONE-DIMENSIONAL STRUCTURES

Chains were the commonest type of structure formed from the interactions of $K_2[Ru(phen)(CN)_4]$ or $Na_2[Os(phen)(CN)_4]$ complexes with Ln(III) cations in aqueous media under mild conditions. These potentially tetradentate Ru(II) or Os(II) building blocks used two, three or four CN groups to connect lanthanide atoms into one-dimensional chains with different topologies including linear, ladder, tube-like chains and chains of squares as shown schematically in Figure 1.

As mentioned above different factors affect the type of chain formed. For example, using 2,2':6',2''-terpyridine as additional N-donor ligand, and in the presence of K_2CO_3, positively charged single-stranded chains are formed as shown in Figure 2. In such chains two cyanide groups from the $[Ru(phen)(CN)_4]^{2-}$ dianion coordinate to two $[Pr(terpy)(H_2O)_4]^{3+}$ cations in $\{[Ru(phen)(CN)_4]_3[Pr(terpy)(H_2O)_4]_2 \cdot 20H_2O\}_n$ (**1**) or to $[Nd(terpy)(H_2O)_4]^{3+}$ cations in $\{[Ru(phen)(CN)_4]_3[Nd(terpy)(H_2O)_4]_2 \cdot 19H_2O\}_n$ (**2**), forming cyanide-bridged $\{-Ru-CN-Ln-CN-\}^{2+}_n$ chains (Figure 1a). The Ru···Pr(Nd) separations alternate between 5.39(5.38) and 5.67(5.65) Å in the chain. Additional "free" $[Ru(phen)(CN)_4]^{2-}$ dianions connect these single stranded $\{-Ru-CN-Ln-CN-\}^{2+}_n$ chains by hydrogen bonds with Ln(III)-bound water ligands.

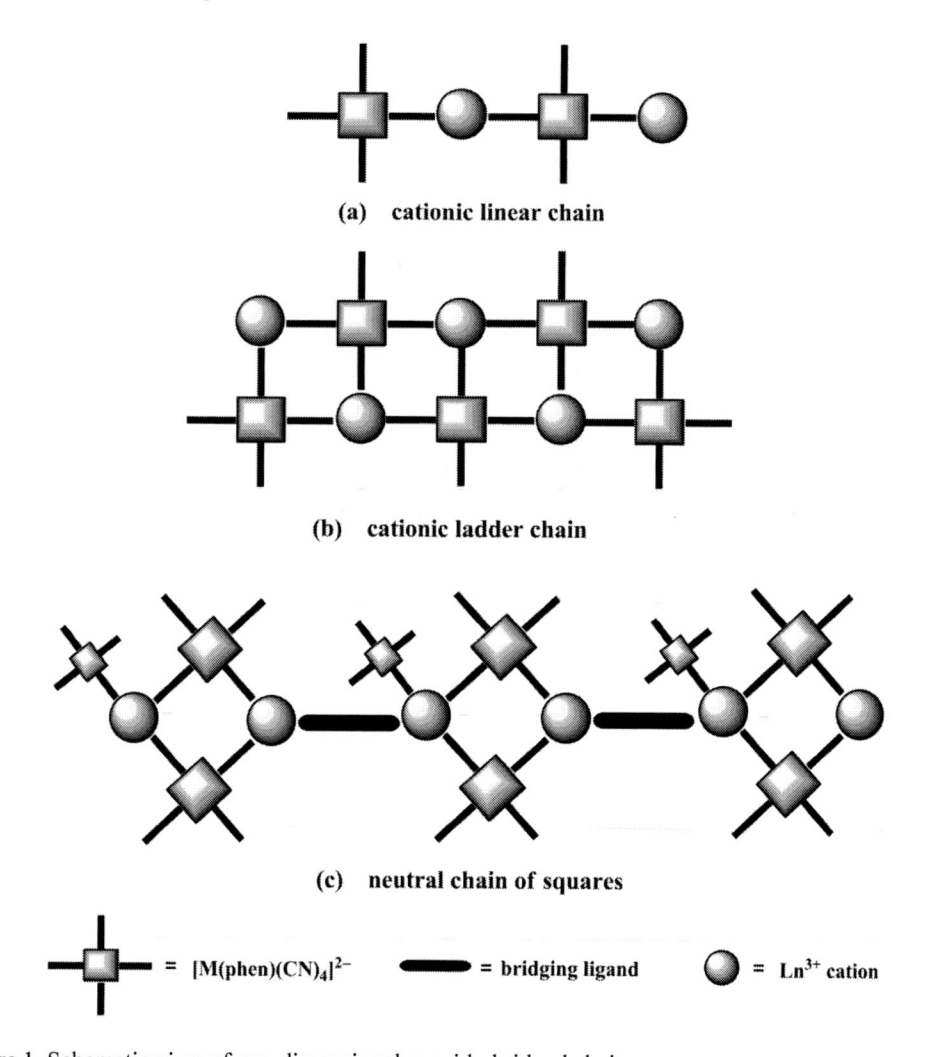

(a)   cationic linear chain

(b)   cationic ladder chain

(c)   neutral chain of squares

= $[M(phen)(CN)_4]^{2-}$ = bridging ligand   = Ln^{3+} cation

Figure 1. Schematic view of one-dimensional cyanide-bridged chains.

Table 1.

N	Formulae	Dimensionality		Reaction	Ref.
1	$\{[Ru(phen)(CN)_4]_3[Pr(terpy)(H_2O)_4]_2 \cdot 20H_2O\}_n$	1D	cationic chains	$Pr(NO_3)_3 \cdot 6H_2O$ + terpy + $K_2[Ru(phen)(CN)_4]$ + K_2CO_3 (ratio 2 : 2 : 3) in H_2O/MeOH	[13]
2	$\{[Ru(phen)(CN)_4]_3[Nd(terpy)(H_2O)_4]_2 \cdot 19H_2O\}_n$	1D	cationic chains	$Nd(NO_3)_3 \cdot 6H_2O$ + terpy + $K_2[Ru(phen)(CN)_4]$ + K_2CO_3 (ratio 2 : 2 : 3) in H_2O/MeOH	[15]
3	$\{[Ru(phen)(CN)_4]_3[Pr(terpy)(H_2O)_3]_2 \cdot 12.5H_2O\}_n$	1D	ladder	$Pr(NO_3)_3 \cdot 6H_2O$ + terpy + $K_2[Ru(phen)(CN)_4]$ (ratio 2 : 2 : 3) in H_2O/MeOH	[13]
4	$\{[Ru(phen)(CN)_4]_3[Nd(terpy)(H_2O)_3]_2 \cdot 9.5H_2O\}_n$	1D	ladder	$Nd(NO_3)_3 \cdot 6H_2O$ + terpy + $K_2[Ru(phen)(CN)_4]$ (ratio 2 : 2 : 3) in H_2O/MeOH	[13]
5	$\{[Os(phen)(CN)_4]_3[Pr(terpy)(H_2O)_3]_2 \cdot MeOH \cdot 14H_2O\}_n$	1D	ladder	$Pr(NO_3)_3 \cdot 6H_2O$ + terpy + $Na_2[Os(phen)(CN)_4]$ (ratio 2 : 2 : 3) in H_2O/MeOH	unpublished
6	$\{[Os(phen)(CN)_4]_3[Nd(terpy)(H_2O)_3]_2 \cdot 9.5H_2O\}_n$	1D	ladder	$Nd(NO_3)_3 \cdot 6H_2O$ + terpy+ $Na_2[Os(phen)(CN)_4]$ (ratio 2 : 2 : 3) in H_2O/MeOH	unpublished
7	$\{[Os(phen)(CN)_4]_3[Pr(phen)(H_2O)_2]_2 \cdot 11.5H_2O\}_n$	1D	tube	$PrCl_3 \cdot 6H_2O$ + phen + $Na_2[Os(phen)(CN)_4]$ (ratio 1 : 2 : 1.5) in H_2O/MeOH	[16]
8	$\{[Os(phen)(CN)_4]_3[Nd(phen)(H_2O)_2]_2 \cdot 10.5H_2O\}_n$	1D	tube	$NdCl_3 \cdot 6H_2O$ + phen + $Na_2[Os(phen)(CN)_4]$ (ratio 1 : 2 : 1.5) in H_2O/MeOH	[16]
9	$\{[Ru(phen)(CN)_4]_3[Er_2(bpm)(H_2O)_7] \cdot MeOH \cdot 15.5H_2O\}_n$	1D	chains of squares	$ErCl_3 \cdot 6H_2O$ + bpm + $K_2[Ru(phen)(CN)_4]$ (ratio 1 : 2 : 1.5) in H_2O/MeOH	[15]
10	$\{[Ru(phen)(CN)_4]_3[Yb_2(bpm)(H_2O)_7] \cdot 19H_2O\}_n$	1D	chains of squares	$YbCl_3 \cdot 6H_2O$ + bpm + $K_2[Ru(phen)(CN)_4]$ (ratio 1 : 2 : 1.5) in H_2O/MeOH	[15]
11	$\{[Os(phen)(CN)_4]_3[Er_2(bpm)(H_2O)_7] \cdot MeOH \cdot 15H_2O\}_n$	1D	chains of squares	$Er(NO_3)_3 \cdot 5H_2O$ + bpm + $Na_2[Os(phen)(CN)_4]$ (ratio 1 : 2 : 1.5) in H_2O/MeOH	[16]
12	$\{[Os(phen)(CN)_4]_3[Yb_2(bpm)(H_2O)_7] \cdot MeOH \cdot 13H_2O\}_n$	1D	chains of squares	$Yb(NO_3)_3 \cdot 5H_2O$ + bpm + $Na_2[Os(phen)(CN)_4]$ (ratio 1 : 2 : 1.5) in H_2O/MeOH	[16]
13	$\{[Ru(phen)(CN)_4]_3[Gd(phen)(H_2O)_3]_2 \cdot 6H_2O\}_n$	2D	honeycomb layer	$Gd(NO_3)_3 \cdot 6H_2O$ + phen + $K_2[Ru(phen)(CN)_4]$ (ratio 2 : 3 : 3) in H_2O	[12]
14	$\{[Ru(phen)(CN)_4]_3[Nd(phen)(H_2O)_3]_2 \cdot 2MeOH \cdot 12H_2O\}_n$	2D	honeycomb layer	$NdCl_3 \cdot 6H_2O$ + phen + $K_2[Ru(phen)(CN)_4]$ (ratio 2 : 2 : 3) in H_2O/MeOH	[15]
15	$\{[Ru(phen)(CN)_4]_3[Er(phen)(H_2O)_3]_2 \cdot 14H_2O\}_n$	2D	honeycomb layer	$Er(NO_3)_3 \cdot 5H_2O$ + phen + $K_2[Ru(phen)(CN)_4]$ (ratio 2 : 2 : 3) in H_2O/acetone	[15]
16	$\{[Ru(phen)(CN)_4]_3[Yb(phen)(H_2O)_3]_2 \cdot 14H_2O\}_n$	2D	honeycomb layer	$Yb(NO_3)_3 \cdot 5H_2O$ + phen+ $K_2[Ru(phen)(CN)_4]$ (2 : 2 : 3 ratio) in H_2O/MeCN	[15]
17	$\{[Os(phen)(CN)_4]_6[Pr_4(H_2O)_{13}(bpm)_2] \cdot MeOH \cdot 25H_2O\}_n$	2D	cellular layer	$Pr(NO_3)_3 \cdot 6H_2O$ + bpm+ $Na_2[Os(phen)(CN)_4]$ (ratio 2 : 1 : 3) in H_2O/MeOH	[16]
18	$\{[Os(phen)(CN)_4]_6[Nd_4(H_2O)_{12}(MeOH)(bpm)_2] \cdot 6MeOH \cdot 19.5H_2O)\}_n$	2D	cellular layer	$Nd(NO_3)_3 \cdot 6H_2O$ + bpm+ $Na_2[Os(phen)(CN)_4]$ (ratio 2 : 1 : 3) in H_2O/MeOH	[16]
19	$\{[Os(phen)(CN)_4]_{1.5}[Gd(H_2O)_4(MeOH)] \cdot 4H_2O\}_n$	2D	sandwich layer	$Gd(NO_3)_3 \cdot 6H_2O$ + $Na_2[Os(phen)(CN)_4]$ (ratio 2 : 3) in H_2O/MeOH	[16]

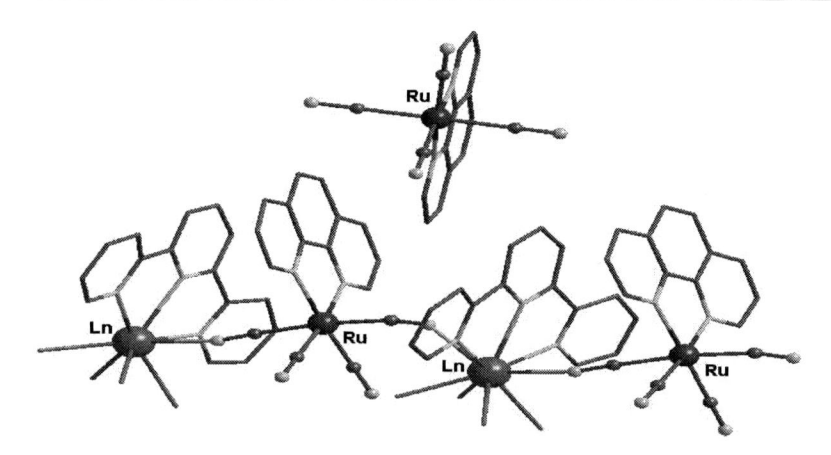

Figure 2. View of the cationic {Ru(phen)–CN–Ln(terpy)}$_n$ chain in coordination polymers **1** and **2** (Ln = Pr^{3+} and Nd^{3+}). Hydrogen atoms and water molecules are omitted for clarity. C and N atoms of the cyanide groups are shown as black and grey balls, respectively.

It was found that under more acidic conditions [the same combination of starting materials but without K$_2$CO$_3$] other types of one-dimensional coordination polymer were formed. Coordination polymers of formulae {[M(phen)(CN)$_4$]$_3$[Ln(terpy)(H$_2$O)$_3$]$_2$ · n solvent}$_n$ where (M = Ru: Ln = Pr (**3**), Nd (**4**); M = Os: Ln = Pr (**5**), Nd (**6**)) (Table 1), shown in Figure 3, have a ladder-like chain structure. In these ladder-like chains the three cyanide groups from a [M(phen)(CN)$_4$]$^{2-}$ dianion coordinate to three [Ln(terpy)(H$_2$O)$_3$]$^{3+}$ cations. Each ladder consists of two strands based on alternating M(II) and Ln(III) centers connected by a cyanide bridge [M···Ln separations in the range of 5.39 - 5.79 Å] to give a chain similar to **1** and **2**. However in contrast to the formation of only single-stranded chains in **1** and **2**, two such parallel chains are connected in **3–6** by additional cyanide cross-pieces (the "rungs" of the ladder) which bridge M(II) and Ln(III) centers (Figure 1b). Again, these ladder chains are overall cationic and "free" [M(phen)(CN)$_4$]$^{2-}$ units provide the charge balance; these "free" [M(phen)(CN)$_4$]$^{2-}$ units are hydrogen-bonded to lattice water molecules forming a 3D network. The syntheses and structures of these coordination polymers provide excellent examples how small variations in reaction conditions can be used to generate families of chain coordination polymers with different topologies.

Using excess 1,10′-phenanthroline in combination with Na$_2$[Os(phen)(CN)$_4$] and a Ln(III) chloride (Ln = Pr or Nd) in methanol-water solutions afforded one-dimensional coordination polymers of the formula {[Os(phen)(CN)$_4$]$_3$[Ln(phen)(H$_2$O)$_2$]$_2$ · nH$_2$O}$_n$ (where Ln = Nd (**7**), n = 11.5; Ln = Pr (**8**), n = 10.5) (Table 1) consisting of an unusual cationic "tube" polymeric structure. Two [Os(phen)(CN)$_4$]$^{2-}$ units and the two [Ln(phen)(H$_2$O)$_2$]$^{3+}$ units are connected via cyanide bridges, forming a one-dimensional "tube" with an approx-imately square cross section as shown in Figure 4. Each square consists of a 12-membered Os$_2$Ln$_2$(μ-CN)$_4$ ring; these are stacked vertically via additional cyanide bridges to give an infinite linear array of cyanide-bridged cubes in which metal types alternate at the corners, similar to a column extracted from a Prussian Blue-type structure. Each [Os(phen)(CN)$_4$]$^{2-}$ unit in this tubular assembly uses all four of its cyanide ligands to connect to different Ln(III) centers. The remaining [Os(phen)(CN)$_4$]$^{2-}$ unit acts as an isolated counterion, with all four of its cyanide groups forming CN···HOH hydrogen bonds with lattice water molecules, so the resulting coordination network is neutral.

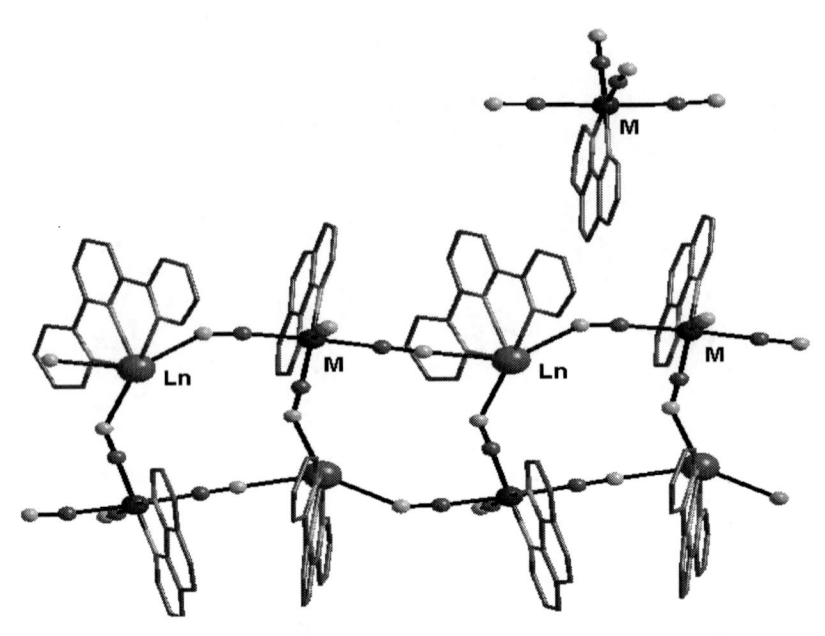

Figure 3. View of the cationic ladder chain in coordination polymers 3–6 (M = Ru^{2+} and Os^{2+}; Ln = Pr^{3+} and Nd^{3+}). Hydrogen atoms, water and methanol molecules are omitted for clarity. C and N atoms of the cyanide groups are shown as black and grey balls, respectively.

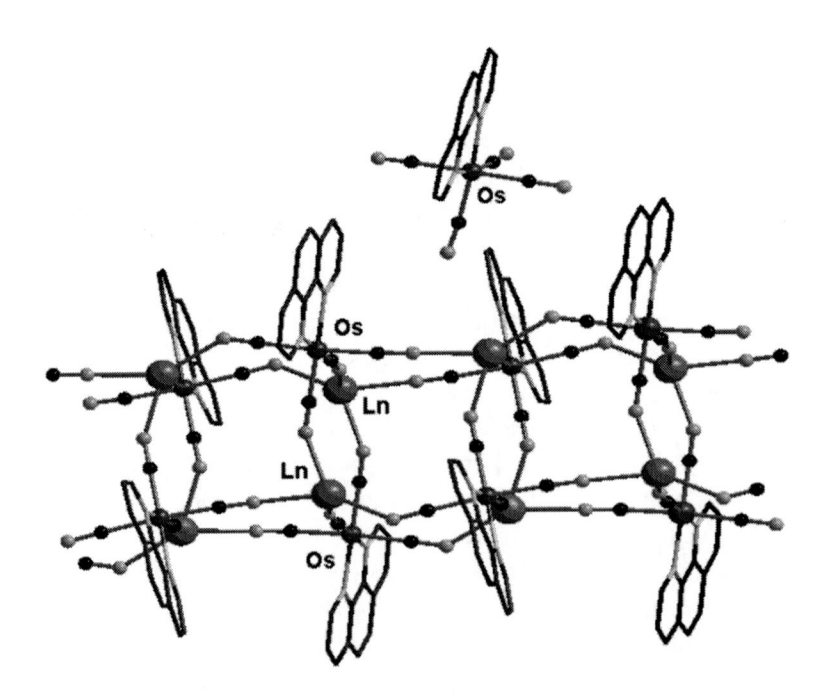

Figure 4. View of the cationic tube-like chain in coordination polymers 7 and 8 (Ln = Pr^{3+} and Nd^{3+}). Hydrogen atoms, water molecules and phen ligands coordinated to Ln(III) ions are omitted for clarity. C and N atoms of the cyanide groups are shown as black and grey balls, respectively.

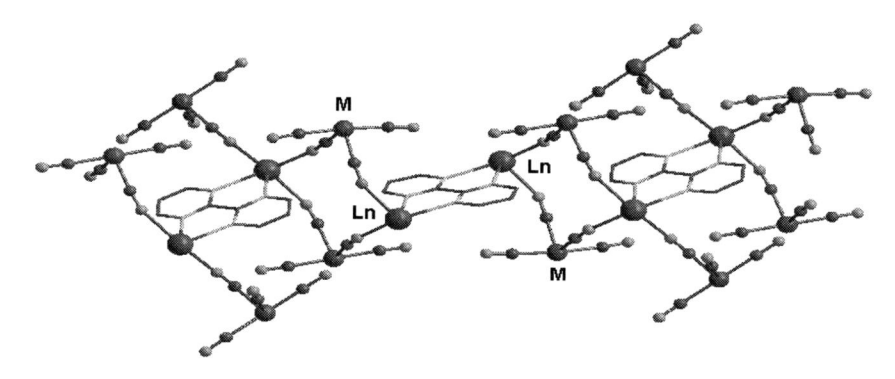

Figure 5. View of the "chain of squares" in coordination polymers **9–12** (M = Ru^{2+} and Os^{2+}; Ln = Er^{3+} and Yb^{3+}). Hydrogen atoms, water and methanol molecules and phen ligands coordinated to Ln(III) ions are omitted for clarity. C and N atoms of the cyanide groups are shown as black and grey balls, respectively.

A different type of 1D chain – a chain of squares – was formed from the crystallization of $K_2[Ru(phen)(CN)_4]$ or $Na_2[Os(phen)(CN)_4]$ with Ln(III) chloride or nitrate salts in the presence of the bis-bidentate 2,2′-bipyrimidine ligand (bpm) in aqueous MeOH. This one-dimensional "chain of squares" is exemplified by the structure of $\{[M(phen)(CN)_4]_3[Ln_2(H_2O)_7(bpm)] \cdot n$ solvent$\}_n$ (M = Ru: Ln = Er (**9**), Yb (**10**); M = Os: Ln = Er (**11**), Yb (**12**)) (Table 1), which is shown in Figure 5. The chain consists of square $M_2Ln_2(\mu\text{-CN})_4$ units that incorporate two $[M(phen)(CN)_4]^{2-}$ units and two Ln(III) ions. Each $[M(phen)(CN)_4]^{2-}$ unit coordinates to the two Ln(III) atoms by the pair of cyanide ligands. A bridging bpm ligand connects adjacent such squares to form an infinite one-dimensional chain (Figure 1c). An additional $[M(phen)(CN)_4]^{2-}$ unit is pendent from the backbone of the chain, thus the repeat unit of the chain is a pentanuclear M_3Ln_2 fragment and the chains are neutral overall.

TWO-DIMENSIONAL STRUCTURES

From the interactions of $K_2[Ru(phen)(CN)_4]$ or $Na_2[Os(phen)(CN)_4]$ complexes with Ln(III) cations arise the formation of cyanide-bridged two-dimensional arrays with different topologies and Figure 6 illustrates some of the observed motifs. The honeycomb network (Figure 6a) is a particularly commonly reported for 2D coordination polymers, but other motifs, found in the synthesized cyanide-bridged coordination polymers (for example the structure shown in Figure 6b) also occur.

Similar to the syntheses of one-dimensional cyanide-bridged chains **1–12**, the preparations of 2D coordination polymers are relatively easy and consist of a one-pot reaction of the cyanometallate Ru(II) or Os(II) anions, lanthanide(III) cations and bidentate N-containing ligands (phen, bpm) in water or water/organic solvent mixtures (Table 1) followed by slow crystallization.

The combination of $LnCl_3 \cdot 6H_2O$, 1,10′-phenanthroline and $K_2[Ru(phen)(CN)_4]$ in a 2 : 2 : 3 molar ratio afforded a series of two-dimensional sheet-like polymeric complexes $\{[Ru(phen)(CN)_4]_3[Ln(phen)(H_2O)_3]_2 \cdot n$ solvent$\}_n$ (Ln = Gd (**13**), Nd (**14**), Er (**15**), Yb (**16**)) (Table 1).

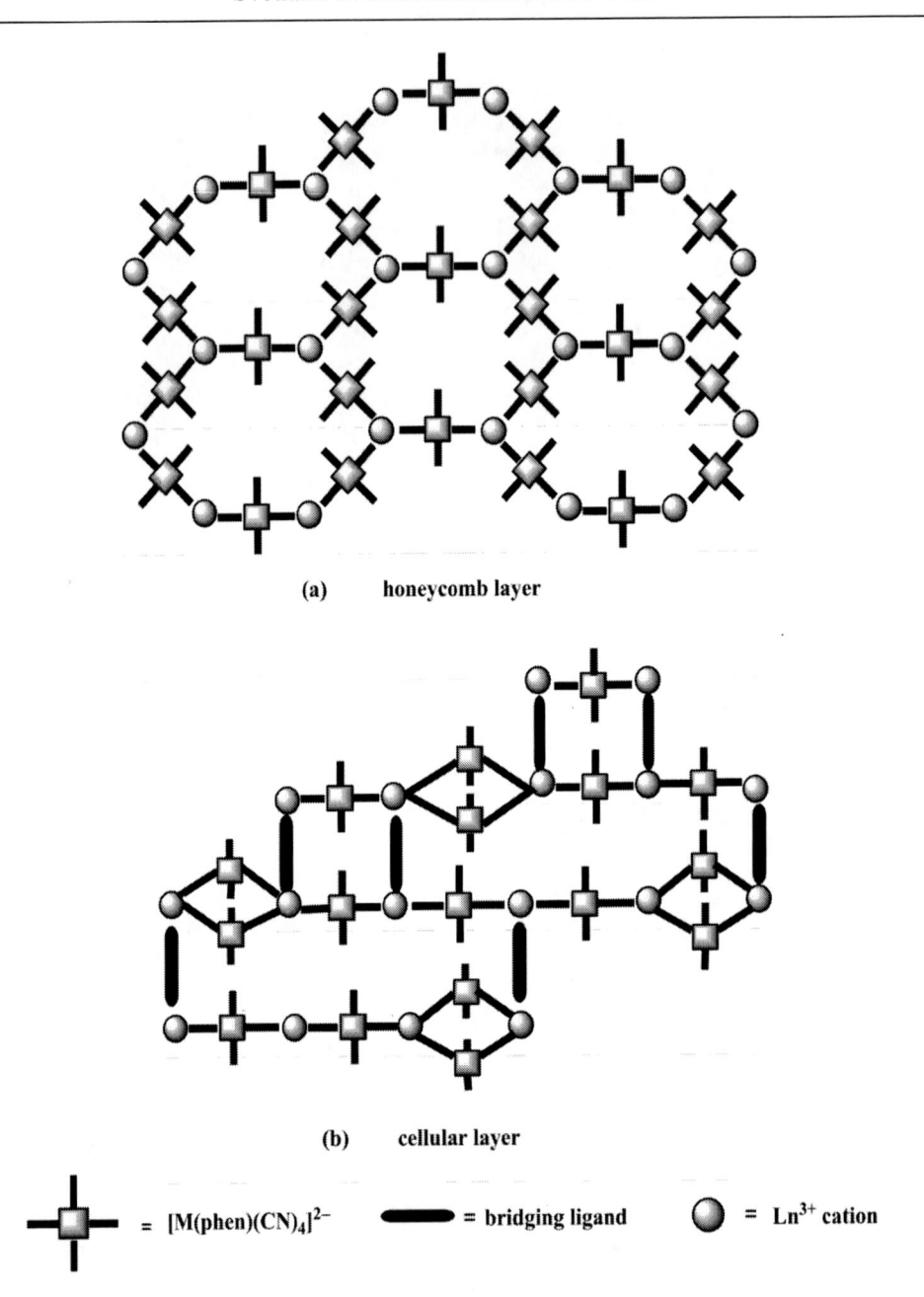

(a) honeycomb layer

(b) cellular layer

= $[M(phen)(CN)_4]^{2-}$ ■ = bridging ligand ● = Ln^{3+} cation

Figure 6. Schematic view of two-dimensional cyanide-bridged layers.

The sheet structure (Figures 7–8) consists of 36-membered $Ln_6Ru_6(\mu\text{-}CN)_{12}$ rings. Each Ru(II) unit is connected to two adjacent Ln(III) centers via the *cis*-related equatorial pair of cyanide ligands that are *trans* to the phenanthroline group; the axial pair of cyanide ligands is not involved in bridging. In contrast each of the Ln(III) centers is connected to three Ru(II) centers via bridging cyanides, and this provides a mechanism for the rings to fuse together, with each Ln(III) centre being at the point where three $Ln_3Ru_3(\mu\text{-}CN)_{12}$ rings meet. The topology of this network is accordingly similar to that of a layer of fused hexagons in graphite, with the Ru atoms in the centre of each edge connected to two vertices, and the Ln atoms at the vertices, connected to three different edges. All of the cyanide groups that are not involved in bridging to Ln(III) are involved in CN···HOH hydrogen-bonding interactions with lattice water molecules.

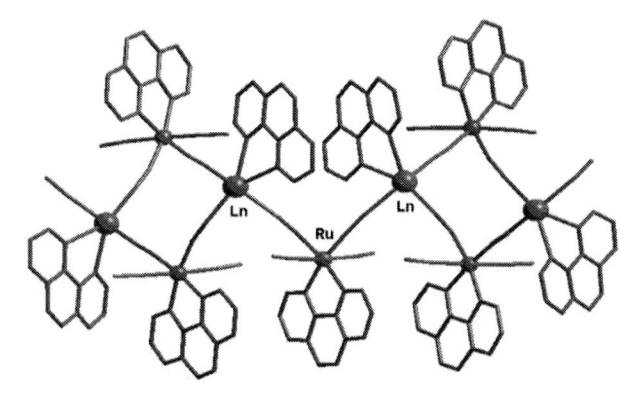

Figure 7. Side view of the honeycomb layer in coordination polymers **13–16** (Ln = Gd^{3+}, Nd^{3+}, Er^{3+} and Yb^{3+}). Hydrogen atoms, water and methanol molecules are omitted for clarity.

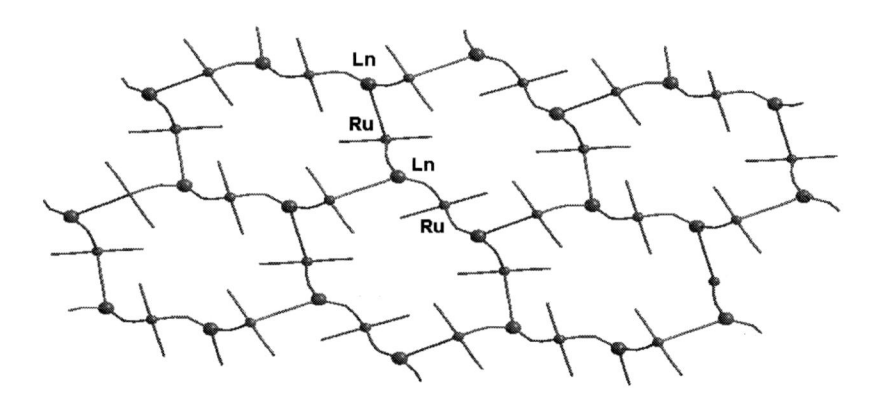

Figure 8. Top view of the honeycomb layer in coordination polymers **13–16** (Ln = Gd^{3+}, Nd^{3+}, Er^{3+} and Yb^{3+}). Hydrogen atoms, water and methanol molecules, and phen ligands are omitted for clarity.

The second type of two-dimensional polymeric structure has been found in the coordination polymers $\{[Os(phen)(CN)_4]_6[Ln_4(H_2O)_{13}(bpm)_2] \cdot n \text{ solvent}\}_n$ (where Ln = Pr (**17**), Nd (**18**)) (Table 1). The 2D coordination polymers **17** and **18** were prepared by the interaction of $Na_2[Os(phen)(CN)_4]$, Ln(III) nitrates and bpm ligands in a molar 3 : 2 : 1 ratio in water-methanol solution. The 2D layer in **17** and **18** has a rather complicated structure (Figure 9). The "top" and "bottom" of each layer consist of the phenanthroline ligands of $[Os(phen)(CN)_4]^{2-}$ units, whose cyanide ligands are directed inwards; the polar center of the layer contains the Ln(III) centers, bridging bpm ligands, and water molecules. The two 2,2'-bipyrimidine ligands each bridge a pair of Ln(III) centers and form portions of heptametallic ring assemblies consisting of the cyclic hexanuclear sequence $\{Os_2Ln_4(\mu\text{-bpm})_2\}$ plus an additional $[Os(phen)(CN)_4]^{2-}$ unit that spans both Ln atoms. These heptanuclear units are linked by smaller $\{Os_2Ln_2(\mu\text{-CN})_4\}$ tetranuclear rings. The result is an alternating one dimensional sequence of heptanuclear and tetranuclear rings. The remaining [Os(phen)

$(CN)_4]^{2-}$ units cross-link these one-dimensional chains via cyanide bridges to two Ln(III) ions from adjacent chains.

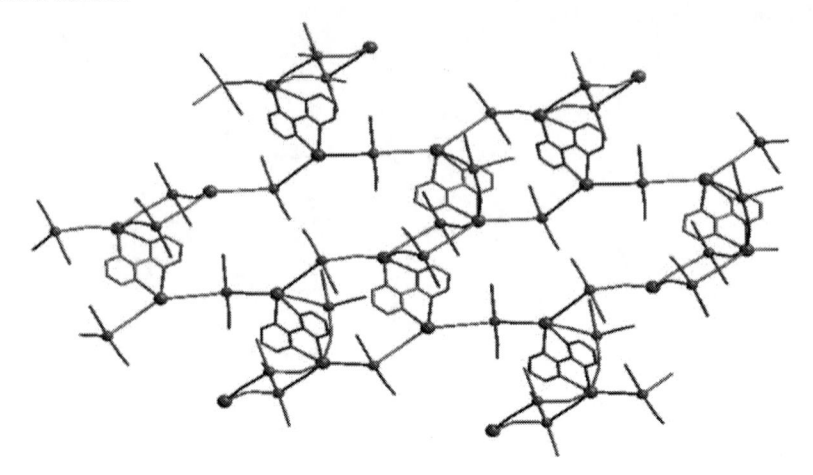

Figure 9. Top view of the layer in coordination polymers 17 and 18 (Ln = Pr^{3+} and Nd^{3+}). Hydrogen atoms, phen ligands coordinated to Os(II) ions, water and methanol molecules are omitted for clarity.

The two-dimensional polymer $\{[Os(phen)(CN)_4]_{1.5}[Gd(H_2O)_4(MeOH)] \cdot 4H_2O\}_n$ (19) was prepared by the reaction of an aqueous solution of $Na_2[Os(phen)(CN)_4]$ with $Gd(NO_3)_3 \cdot 6H_2O$ in methanol (Table 1). The structure is shown in Figures 10 and 11 and consists of two-dimensional sheets, each of which has a sandwich-like structure. $[Os(phen)(CN)_4]^{2-}$ units at the "top" and "bottom" of each sheet (the layers of bread in the sandwich) have their cyanide groups directed toward the center of the sheet, where they interact with Gd(III) ions (the filling of the sandwich) by direct coordination and with water molecules by CN⋯HOH hydrogen bonds. In each $[Os(phen)(CN)_4]^{2-}$ unit, the two cyanides in the same plane as the phen ligand are coordinated to different Gd(III) centers, whereas the two axial (out-of-plane) cyanides are hydrogen-bonded to lattice water molecules with typical nonbonded N⋯O separations of 2.8 - 2.9 Å.

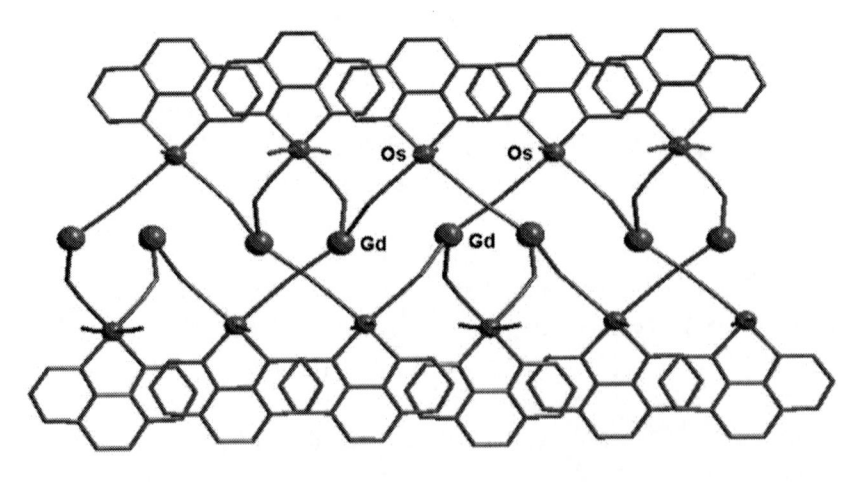

Figure 10. Side view of a sandwich-like layer in coordination polymer **19**. Hydrogen atoms, water and methanol molecules are omitted for clarity.

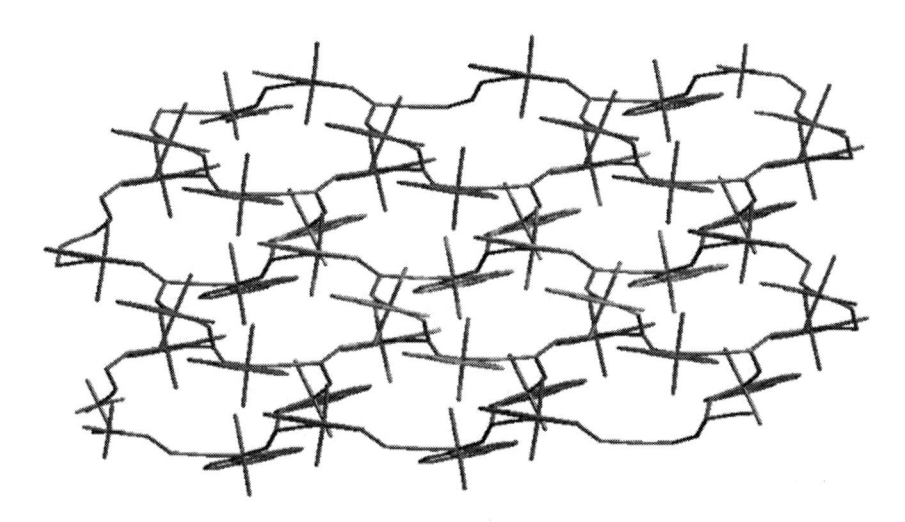

Figure 11. Top view of a sandwich-like layer in coordination polymer **19**. Hydrogen atoms, water and methanol molecules are omitted for clarity.

In general when the Ln(III) ion is a near-infrared red emitter with a low-energy luminescence excited state [such as Yb(III) or Nd(III)], absorption of light by the Ru(II) or Os(II) polypyridine chromophore is rapidly followed by luminescence from the lanthanide(III) ion at the characteristic wavelengths of 980 nm [for Yb(III)] and 1060 nm [for Nd(III)]. This occurs because the MLCT excited state of the Ru(II) or Os(II) unit has an energy sufficiently high for energy-transfer to the emissive excited state of the lanthanide ion to be endergonic. Thus the excited state of the Ru(II)/Os(II) unit is quenched, with loss of the characteristic luminescence, and sensitized Ln(III)-based emission appears.

The efficiency with which energy is transferred to the Ln(III) ion is more or less independent of structural type because the M···Ln distances across the cyanide bridge are almost invariant and this distance is what dominates the energy-transfer rate. However, energy-transfer the rate varies with different lanthanide ions across the same distance – it is much faster to Nd(III) than to Yb(III) – because of better donor/acceptor spectroscopic overlap arising from the high density of f-f states on Nd(III) which can act as energy acceptors.

Blocking some of the coordination sites on the Ln(III) ions using ancillary polypyridine ligands such as bpm, phen or terpy results in slightly improved luminescence by exclusion of water molecules from the Ln(III) coordination sphere but the lifetimes of Ln(III)-based luminescence remain quite short in all cases. Nevertheless these systems are interesting for generating near-infrared luminescence from Ln(III) ions using visible-light excitation of Ru(II) or Os(II) units.

CONCLUSION AND PERSPECTIVES

In summary, an extensive series of luminescent M–CN–Ln coordination polymers has been synthesized based on $[M(phen)(CN)_4]^{2-}$ (M = Ru, Os) cyanometallate building blocks, which have highly desirable photophysical properties such as a long-lived, luminescent excited state and the important structural feature of four externally-directed CN donor groups to bind to Ln(III) cations to form the infinite networks. Depending on the different factors such as precursor anion, lanthanide cation, molar ratio, the nature of the ancillary polypyridyl ligands, and solvents used all of which play important roles in the formation of the final structure of coordination polymers, different types of 1D and 2D assembly arise. For instance, the flexibility of the coordination sphere around lanthanide centers [Ln(III) centre was coordinated to typically three or four cyanide ligands from different M(II) units, some water molecules, and possibly a polypyridine co-ligand] meant that a wide variety of structural types was observed. Different structural types include one-dimensional chains [linear, ladder, tube-like chains and chains of squares] and the two-dimensional honeycombed sheet, cellular and sandwich-like layers.

Photophysical analysis of these coordination polymers showed that in nearly every case laser irradiation with visible light [which is absorbed selectively by the transition metal fragment] results in transfer of the energy to, and subsequent luminescence from, the Ln centers [especially Yb and Nd, which have characteristic long-wavelength luminescence; 980 nm for Yb, 880 and 1060 nm for Nd]. Thus the compounds, in addition to having unusual types of new structure, also show the desired property of allowing a transition metal ion to "sensitize" luminescence from a nearby Ln(III) ion.

ACKNOWLEDGEMENT

We thank the European Commission for a Marie-Curie post-doctoral fellowship to S. G. B. (contract M1F1-CT-2005-513860).

REFERENCES

[1] Batten, S.R.; Neville, S.M.; Turner, D.R. Coordination Polymers: Design, Analysis and Application; Publisher: Royal Society of Chemistry, Cambridge; 2009; pp. 1–471.

[2] Kitagawa, S.; Kitaura, R.; Noro, S.-I. Angew. Chem. Int. Ed. 2004, 43, 2334–2375. (b) Lin, W.; Rieter, W.J.; Taylor, K.M. Angew. Chem. Int. Ed. 2009, 48, 650–658.

[3] Ma, L.; Abney, C.; Lin, W. Chem. Soc. Rev. 2009, 38, 1248–1256. (b) Lee, J.Y.; Vaidhyanathan, R.; Taylor, J.M. Chem. Soc. Rev. 2009, 38, 1450–1459.

[4] Yaghi, O.M.; Li, Q. MRS Bulletin 2009, 34, 682–690.

[5] Kitagawa, S.; Kitaura, R.; Noro S.-i. Angew. Chem. Int. Ed. 2004, 43, 2334–2375.

[6] Férey, G. Chem. Soc. Rev. 2008, 37, 191–241. (b) Murray, L.J.; Dinca, M.; Long, J.R. Chem. Soc. Rev. 2009, 38, 1294–1314. (c) Barman, S.; Furukawa, H.; Blacque, O.; Venkatesan, K.; Yaghi, O.M.; Berke, H. Chem. Commun. 2010, 46, 7981–7983. (d) Furukawa, H.; Ko, N.; Go, Y.B.; Aratani, N.; Choi, S.B.; Choi, E.; Yazaydin, A.O.; Snurr, R.Q.; O'Keeffe, M.; Kim, J.; Yaghi, O.M. Science, 2010, 239, 424–428.

[7]   Janiak, Ch. Dalton Trans. 2003, 2781–2804. (b) Allendorf, M.D.; Bauer, C.A.; Bhakta, R.K.; Houk, J.T. Chem. Soc. Rev. 2009, 38, 1330–1352.

[8]   Tranchemontagne, D.J.; Mendoza-Cortes, J.L.; O'Keeffe, M.; Yaghi, M. Chem. Soc. Rev. 2009, 38, 1257–1283. (b) Perry, J.J.; Perman, J.A.; Zaworotko, M.J. Chem. Soc. Rev. 2009, 38, 1400–1417.

[9]   Klink, S.I.; Keizer, H.; van Veggel, F.C.J.M. Angew. Chem., Int. Ed. 2000, 39, 4319–4321. (b) Shavaleev, N.M.; Moorcraft, L.P.; Pope, S.J.A.; Bell, Z.R.; Faulkner. S.; Ward, M.D. Chem.–Eur. J. 2003, 9, 5283–5291. (c) Shavaleev, N.M.; Accorsi, G.; Virgili, D.; Bell, Z.R.; Lazarides, T.; Calogero, G.; Armaroli N.; Ward, M.D. Inorg. Chem. 2005, 44, 61–72. (d) Pope, S.J.A.; Coe, B.J.; Faulkner S.; Laye, R.H. Dalton Trans. 2005, 1482–1490.

[10] Ward, M.D. Dalton Trans. 2010, 39, 8851–8867.

[11] Miller, T.A.; Jeffery, J.C.; Ward, M.D.; Adams, H.; Pope, S.J.A.; Faulkner, S. Dalton Trans. 2004, 1524–1526. (b) Davies, G.M.; Pope, S.J.A; Adams, H.; Faulkner, S.; Ward, M.D. Inorg. Chem. 2005, 44, 4656–4665. (c) Herrera, J.M.; Ward, M. D.; Adams, H.; Pope S.J.A.; Faulkner, S. Chem. Commun. 2006, 1851–1853. (d) Adams, H.; Alsindi, W.Z.; Davies, G.M.; Duriska, M.B.; Easun, T.L.; Fenton, H.E.; Herrera, J.M.; George, M.W.; Ronayne, K.L.; Sun, X.Z.; Towrie M.; Ward, M.D. Dalton Trans. 2006, 39–50.

[12] Herrera, J.-M.; Baca, S.G.; Adams, H.; Ward, M.D. Polyhedron 2006, 25, 869–875.

[13] Baca, S.G.; Adams, H.; Ward, M.D. CrystEngComm. 2006, 8, 635–639.

[14] Baca, S.G.; Adams, H.; Grange, Ch.S.; Smith, A.P.; Sazanovich I.; Ward, M.D. Inorg. Chem. 2007, 46, 9779–9789.

[15] Baca, S.G.; Adams, H.; Sykes, D.; Faulkner, S.; Ward, M.D. Dalton. Trans. 2007, 2419–2430.

[16] Baca, S.G.; Pope, S.; Adams, H.; Ward, M.D. Inorg. Chem. 2008, 47, 3736–3747.

In: Polymer Synthesis
Editor: E. Kowsari

Chapter 3

POLYMERIZED IONIC LIQUIDS: SYNTHESIS AND APPLICATIONS

*E. Kowsari**

Department of Chemistry, Amirkabir University of Technology, No. 424,
Tehran, Iran

ABSTRACT

Ionic liquids have attracted extensive research interest in recent years as environmental benign solvents due to their favorable properties like non-inflammability, negligible vapour pressure, reusability and high thermal stability. Poly(ionic liquid)s, the polymers made from ionic liquid monomers, have received much research interests for their potential applications such as gas separation materials, catalytic membranes, polymer electrolytes, and ionic conductive materials. Poly(ionic liquid)s contain anion cation pairs and therefore have a relatively high density of strong dipoles, which makes them promising candidates for different applications. Their physical properties can be tuned to specific design criteria by adjusting their chemical structure including the cation, anion, or their combination. Recent developments of the synthesis and applications of poly(ionic liquid)s containing either imidazolium or non imidazolium such as ammonium cations have been highlighted in this review chapter.

1. INTRODUCTION

Ionic liquids are defined as pure compounds, consisting only of cations and anions (i.e., salts), which melt at or below 100 °C [1]. Many are liquid at 25 °C (and are sometimes called room temperature ionic liquids) but, as this is a some what arbitrary definition all ionic liquids are considered, along with the related higher melting salts, where these shed light on the mechanism of action or reaction of ionic liquids. Structures and abbreviations for

---

* Department of Chemistry, Amirkabir University of Technology, No. 424, Hafez Avenue, 1591634311, Tehran, Iran, E-mail address: kowsarie@aut.ac.ir. Corresponding author: Fax: +98 (21)64542762, Tel.: +98 (21) 64542769.

commonly occurring cations and anions are provided in Fig 1.Ionic liquids have interesting advantages such extremely low vapor pressure, excellent thermal stability, reusability, talent to dissolves many organic and inorganic substrates [2]. The large number of cations and anions allow a wide range of physical and chemical characteristics to be achieved, including volatile and involatile systems, and thus the terms "designer" and "task-specific" ionic liquids have been developed [3,4]. This allows not only control over processing of the reaction but also control over solvent-solute interactions.

Ionic liquid technology when used in place of classical organic solvents, offers a new and environmentally benign approach toward modern synthetic chemistry [5-7].

Ionic liquids are used not only as solvents in chemical syntheses but also as solvents in extraction processes or as electrolytes. An important property of ionic liquids that stimulates interest in using them in the context of so-called green chemistry is their essentially zero vapor pressure. Thus, they are nonvolatile (noncontaminating) liquids and therefore are considered alternatives to, for example, supercritical CO_2 [8].

The design of new advanced materials with high-ionic conductivity has been explored by the formation of monomers incorporating an ionic liquid part. Polymerization of monomers that are ionic liquids is another strategy to prepare ionic gels. Ohno's group synthesized a series of polymer electrolytes by polymerization of ionic liquid-based monomers [9-15].

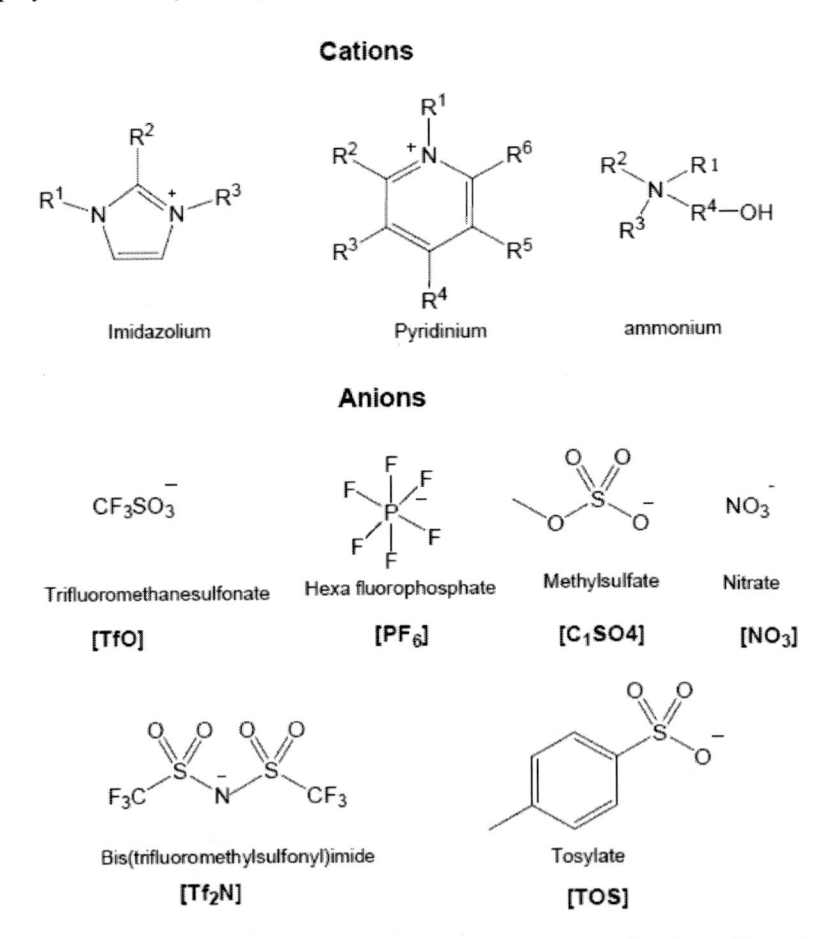

Figure 1. Structures and abbreviations for commonly occurring cations and anions of ionic liquids.

In general, a polymerizable vinyl group was covalently introduced on the cation or (and) anion moiety of an ionic liquid. Polymerization at suitable temperature yielded polymerized (or cross-linked) ionic gels [15, 16]. A variety of polymerizable ionic liquid systems such as polycation-type ionic liquids [12], polyanion-type ionic liquids [11], copolymer [11], and poly(zwitterion) [12] have recently be reported.

Gold-nanoparticle-containing poly ionic liquid composites were synthesized in a single step by UV irradiation of a metal-ion-precursor-doped, self-assembled polymerizable ionic liquid gel, 1-decyl-3-vinylimidazoliumchloride. Most microwave-absorbing polymer composites are not optically transparent, which poses a problem for many applications. Shen et al. [17] recently reported that poly(ionic liquid)s (tetrachloroferrate (P[VBBI][FeCl$_4$]), poly[1-(p-vinylbenzyl)-3-butylimidazolium o-benzoicsulphimide] (P[VBBI][Sac]), and poly[p-vinylbenzyltrimethylammonium tetrafluoroborate] (P[VBTMA][BF$_4$])) have dielectric constants from 3.7 to 5.3 and dielectric loss factors from 0.18 to 0.37, which are much higher than those of other known polymers, due to the high dipole concentrations of poly(ionic liquid)s. Poly(ionic liquid)s can also be used for CO_2 sorption. Radosz et al. recently demonstrated that the CO_2 solubility in poly(*p*vinylbenzyltrimethyl ammonium tetrafluoroborate) is much higher than that in poly(1-(*p*-vinylbenzyl)-3-methyl-imidazolium tetrafluoroborate), which points to a cation-type effect on the CO_2 sorption in polymerized ionic liquids [18]. In addition, the number of polymerizable ionic liquids is steadily increasing, and ionic liquid polymers of polymerizable ionic liquid monomers have been produced as exotic polyelectrolytess. In This review chapter, basic design of poly(ionic l;iquid)s and their unique characteristics are introduced

2. DISSCUSSION

2.1. Synthesis of Polymeric Imidazole–Based Ionic Liquids for Advanced Materials

Imidazolium-based ionic liquids and ionic liquid monomers are becoming increasingly popular in a variety of areas including, proton-conducting hybrid membranes, solid state electrolytes, ionic gels, optically transparent microwave-absorbing materials. This section reviews some of the more recent advances that are associated with Imidazolium-based ionic liquids. The structures of imidazole–based ionic liquids monomers are shown in Figure 2

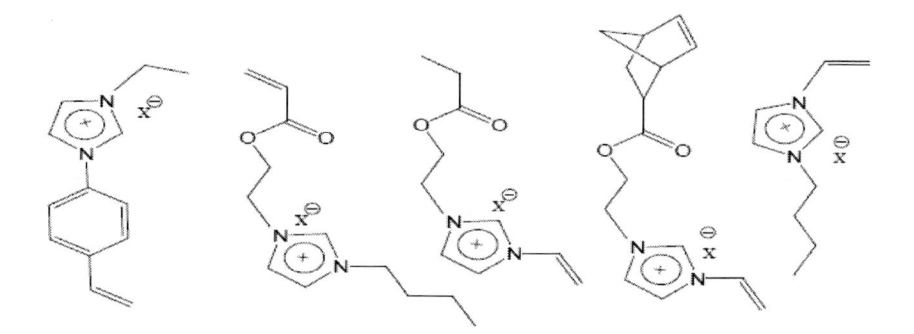

Figure 2. Imidazole–based ionic liquids monomers.

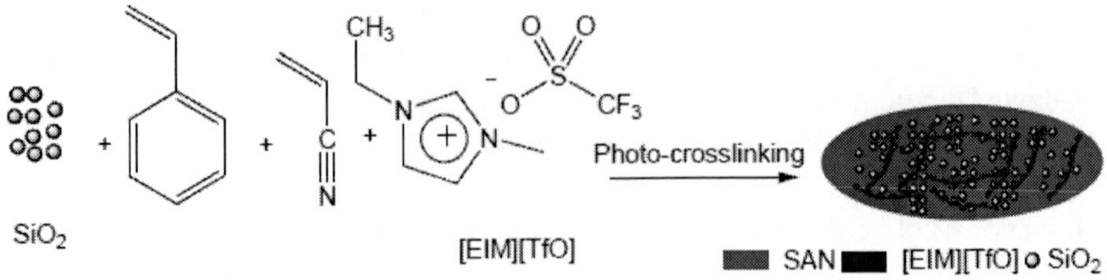

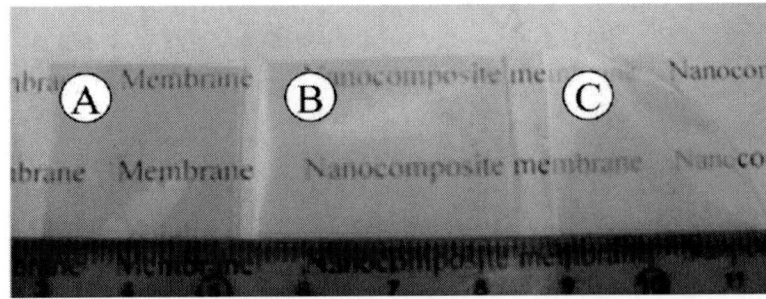

Figure 3. Reaction Scheme for the Preparation of PIL-Based Hybrid Membranes.

A new type of proton-conducting hybrid membranes were prepared by in situ cross-linking of a mixture of polymerizable oils containing protic ionic liquids (PILs) and silica nanoparticles or mesoporous silica nanospheres by B. Lin and coworkers [19].

The hybrid membranes were prepared via in situ photo crosslinking of a mixture of styrene, acrylonitrile, [EIm][TfO], and various amounts of silica fillers in a glass mold (Figure 3).

Figure 4 shows the photographs of produced PIL-based polymeric composite membranes without (Figure 4A) and containing silica fillers (Figure 4B, C).

The resultant hybrid membranes are semitransparent, flexible, and show good thermal stability, good and tunable mechanical properties. Compared with silica nanoparticles, mesoporous silica nanospheres is more effective in enhancing the conductivity and in preventing the release of ionic liquid component from the composite membranes. Under anhydrous conditions, the produced hybrid membranes show proton conductivity up to the order of 1×10^{-2} S/cm at 160 °C. These properties make this type of PIL-based hybrid membranes suitable for high-temperature polymer electrolyte membrane fuel cells.

Ionic conductivity in new polymerized ionic liquids is of great interest as it applies to solid state electrolytes for electrochemical and electromechanical applications, a new ionic liquid monomer was synthesized and polymerized into random copolymers and their ionic conductivity and structure were investigated as a function of copolymer composition by H. Chen and coworkers [20].

The imidazolium monomer was synthesized by a three-step method as shown in Figure5.

Figure 4. Photographs of PIL-based polymer hybrid membranes: (A) without silica fillers, (B) containing 4 wt % of 130 nm silica particles and (C) containing 4 wt %of mesoporous silica nanospheres.

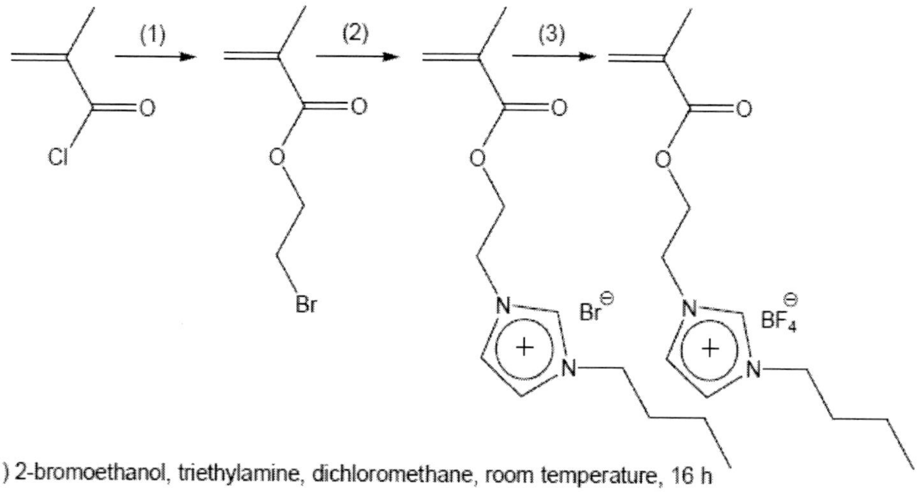

(1) 2-bromoethanol, triethylamine, dichloromethane, room temperature, 16 h

(2) 1-butylimidazole, 40 °C, 24 h

(3) NaBF$_4$, acetonitrile, room temperature, 48 h

(Reproduced from Chen, H., Choi, J. H., Salas-de la Cruz, D., Winey,K. I., Elabd, Y. A. *Macromolecules* 2009, *42*, 4809 , Copyright (2009), with permeation from American Chemical Society)).

Figure 5. Synthesis of Imidazolium-Containing Monomer, MEBIm-BF$_4$

The copolymers were synthesized with various concentrations of hexyl methacrylate by the same free-radical polymerization (structures shown in Figure 6).

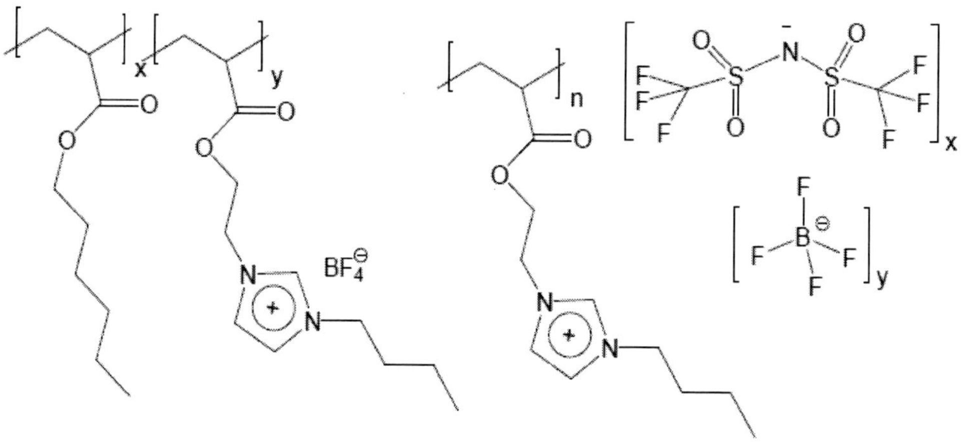

Poly(HMA-*co*-MEBIM-BF$_4$ Poly(MEBIm-TFSI-*co*-MEBIm-BF$_4$)

(Reproduced from Chen, H., Choi, J. H., Salas-de la Cruz, D., Winey,K. I., Elabd, Y. A. *Macromolecules* 2009, *42*, 4809, Copyright (2009), with permeation from American Chemical Society).

Figure 6. Structures of Imidazolium Copolymers.

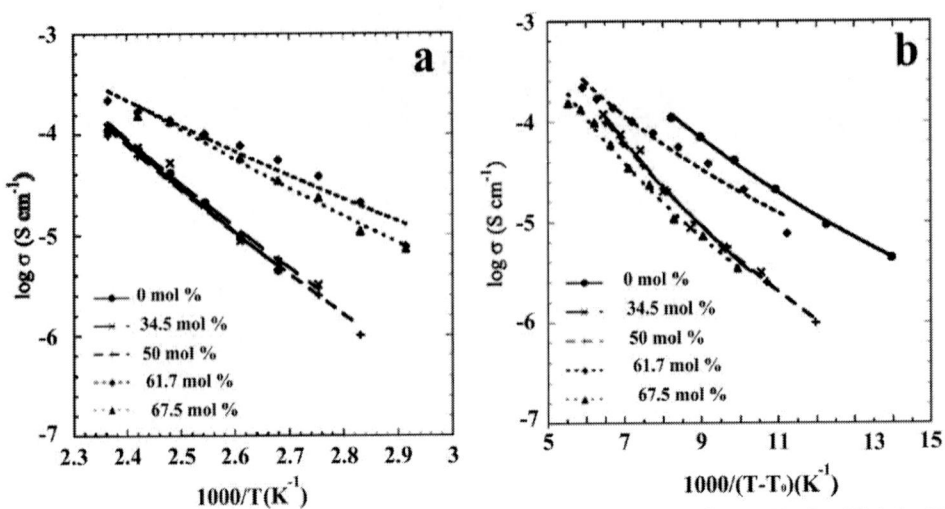

(Reproduced from Chen, H., Choi, J. H., Salas-de la Cruz, D., Winey, K. I., Elabd, Y. A. *Macromolecules* 2009, *42*, 4809, Copyright (2009), with permeation from American Chemical Society)).

Figure 7. Temperature dependence of ionic conductivity for poly(HMA-co-MEBIm-BF4): (a) regression to Ahrrhenius equation; (b) regression to VFT equation. Numbers on graphs correspond to mol % HMA.

In the nonionic-ionic copolymer, the ionic conductivity increased by over an order of magnitude with increasing HMA composition, even though the overall charge content decreased, because the addition of HMA significantly lowered the glass transition temperature. Figure 7 shows a regression of the poly(HMA-co-MEBIm-BF4) copolymer conductivity-temperature data to the Arrhenius equation, $\sigma = \sigma_0 \exp(E_a/RT)$, and the Vogel-Tamman- Fulcher (VFT) equation, $\sigma = \sigma_0 \exp[-B/(T - T_0)]$, where E_a (kJ/mol) is the activation energy and T_0 (K) is the temperature at which the polymer relaxation time becomes infinite or where the ion mobility goes to zero.[4,36] T_0 has been referred to as the equilibrium or true glass transition temperature and is usually approximately 50 K below the measured glass transition temperature.

The synthesis of well-defined poly (ionic liquid) brushes with tunable wettability using surface initiated atom transfer radical polymerization (ATRP) was reported by X. He and coworkers [21]. To graft polymer chains from the surface, a uniform and dense initiator layer on the silicon substrate is important. The immobilization of ATRP initiator on a flat silicate substrate is illustrated in Figure 8.

Kinetic studies revealed a linear increase in polymer film thickness with reaction time, indicating that chain growth from the surface was a controlled process with a "living" characteristic. Furthermore, the surface of poly (ionic liquid) brushes with tunable wettability, reversible switching between hydrophilicity and hydrophobicity can be easily achieved by exchanging their counteranions.

The solution properties and electrospinning of a polymerized ionic liquid was explored by Chen and coworkers [22]. Polymerized ionic liquid poly(MEBIm-BF4) was synthesized by a conventional free-radical polymerization (Figure 9).

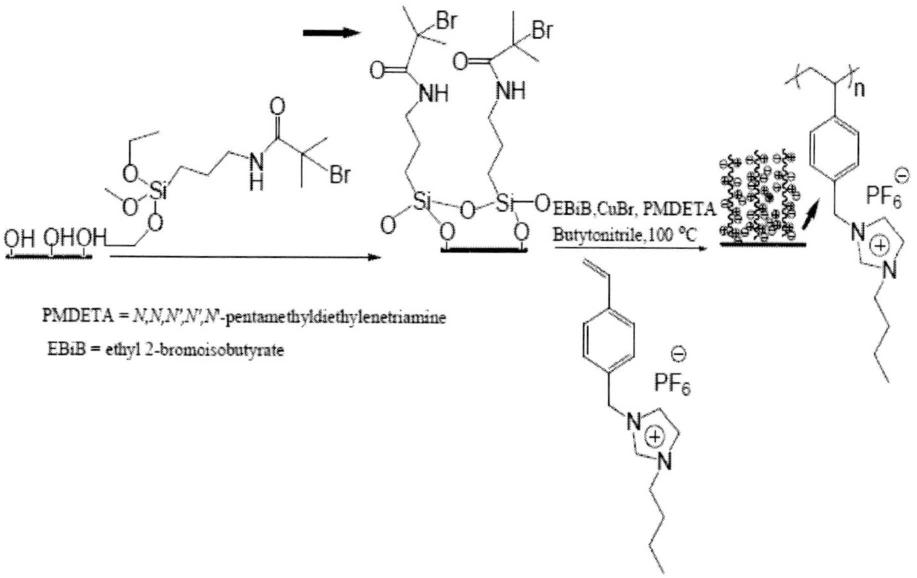

(Reproduced from He, X., Yang,W., Pei, X. *Macromolecules* 2008, *41*, 4615, Copyright (2008), with permeation from American Chemical Society)).

Figure 8. Scheme for the preparation of PVBIm-PF6 brushes via surface-initiated ATRP polymerization.

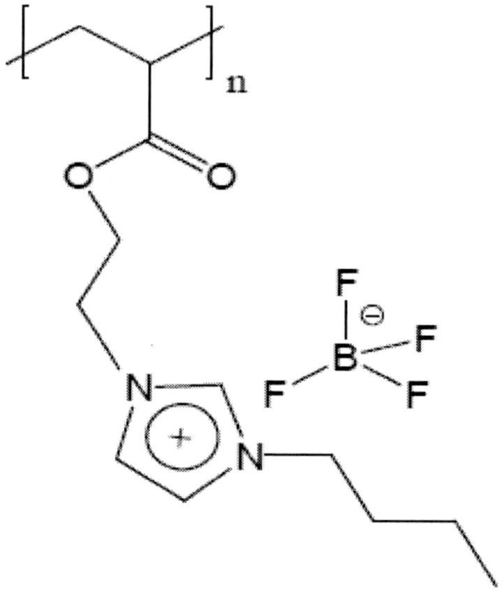

(Reproduced from Chen, H., Elabd, Y. A. *Macromolecules* 2009, *42*, 3368, Copyright (2009), with permeation from American Chemical Society).

Figure 9. Structure of Polymerized Ionic Liquid Poly(MEBIm-BF$_4$) .

The electrospun fibers were collected on aluminum foil and their morphology is shown in Fig 10. Despite a steady polymerjet, only beads were collected from DMF solution. Beaded fiberswere observed from the 1/1 MeCN/DMF cosolvent. A further increase in MeCN content in the solution (3/1 MeCN/DMF)produced defect free fibers.

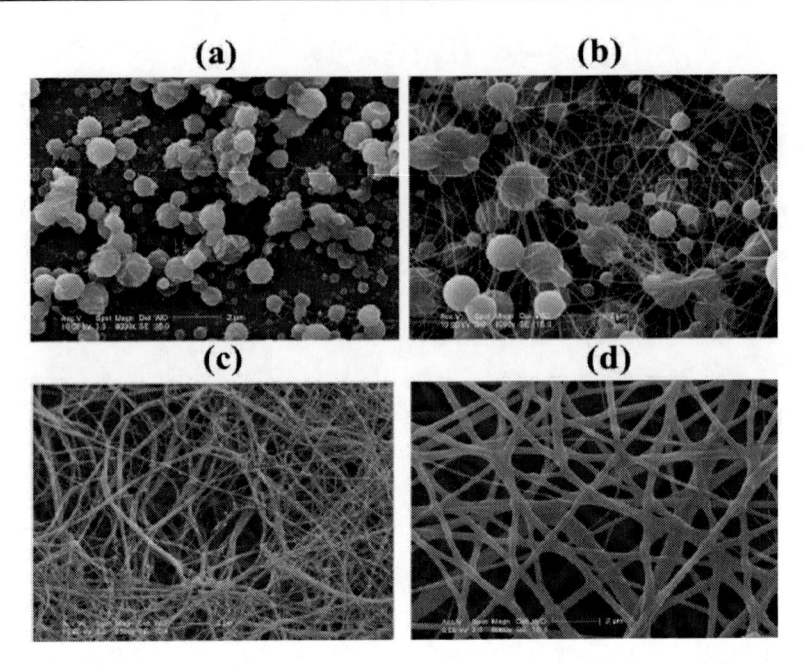

(Reproduced from Chen, H., Elabd, Y. A. *Macromolecules* 2009, *42*, 3368, Copyright (2009), with permeation from American Chemical Society).

Figure 10. Solvent effect on the electrospinning of poly(MEBIm-BF4) (10 wt % solution): (a) DMF; (b) 1/1 MeCN/DMF; (c) 3/1 MeCN/ DMF; (d) MeCN.

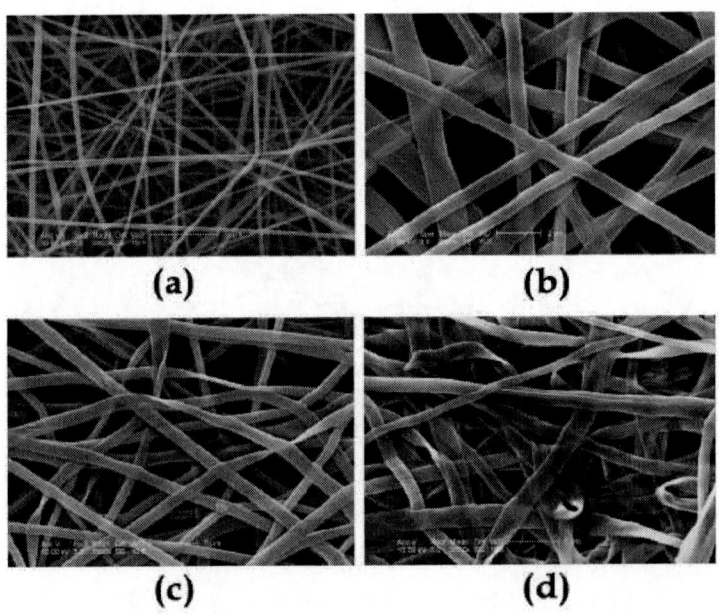

(Reproduced from Chen, H., Elabd, Y. A. *Macromolecules* 2009, *42*, 3368-3373, Copyright (2009), with permeation from American Chemical Society).

Figure 11. Field emission scanning electron microscope images of electrospun Nafion-PAA-BMIm-BF4 blend at ionic liquid weight fraction of (a) 0%, (b) 10%, (c) 20%, (d) 30%. The weight ratio of Nafion:PAA is 3:2 at a 10 wt % total polymer concentration.

Compared to other polyelectrolyte solutions, this polymerized ionic liquid solution exhibits similar viscosity scaling relationships in the semidilute unentangled and semidilute entangled regimes. Due to high solution conductivities, electrospinning produces fibers approximately an order of magnitude smaller than neutral polymers at equivalent normalized solution concentrations.

Figure 11 shows the morphology of the fibers at various ionic liquid contents. Instead of reduced fiber sizes, the existence of ionic liquid results in larger fibers with a ribbon structure. This can be attributed to the nonvolatility of ionic liquid that hinders the solidification of fibers to smaller sizes. With the increase of ionic liquid content, more ribbons were observed in the fiber mat (Figure 11).

The polymerizations of *N*-vinylimidazolium salts, 1-(3-phenylpropyl)-3-vinylimidazolium bromide (PVI-Br), 1-(6-ethoxycarbonylhexyl)-3-vinylimidazolium bromide (EHVI-Br), and 1-(2-ethoxyethyl)-3-vinylimidazolium bromide (EtOEVI-Br), were performed with reversible addition-fragmentation chain transfer (RAFT)/macromolecular design via interchange of xanthate (MADIX) process by H. Mori and coworkers. [23]

In this study, they focused on N-vinylimidazolium salts having different substituent groups, 1-(3-phenylpropyl)-3-vinylimidazoliumbromide (PVI-Br), 1-(6-ethoxycarbonylhexyl)-3- vinylimidazolium bromide (EHVI-Br), and 1-(2-ethoxyethyl)- 3-vinylimidazolium bromide (EtOEVI-Br), as shown in Figure 12.

Two xanthate-type chain transfer agents (CTAs), *O*-ethyl-*S*-(1-phenylethyl) dithiocarbonate (CTA 1) and *O*-ethyl-*S*-(1-ethoxycarbonyl) ethyldithiocarbonate (CTA 2), proved efficient for obtaining poly(PVI-Br)s and (EHVI-Br)s with relatively low polydispersities (Mw/Mn<1.4). Poly(EtOEVI-Br)s with moderate molecular weight distributions (Mw/Mn= 1.5-1.6) were also obtained under the same conditions.(Table 1)

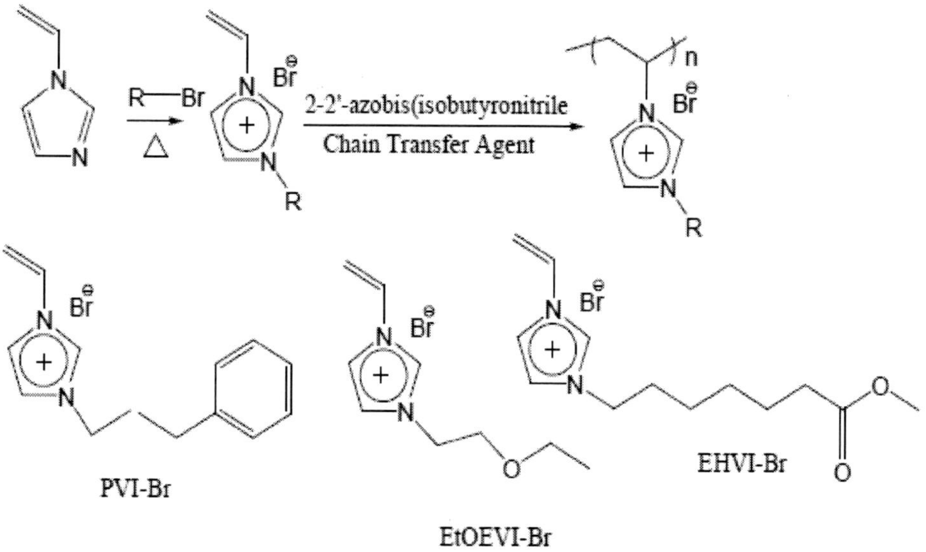

Figure 12. Synthesis of Poly(*N*-vinylimidazolium Salt)s and Structures of *N*-Vinylimidazolium Salts.

Table 1. Polymerization of N-Vinylimidazolium Salts Using 2,2'- Azobis (isobutyronitrile) (AIBN) in *N,N'*-Dimethylformamide (DMF) at 60 °C for 20 h[a]

Run monomer[b]	CTA[c]	conv[d] (%)	Mn (g/mol) theory[e]	SEC[f]	$M_w/M_n{}^{f}$ (SEC)
1[g]	PVI-Br		>99		51900	1.89
2		CTA1	72	10600	25300	1.32
3		CTA2	77	11500	25600	1.39
4 [g, h]	EtOEVI-Br		55		42000	2.23
5		CTA1	62	7900	22200	1.53
6		CTA2	65	8200	23600	1.56
7[g]	EHVI-Br		94		195100	1.87
8		CTA1	46	8500	11000	1.26
9		CTA2	50	9200	12100	1.27

[a] [M]/[CTA]/[AIBN] = 100/2/1, monomer concentration = 0.9 M. [b] PVI-Br = 1-(3-phenylpropyl)-3-vinylimidazolium bromide, EtOEVIBr = 1-(2-ethoxyethyl)-3-vinylimidazolium bromide, and EHVI-Br =1-(6-ethoxycarbonylhexyl)-3-vinylimidazolium bromide. [c] CTA 1 = Oethyl- S-(1-phenylethyl) dithiocarbonate and CTA 2 = O-ethyl-S-(ethoxycarbonyl) ethyldithiocarbonate. [d] Calculated by 1H NMR in CDCl$_3$. [e] $M_{n(theory)}$ = M$_W$ of monomer × [M]/[CTA] × conv + M$_W$ of CTA. [f] Measured by SEC using poly(ethylene oxide) standards in H$_2$O/acetonitrile (50/50 vol % containing 0.05 M NaNO$_3$). [g] [M]/[AIBN]=100/1. [h] Polymerization for 2 h. (Reproduced from Mori, H., Yahagi, M., Endo, T. *Macromolecules* 2009, *42*, 8082, Copyright (2009), with permeation from American Chemical Society).

The NMR technique was used to determine the chain-end structure and absolute molecular weights of poly(PVI-Br)s.

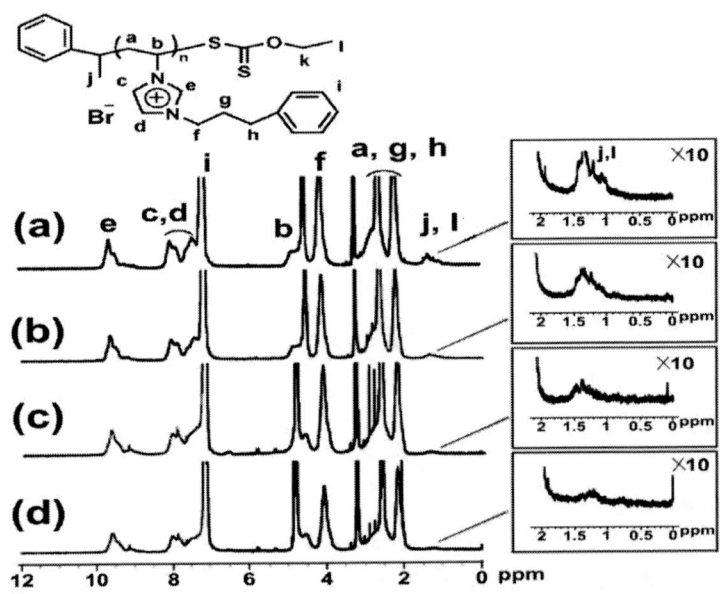

(Reproduced from Mori, H., Yahagi, M., Endo, T. *Macromolecules* 2009, *42*, 8082, Copyright (2009), with permeation from American Chemical Society)).

Figure 13. ^{1}H NMR spectra (CD$_3$OD) of poly(PVI-Br)s obtained at [PVI-Br]/[CTA 1] = (a) 10, (b) 50, (c) 100, and (d) 150.

The ^{1}HNMRspectra of the poly(PVI-Br)s obtained at different $[PVI\text{-}Br]_0/[CTA\ 1]_0$ ratios is presented in Figure 13.

Controlled character of the polymerization of PVI-Br was confirmed by the molecular weight controlled by the monomer/CTA molar ratio, a linear increase in the number-average molecular weight (Mn) with conversion, and the ability to extend the chain by a second addition of monomer.

The dielectric relaxation behavior of polymerized ionic liquid, poly(1-ethyl-3-vinylimidazolium bis(trifluoromethanesulfonylimide)) (PC2VITFSI) over a wide frequency, 10 mHz to 2 MHz, and temperature range, -90 to +90 °C was investigated by K. Nakamura [24] Figure 14. shows chemical structure of poly(1-ethyl-3-vinylimidazolium bis-(trifluoromethanesulfonylimide)) (PC2VITFSI).

(Reproduced from Nakamura, K., Saiwaki,T., Fukao, K. *Macromolecules* 2010, *43*, 6092, Copyright (2010), with permeation from American Chemical Society).

Figure 14. Chemical structure of poly(1-ethyl-3-vinylimidazolium bis- (trifluoromethanesul-fonylimide)) (PC2VITFSI).

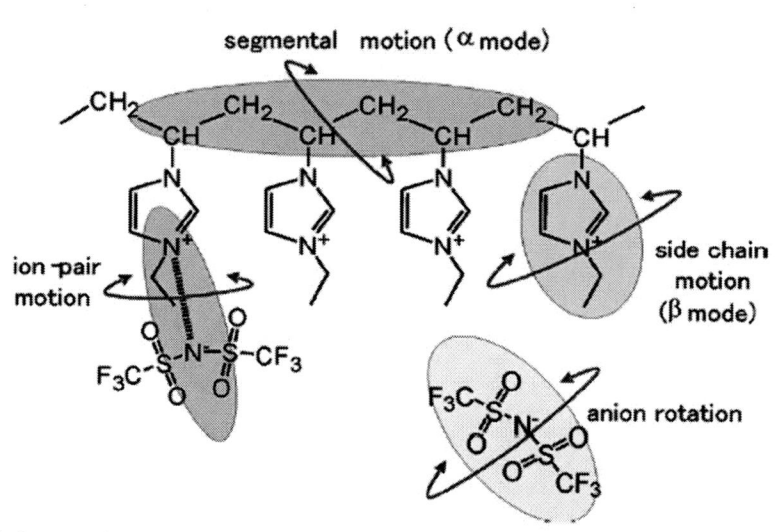

(Reproduced from Nakamura, K., Saiwaki,T., Fukao, K. *Macromolecules* 2010, *43*, 6092–6098 , Copyright (2010), with permeation from American Chemical Society)).

Figure 15. Schematic representation of the candidates for relaxation modes observed in PC2VITFSI.

Three relaxation modes including that of electrode polarization were observed. Relaxation times of these relaxation modes and the specific direct current conductivity showed an Arrhenius-type temperature dependence. They believe that less fragile behavior is the essential property for the polymerized ionic liquids and suppose that the charge transport mechanismis achieved by an anion relay process, even at temperatures below the glass transition temperature; this is facilitated by the interaction between one and another ion-pair through the rotational motion of side chains on the axis of the polymer main chain. The slower relaxation mode is assigned to the segmental relaxation mode of the polymer because the magnitude of its dielectric strength is similar to that of segmental relaxation mode observed in an ionomer system. The faster relaxation mode is attributed to the rotational relaxation mode of the polymer side chain Synthesis of polymeric non imidazole–based Ionic Liquids for advanced materials (Figure 15).

2.2. Synthesis and Applications of Polymeric Non Imidazole–Based Ionic Liquids

A novel ionic polymer, poly(1,1,3,3-tetramethylguanidine acrylate) (PTMGA), was synthesized, and its SO_2 absorption and desorption properties were studied by D. An and coworkers [25]. 1,1,3,3-Tetramethylguanidine acrylate, a polymerizable ionic liquid (IL), was prepared via neutralization of 1,1,3,3-tetramethylguanidine and acrylic acid. PTMGA was then synthesized via free radical polymerization of TMGA. (Figure 16)

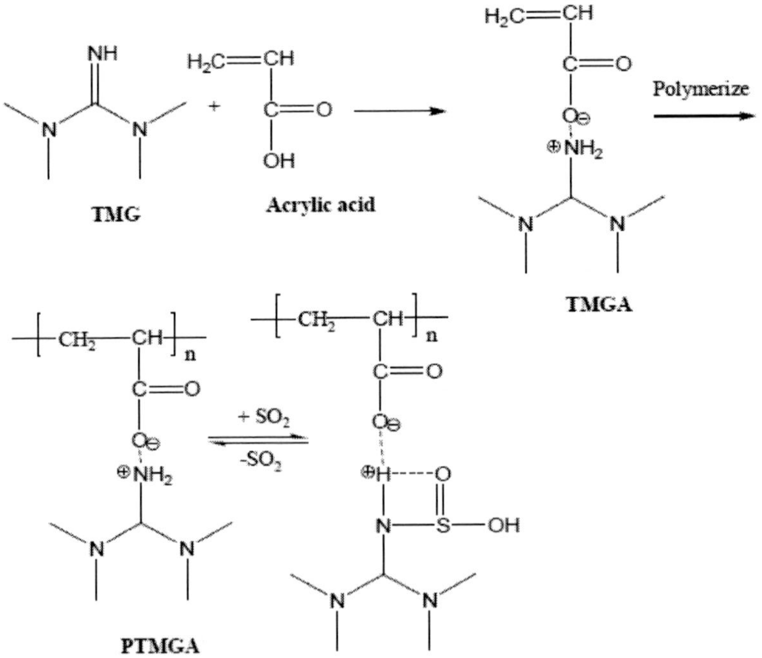

(Reproduced from An,D., Wu, L., Li, B. G., Zhu, S. *Macromolecules* 2007, *40*, 3388, Copyright (2007), with permeation from American Chemical Society).

Figure 16. Synthesis of TMGA and PTMGA Using TMG and AA) as Starting Materials and possible chemical absorption mechanism of PTMGA.

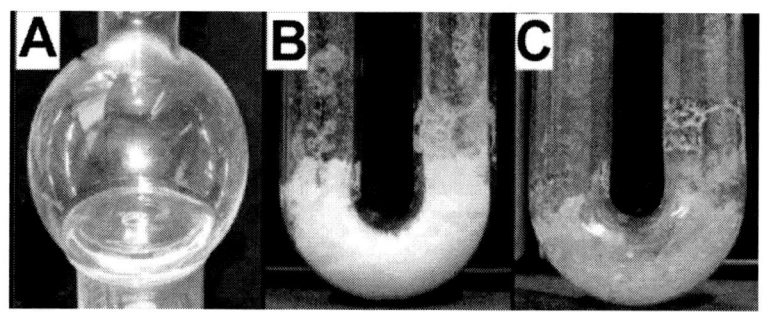

(Reproduced from An,D., Wu, L., Li, B. G., Zhu, S. *Macromolecules* 2007, *40*, 3388, Copyright (2007), with permeation from American Chemical Society).

Figure 17. Appearance of (A) monomer TMGA, (B) polymer PTMGA, and (C) SO_2- saturated PTMGA sample (absorbed at 50 °C).

The polymer adsorbed SO_2 with high selectivity, capacity, and rate. The absorption capacity and rate of PTMGA were significantly higher than the monomer. Appearance of monomer, TMGA, polymer PTMGA, and SO_2- saturated PTMGA sample was shown in Figure 17.

The SO_2 absorbed at a relatively low temperature was effectively desorbed at higher temperatures and/or under vacuum. A complete desorption was achieved at 140 °C, as shown in Figure 18. However, the desorption rate was low under an atmospheric pressure. It took 8-10 h to reach the desorption equilibrium even at 125-140 °C. The absorption/desorption process could be repeatedly operated, and thus the polymer was reused. Under a typical operation condition, about 0.3 g of SO_2 per gram of polymer was separated in each cycle. The PTMGA material showed a good potential as solid-state SO_2 absorbent for applications in purification of SO_2-containing gas such as fuel gas desulfurization.

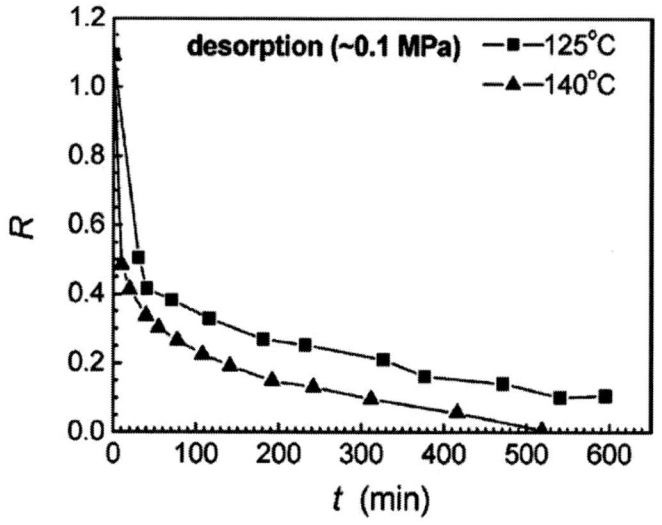

(Reproduced from An,D., Wu, L., Li, B. G., Zhu, S. *Macromolecules* 2007, *40*, 3388, Copyright (2007), with permeation from American Chemical Society).

Figure 18. Desorption of SO2-saturated PTMGA (absorbed at atmospheric pressure and 50 □C) under atmospheric pressure and at high temperatures: 125 and 140 °C.

The use of polymerizable ionic liquids for application to polymer light-emitting electrochemical cells was demonstrated by L. V. Kosilkin and coworkers [26].

They describe the synthesis of novel ionic materials for use in an LEC device structure that combine the good processing properties of ionic liquids (such as good solubility in conventional organic solvents and good miscibility with emissive materials) with the ability of ion-pair monomers to form covalent bonds to achieve ion immobilization. The structures of the ionic liquids developed here are presented in Figure 19.

Devices employing PILs had uniformfilm morphologies and showed diode-like behavior. In addition, brightness and turn-on times were improved by an order of magnitude compared to the ion-pair monomer based devices reported earlier. Further studies for optimizing PIL structure to improve device lifetime and stability are currently underway.

A method for creating high performance fixed junction LECs can lead to advances in low-cost and low power consumption solid-statelighting and photovoltaic devices. In addition, the materials developed here may have further utility for additional applications in which control over ionic mobility or electrochemical doping is critical.

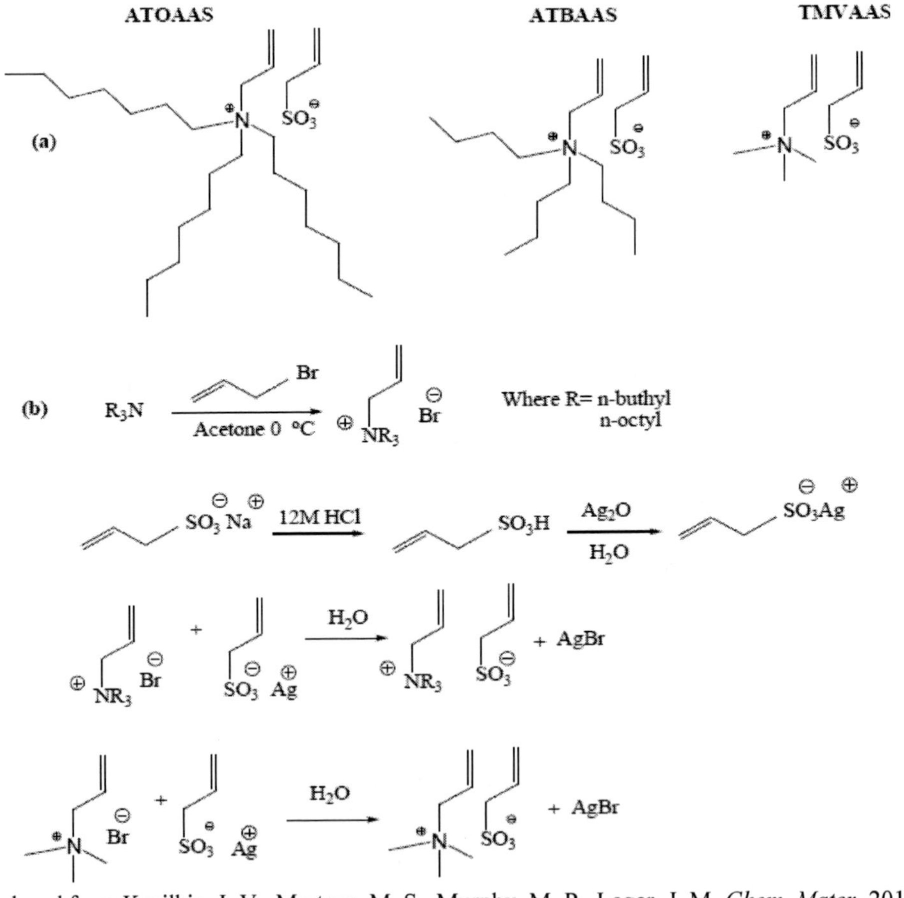

(Reproduced from Kosilkin, I. V., Martens, M. S., Murphy, M. P., Leger, J. M. *Chem. Mater.* 2010, *22*, 4838, Copyright (2010), with permeation from American Chemical Society).

Figure 19. (a) Chemical structures of materials synthesized in this study. (b) General synthetic.

A new strategy based on ionic self-assembly technology was provided for design of photosensitive material as liquid crystals (LC) alignment layer by S. Xiao [27]. The complex material was constructed by the coupling of poly(ionic liquid) and photosensitive unit azobenzene dye methyl orange. The poly(ionic liquid) and the Complex of poly(ionic liquid) and Methyl orange was prepared according to Figure 20

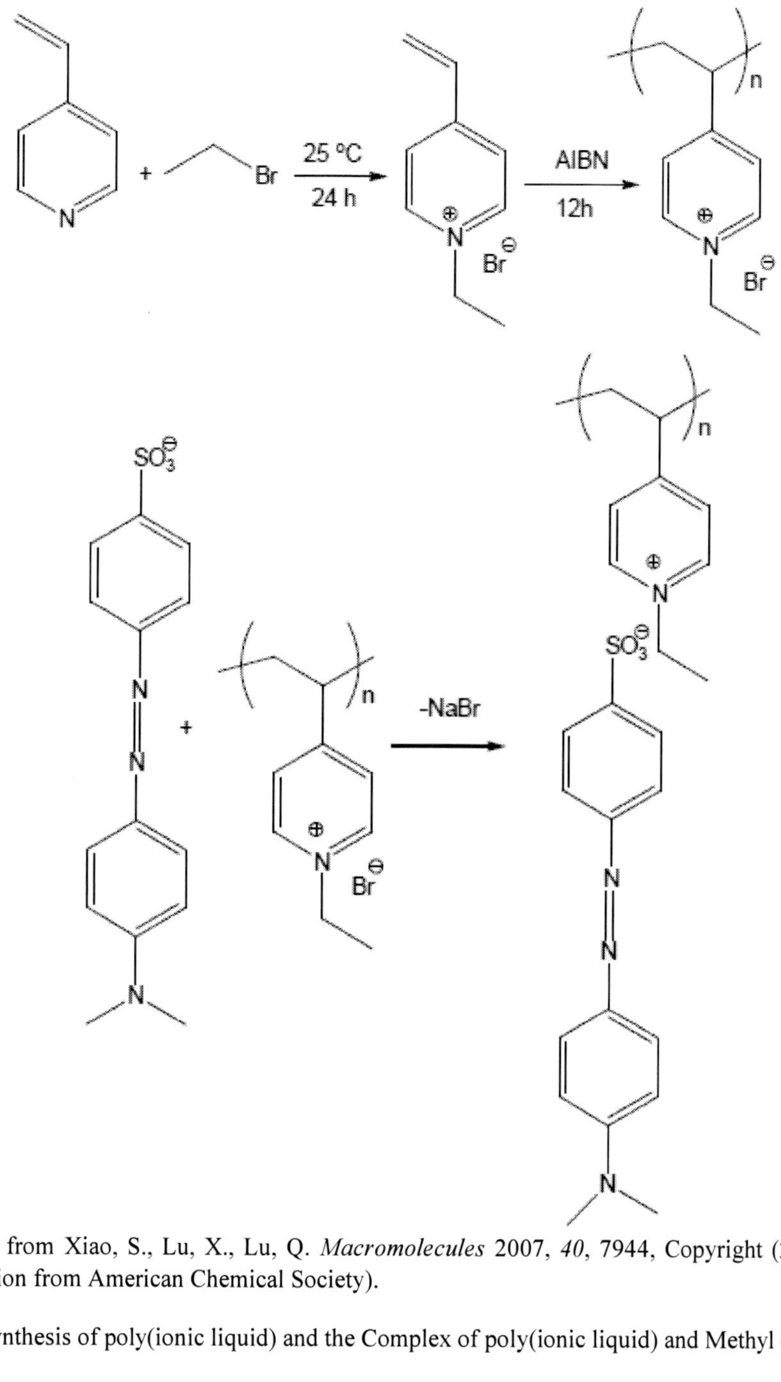

(Reproduced from Xiao, S., Lu, X., Lu, Q. *Macromolecules* 2007, *40*, 7944, Copyright (2007), with permeation from American Chemical Society).

Figure 20. Synthesis of poly(ionic liquid) and the Complex of poly(ionic liquid) and Methyl orange.

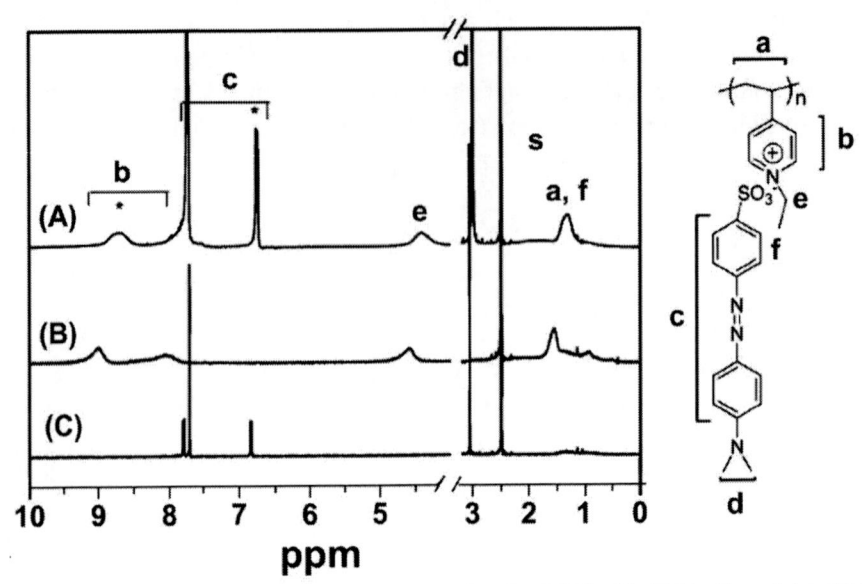

(Reproduced from Xiao, S., Lu, X., Lu, Q. *Macromolecules* 2007, *40*, 7944, Copyright (2007), with permeation from American Chemical Society).

Figure 21. ^{1}H NMR spectra of (A) PILMO, (B) PIL, and (C) MO.

The structure, phase behavior and photoresponse were examined by a variety of techniques including FTIR, NMR, thermal analysis, polarized optical microscopy, X-ray diffraction, small-angle X-ray scattering, and birefringence measurements.

^{1}H NMR spectra of PILMO complexes and two building blocks in DMSO-d_6. were shown in Figure 21.

Highly ordered lamellar nanostructure and photosensitive character were confirmed. Schematic representation of the layered architecture obtained from PILMO complexes was shown in Figure 22.

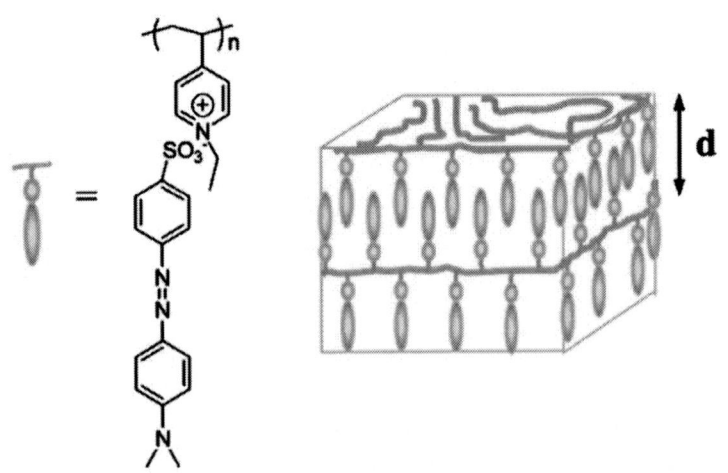

(Reproduced from Xiao, S., Lu, X., Lu, Q. *Macromolecules* 2007, *40*, 7944, Copyright (2007), with permeation from American Chemical Society).

Figure 22. Schematic representation of the layered architecture obtained from PILMO complexes.

Under the irradiation of pulsed UV laser with certain fluences, a pronounced optical anisotropic surface with the preferred direction perpendicular to the pulsed polarization or regular periodic grooves microstructure surface parallel to the pulsed polarization was obtained.

2.3. Ionic Liquid Polymer Composites

A new design of a lithium gel-polymer battery, fabricated with a (LILP) composite consisting of a lithium salt dissolved in an ionic liquid (binary Li-IL) and an ultra high molecular weight ionic liquid polymer (ILP) by T. Sato and co worker [28]. This polymer, with a Mw of over a million, was prepared by the bulk radical polymerization of a novel ionic liquid monomer, *N,N*-diethyl-*N*-(2-methacryloylethyl)-*N*-methylammonium bis(trifluoromethylsulfonyl)imide (DEMM-TFSI). The polymer could form a binary Li-IL solid at a concentration of only 5 wt%. They selected high power-active electrode materials, and combined them with the LILP system. The demonstration vapor-free cell had a higher discharge performance than a conventional lithium polymer battery: at 40 ∘C, it retained 83% of its discharge capacity at a 3C current, and relatively good cycle performance. This is the first report of to knowledge that a lithium ion cell with a LILP system performed, in terms of cell performance and cycle durability, at a level of practical utility.

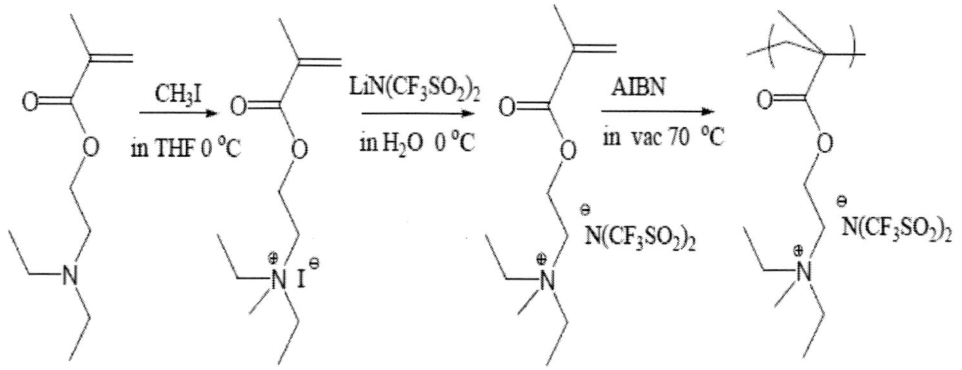

(Reproduced from Sato, T., Marukane, S., Narutomi,T., Akao, T. *J Power Sources* 2007,*164*, 390, Copyright (2007), with permeation from Elsevier).

Figure 23. Synthesis of ionic liquid monomer (DEMM-TFSI) and poly(DEMM-TFSI).

CONCLUSION

This chapter reviews some of the more recent advances that are associated with poly(ionic liquid)s. Polymerized ionic liquids are macromolecules synthesized from polymerizing ionic liquid monomers This class of smart polymers has already found applications in conducting polymer coatings, anion detectors, lithium-polymer batteries, gene-delivery vectors, gas separation membranes, electrospinable polymers and dye sensitized solar cells.

REFERENCES

[1] Anderson, J. L.; Armstrong, D. W.; Wei, G.-T. Anal. Chem. 2006, 78, 2893.

[2] Welton, T.; Chem. Rev. 1999; 99, 2071.

[3] Dupont, J.; de Souza R. F.; Suarez P A Z. Chem. Rev. 2002, 102, 3667.

[4] Freemantle, M. Chem. Eng. News 1998, 76, 32.

[5] Davis, J. H. Chem. Lett. 2004, 33, 1072.

[6] Van Rantwijk, F.; Sheldon, R A. Chem. Rev. 2007, 107, 2757.

[7] Wasserscheid, P.; Keim, W. Angew. Chem. Int Ed. 2000; 39, 3772.

[8] Pârvulescu, V. I.; Hardacre, C. Chem. Rev. 2007, 107, 2615-2665.

[9] Kubisa, P. Prog. Polym. Sci., 2004, 29,1, 3.

[10] Hirao, M.; Ito, K.; Ohno, H.. Electrochim. Acta 2000, 45, 1291.

[11] Ohno, H.; Ito, K. Chem. Lett. 1998,751.

[12] Yoshizawa, M.; Ogihara, W.; Ohno, H. Polym. Adv. Technol. 2002;13,589.

[13] Yoshizawa, M.; Hirao, M.; Ito, K.; Ohno, H. J. Mater Chem. 2001, 11,1057.

[14] Ogihara,W.;Washiro, S.;Nakajima, H.;Ohno, H. Electrochim. Acta 2006, 51,2614.

[15] Nakajima H, Ohno, H. Polymer 2005,46,11499..

[16] Ohno, H. Macromol. Symp. 2007, 249,551.

[17] Muldoon, M. J.; Gordon, C. M. J. Polym. Sci. Part A Polym. Chem. 2004, 42, 3865.

[18] Tang, J.; Radosz, M.; Shen, Y. Macromolecules 2008, 41, 493.

[19] Blasig, A.; Tang, J.; Hu, X.; Tan, S. P.; Shen, Y.; Radosz, M. Ind. Eng. Chem. Res. 2007, 46, 5542.

[20] Lin, B.; Cheng, S.; Qiu, L.; Yan, F.; , Shang, S.; Lu, J. Chem. Mater. 2010, 22, 1807.

[21] Chen, H.; Choi, J. H.; Salas-de la Cruz, D.; Winey, K. I.; Elabd, Y. A. Macromolecules 2009, 42, 4809.

[22] He, X.; Yang,W.; Pei, X. Macromolecules 2008, 41, 4615.

[23] Chen, H.; Elabd, Y. A. Macromolecules 2009, 42, 3368.

[24] Mori, H.;Yahagi, M., Endo, T. Macromolecules 2009, 42, 8082.

[25] Nakamura, K.; Saiwaki,;T.; Fukao, K. Macromolecules 2010, 43, 6092.

[26] An,D.; Wu, L., Li, B. G.; Zhu, S. Macromolecules 2007, 40, 3388.

[27] Kosilkin,;I. V.; Martens, M. S.; Murphy, M. P.; Leger, J.M. Chem. Mater. 2010, 22, 4838.

[28] Xiao, S.; Lu, X.; Lu, Q. Macromolecules 2007, 40, 7944.

[29] Sato, T.; Marukane, S.; Narutomi, T.; Akao, T. J. Power Sources 2007,164, 390.

In: Polymer Synthesis
Editor: E. Kowsari

ISBN 978-1-61324-672-6
© 2012 Nova Science Publishers, Inc.

Chapter 4

RECENT ADVANCES IN FLUORINATED POLYIMIDES: SYNTHESIS AND PROPERTIES

Susanta Banerjee,* Suman Kumar Sen,
Anindita Ghosh and Barnali Dasgupta
Materials Science Center, Indian Institute of Technology,
Kharagpur, India

ABSTRACT

Aromatic polyimides constitute one of the most important classes of high-performance polymers exhibited a number of outstanding properties, such as high thermal and mechanical properties, high optical transparencies along with chemical and solvent resistance. These excellent combinations of properties make them suitable for a wide range of applications, from engineering plastics in aerospace industries to membranes for fuel-cell applications and gas or solvent separation. Incorporation of fluorine in high performance polymers like poly(ether imide)s is a subject of great research as it brings about dramatic improvements in several properties of the polymers. Polymers containing fluorine in the form of pendent trifluoromethyl ($-CF_3$) groups showed increased solubility, lower dielectric constant and water uptake, higher glass transition temperature, higher thermal and thermo oxidative stability, better optical transparency, higher gas-permeability and flame resistance in comparison to their non-fluorinated analogues. Extensive researches have been directed towards synthesis of $-CF_3$ substituted diamine or dianhydride monomers and their polymerization followed by property evaluations of the resulting polymers to understand the structure-property correlation in these polymers. The present article provides a comprehensive review, particularly on $-CF_3$-substituted aromatic polyimides that have been developed in the last decade as low dielectric constant polymers and as membrane based applications like gas separation, pervaporation and fuel cell membranes. A major effort towards development of novel polyimides and their gas transport properties that have been devoted by our group since several years thoroughly covered in this article.

* E-mail: susanta@matsc.iitkgp.ernet.in.

1. General Introduction

The first synthesis of aromatic polyimide was reported by M. T. Bogert and R. R. Renshaw in 1908 [1]. However, the class of aromatic polyimide came into focus since 1950, after successful development of the two step polyimide synthesis by DuPont [2]. This class of polymers possesses a number of outstanding properties such as, excellent thermal, mechanical, and electrical properties which lead to their application in several fields starting from engineering thermoplastic to aerospace and electronic industries as well as for fibers, adhesives and in matrixes for composite material [3-5]. In addition to the above properties polyimides are endowed with high thermo-oxidative stability, chemical and solvent resistive properties leading to many membrane based applications such as fuel cell application, gas separation, pervaporation etc. [6,7]. However, high softening temperature and poor solubility in different organic solvents of these polymers preclude the processing in both melt and solution routes. Several approaches have been taken to circumvent the poor processability of this class of polymer. One successful approach has been executed by General Electrical with an introduction of flexible ether linkage (–O–) [8,9] and isopropylidene [–C(CH$_3$)$_2$–] moiety [10,11] into the polymer backbone resulting the Ultem 1000$^®$ [12,13]. It exhibited excellent thermal stability and good mechanical properties. Various approaches investigated to ease the polyimide processing include addition of bulky side groups or bulky units in the polymer backbone [14,15], or noncoplanar [16] or alicyclic monomers [17] in the main chain. In the 1970s, Korshak and coworkers first introduced bulky cyclic side groups (termed as "cardo groups") in the polymer backbone to make polyimides soluble without compromising the higher glass transitions temperatures (T$_g$) and thermomechanical properties [18]. Incorporation of hexafluoroisopropylidene [-C(CF$_3$)$_2$-] or pendent trifluoromethyl (-CF$_3$) groups is of great interest which increases the free volume of the polyimides, thereby improving various properties like solubility, chemical resistivity, and gas permeabilities without forfeiture of thermal stability [19-22]. Moreover, these groups decrease crystallinity, color, water absorption as well as increase optical transparency, environmental stability, and flame resistance [23-25]. Optical transparency of polyimide membranes is an important property because of its demand in many optoelectronic devices [26]. All the approaches to make processable polyimides as discussed reduces several types of polymer interchain interactions, chain packing and the charge transfer electronic polarization interactions or charge transfer complex (CTC) formation.

2. Polyimide Synthesis

Polyimides are class of condensation polymers and generally prepared from organic diamines and organic tetracarboxylic acid dianhydrides. There are mainly two synthetic routes for polyimides preparation namely, i) one step and ii) two step polymerization method. Scheme 1 shows a generalized synthesis of polyimide from an aromatic dianhydride and an aromatic diamine.

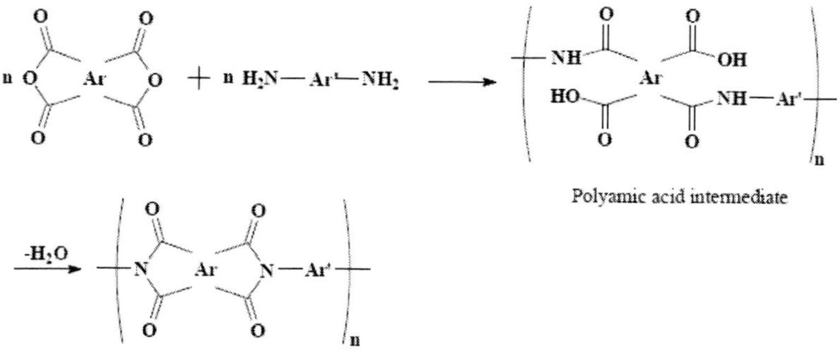

Scheme 1. Synthesis of polyimide.

2.1. One Step Polymerization

In the one-step polymerization method, completely cyclized polyimides are obtained directly from their corresponding equimolar mixture of tetracarboxylic acid dianhydride and diamine in presence of a high boiling tertiary amine (e.g., isoquinoline). Several high boiling solvent are used e.g. m-cresol, p-chlorophenol, α-chloronaphthalene, nitrobenzene, o-dichlorobenzene and, dipolar aprotic amide solvents and their mixtures for the direct conversion. During the progress of the polymerization, the byproduct water that generated is continually removed by azeotropic distillation. This method of direct polymerization is usually used when the final polyimide is soluble in the solvent. However, this method suffers from some enormous drawbacks, such as the use of very toxic carcinogen solvents and a rather low polymerization concentration (usually less than 10% w/v), which hampers direct processing of the polyimides into the final products (e.g., films and fibers) from their polymerization solutions, and a long reaction time (typically, more than 18 h).

2.2. Two Step Polymerization

In the two-step polymerization method, initially the polyamic acid is formed from an equimolar mixture of dianhydride and diamine in a polar aprotic solvent, such as N,N-dimethylacetamide (DMAc) or N,N-dimethylformamide (DMF). The reaction pathway for the formation of poly(amic acid) involving the intermediates is presented in Scheme 2. The reaction mechanism involves the nucleophilic attack of the amino group to the electrophilic carbonyl carbon of the anhydride group. This results in opening of the anhydride ring to form an amic acid group. The formation of the poly(amic acid) is an equilibration reaction where the forward reaction is thought to start with the formation of a charge transfer complex between the dianhydride and the diamine [30]. The susceptibility of the nucleophilic attack increases with increasing the electrophilicity of the dianhydride group. Thus, the reactivity of the dianhydride monomer has been correlated to its electron affinity; higher values indicate greater reactivity of the dianhydride [31]. Electron affinity values for some dianhydrides are listed in Table 1. Strong electron withdrawing groups activate the anhydride toward nucleophilic attack on the anhydride carbonyl. On the other hand, the reactivity of the diamine is related to its basicity. The rate constants for imidization increase as the pK_a of the protonated amine increases. Highly basic amines (e.g. aliphatic amines) may form salts during the initial stages of the reaction, upsetting the stoichiometry and preventing the formation of high molecular weight. Solvent also play an important role in the reaction. Highly polar aprotic solvents used for this type of polymerization form strong hydrogen bonds with the carboxyl group that help in shifting the equilibrium to the forward side i.e., to the amic acid side. The reaction rate is generally faster in more basic solvents. Several other minor important side reactions concurrently proceed with the main reaction. This side reaction may become significant under certain conditions particularly when the main reaction is slow because of low monomer concentration and low monomer reactivity. Presence of water in the reaction system causes lower molecular weight development of poly(amic acid) due to the hydrolysis of the dianhydride moiety [32]. During this step, the self-catalyzed cyclization to form polyimides cannot occur due to the strong interaction between the amic acid and the basic solvent or the larger acylation equilibrium constant [3,6].

Table 1. Electron affinity values for different anhydrides (Data taken from reference 31)

Dianhydride	Name	Electron affinity (ev)
	PMDA	1.90
	DSDA	1.57
	BTDA	1.55
	BPDA	1.38
	ODPA	1.30
	HQDA	1.19
	BPADA	1.12

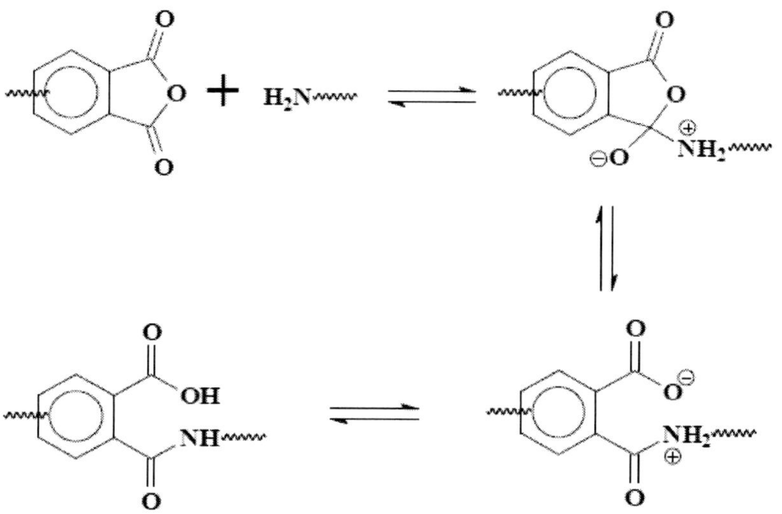

Scheme 2. Reaction pathway in poly(amic acid) synthesis.

In the second step, the polyamic acid is cyclodehydrated at elevated temperatures (thermal imidization) or in presence of a cyclizing agent (chemical imidization). The advantages of this method over the one-step polymerization are the use of less toxic solvents and direct processing of the soluble polyamic acids to form the final polyimide products in the form of films or fibers by thermal imidization. However, the stability of polyamic acid on storage and control of thermal imidization are still important issues [33]. A detail description of thermal imidization and chemical imidization is discussed below.

2.2.1. Thermal Imidization of Poly(Amic Acid)

The most common method for the conversion of the poly(amic acid) to the polyimide is the bulk (or melt) imidization [34-36]. Therefore, thermal imidization method is generally used in industry where poly(amic acid) is heated at ~200-300 °C for a given amount of time to form the imide ring by removing the solvent and the water. At high temperature, irreversible cyclodehydration reaction occurs which leads to a high molecular weight polyimide. In this method, the films of the poly(amic acid)s are often cast from polar aprotic solvents (e.g., N,N-dimethylformamide, N,N-dimethylacetamide etc.) and subsequently dried and imidized. This method is suitable for polyimides in the form of films, coatings and powders in order to allow the diffusion of byproducts and solvent without formation of voids in the final polyimide products. The problem of film cracking as a result of shrinking can be avoided by carefully controlling the curing profile. A typical heating schedule involves a stage below 150 °C, followed by a relative rapid temperature rise to a second stage above the glass transition temperature T_g of the resulting polyimides. The cast films are dried and heated gradually upto 250–350 °C depending on the stability and T_g of the polyimides. Maximum amount of solvent is slowly driven off in the first stage and the imidization occurs in the second stage, where curing and shrinkage is reliable [37]. At the initial stage of the imidization, a small amount of the poly(amic acid) undergoes a reversible reaction with the anhydride and amine instead of forming the imide ring resulting in a lower molecular weight development [39]. Two possible pathways for the imidization are possible during thermal imidization, as shown in Scheme 3 [40].

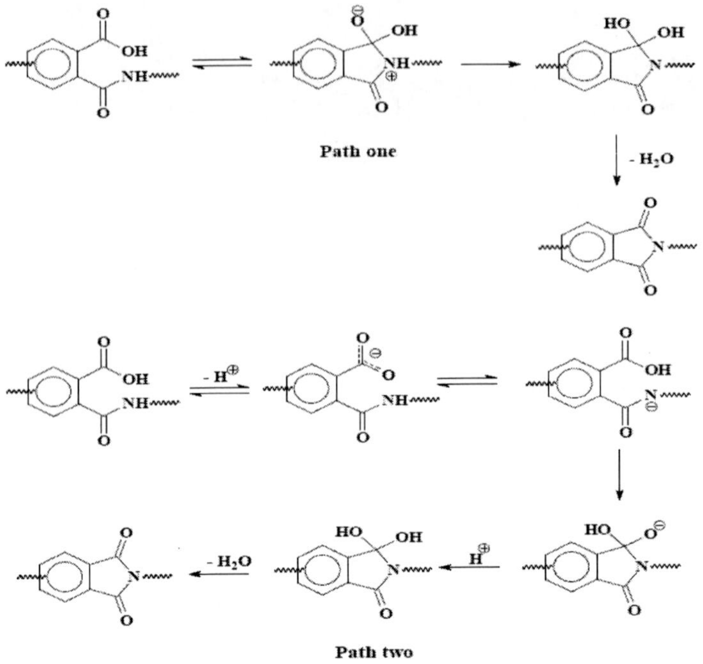

Scheme 3. Two possible pathways in thermal imidization.

2.2.2. Chemical Imidization of Poly(Amic Acid)

Poly(amic acid)s can also be chemically imidized. This is accomplished by using chemical dehydrating agents in combination with basic catalysts [41]. Various reagents have been employed including acetic anhydride, propionic anhydride, and n-butyric anhydride as dehydrating agents and pyridine, triethylamine and isoquinoline as basic catalysts. A reaction pathway for chemical imidization is shown in Scheme 4.

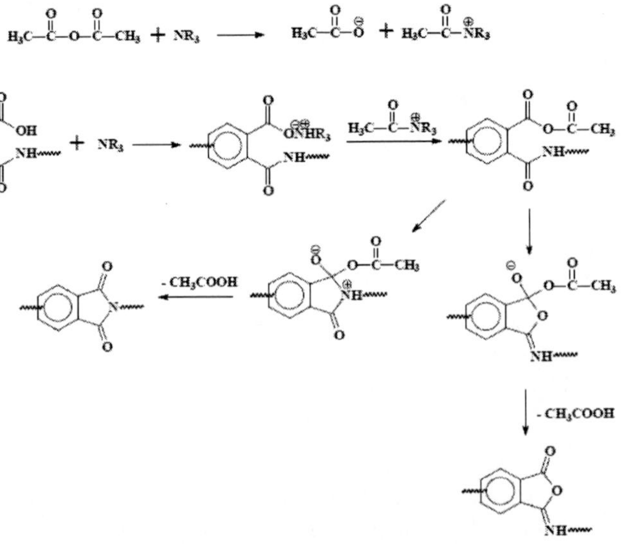

Scheme 4. Reaction pathway in chemical imidization.

3. UNIQUENESS OF FLUORINE

Fluorine containing compounds are unique. The factors which make fluorine different from other halogens are:

1. The low dissociation energy of the fluorine molecule
2. The relatively high strength of bonds formed between fluorine and metallic or non-metallic elements.
3. The relatively small size of the fluorine atom and the fluoride ion.

These factors are to some extent interrelated; it will be convenient to consider them in turn. Considering the bond dissociation energies of the halogens, the following are the experimental values.

$D (F_2)$=37.7 Kcal/ mol $D (Cl_2)$=58.2 Kcal/ mol

$D (Br_2)$=46.1 Kcal/ mol $D (I_2)$=36.1 Kcal/ mol

The ready dissociation of fluorine into atoms is cause of its high reactivity. Though Iodine has almost equally dissociation energy but forms much weaker bonds, so that the overall energy release in the formation of a fluoride is considerably greater (Table 2).

Several authors have discussed the reason for low dissociation energy of fluorine compared to the other halogens. Mulliken pointed out that the energy of the O-O single bond (34 Kcal/mol) is less than that of the S-S bond (63 Kcal/mol). In qualitative terms his view is that bonding for the top row element is normal but in the lower row partial multiple bond character is possible by d-hybridization. Fluorine with a septet of electrons only has a single bond in the molecules whereas oxygen and nitrogen in the molecular form have multiple bonds with much higher energies (225 and 117 Kcal/mol respectively). Cadlow and Coulson have stressed the relatively large electron- electron repulsion in the fluorine molecule is responsible for the low bond dissociation energy of the element compared with the other halogen.

Table 2. Bond energy values

Molecule	Bond	Bond Energy (Kcal/mol)
HF	H-F	135
HCl	H-Cl	103
HBr	H-Br	87
HI	H-I	71
CF_4	C-F	116
CCl_4	C-Cl	78
CBr_4	C-Br	58
CI_4	C-I	57
SiF_4	Si-F	135
$SiCl_4$	Si-Cl	91
SF_6	S-F	68
S_2Cl_2	S-Cl	61

Table 3. Electronegativity values

Atoms	Electro Negativity	Atoms	Electro Negativity	Atoms	Electro Negativity
B	2.04	N	3.04	F	3.98
Al	1.61	P	2.19	Cl	3.16
C	2.55	O	3.44	Br	2.96
Si	1.9	S	2.58	I	2.66

Fluorine is the most electronegative of the chemical elements. The concepts of electro negativity, described by Pauling as "the power of an atom in a molecule to attract electrons to itself" has been expressed in terms of a numerical scale by several authors. These values given in Table 3 are due to Pauling and show both fluorine and oxygen to be more electronegative than the other halogens.

3.1. Effects of Fluorine /Trifluoromethyl Group on Chemical Reactivity

The presences of fluorine and/ or trifluoromethyl groups in a molecule considerably modify chemical properties, biological activity and stability. Therefore, fluorine containing compounds play a significant role in medical and agricultural chemistry as well as material science.

The trifluoromethyl group is the most prominent representative in the series of perfluoroalkyl groups. The Vander Waals radius of trifluoromethyl group is 2.7; the Vander Waal volume 42.6, whereas the corresponding values for a methyl group are 2 and 16.8 [42]. Therefore the sterically demand of trifluoromethyl group is close to that of isopropyl group. The electronegativity of a trifluoromethyl group is similar to oxygen [43]. The presence of trifluoromethyl groups in molecule can alter the reaction behavior of the adjacent functional group. The fluorine atom of trifluoromethyl groups is able to participate in hydrogen bonding as electron pair donors and stabilizes certain conformations [44]. With increasing fluorination the C-C bond length shortens and consequently the bond strength increase [44]. Therefore, introduction of fluorine and of trifluoromethyl group stabilizes molecule towards thermal and metabolic degradation. Another interesting property of trifluoromethyl group is its high lipophilicity [45].

Lipophilicity $F < CF_3 < OCF_3 < SCF_3$

In the context of imidization reaction, fluorine or trifluoromethyl group have a great influence. Polyimide is generally prepared from the reaction of a diamine (AA) with the dianhydride (BB). Although the way in which the -F or $-CF_3$ group affect the basicity of the diamines is quite complex, but it is generally accepted that presence of -F or $-CF_3$ group, due to its electron withdrawing effect or high electronegativity, act to reduce the reactivity of the diamines. However, the position of the -F or $-CF_3$ group in respect to the amine group have an important role to control the reactivity of the amine groups. G. Hougham et al. demonstrated that if the -F or the $-CF_3$ group is present at the meta position in respect to the amine groups than they have much less steric as well as electronic affect on the reactivity of the diamines [46].

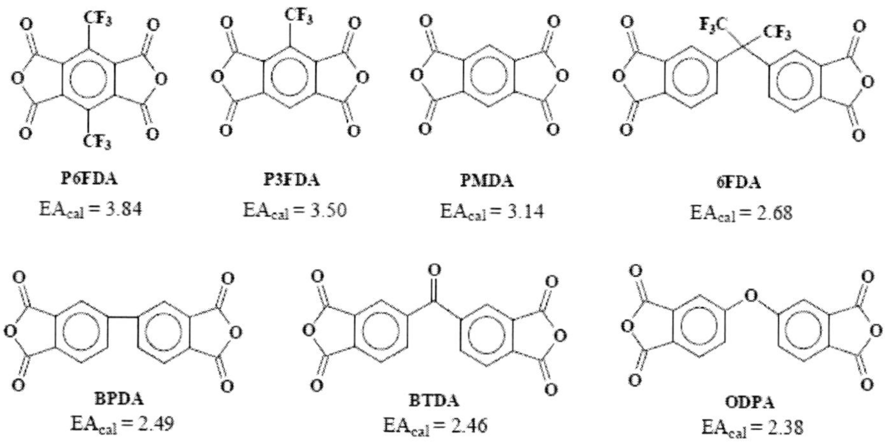

Figure 1. Few dianhydrides with their calculated electron affinity values [47a].

Due to high electronegativity of fluorine or trifluoromethyl group, when these are attached to the benzene ring directly causes an enhancement of reactivity of the dianhydrides. How the reactivity of the anhydride is affected by trifluoromethyl group is nicely presented by Shinji et al. [47a]. The following figure (Figure 1) depicts few dianhydride structures with their calculated electron affinity values. The electron affinity values were calculated by MNDO-PM3 method.

The dianhydrides are arranged in decreasing order of their calculated electron affinity values. From Figure 1, it is seen that when -CF_3 groups are directly attached to the benzene ring the electron affinity of the dianhydrides, therefore, the reactivity of the dianhydrides towards the diamine increases. For example, P6FDA and P3FDA are much more reactive than 6FDA. Even PMDA is much reactive than 6FDA. This observation indicates that the quaternary carbon to which two -CF_3 groups are attached blocks the electronegativity effect of the -CF_3 groups by a significant amount.

3.2. Stability of Trifluoromethyl Groups

Although the trifluoromethyl group is often considered to be chemically inert, it is known to undergo a variety of reactions. The stability of a trifluoromethyl group is very much dependent on its position in a molecule. Trifluoromethyl groups directly bonded to aromatic system undergo hydrolysis only in strong acidic medium [48]. Trifluoromethyl groups directly attached to carbanionic centers are labile and undergo fluoride elimination, sometimes under very mild conditions. The ability to eliminate fluoride ions from trifluoromethyl and perfluoroalkyl groups on treatment with bases or on reduction of trifluoromethyl substituted unsaturated compounds can be applied to generate synthetic valuable fluoro containing building blocks in situ [49]. It is noteworthy that, trifluoromethyl groups attached to certain positions of heterocyclic systems are prone to facile base- induced hydrolysis; for example, the trifluoromethyl group of 2- trifluoromethylimidazole [50] as shown in Scheme 5. Following the same reaction sequence, 5-amino-4-trifluoromethyl-oxazoles can be transformed into 5-amino-4-methyloxazole on treatment with $LiAlH_4$ after a sequence of successive elimination/ addition steps [51].

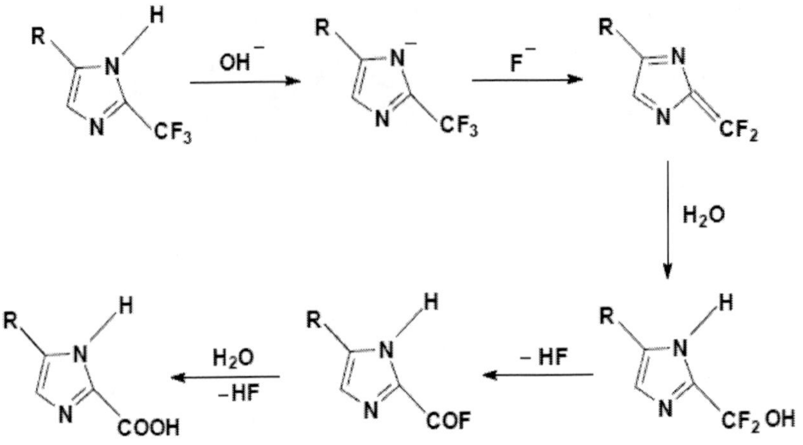

Scheme 5. Base-induced hydrolysis of trifluoromethyl group of 2- trifluoromethylimidazole.

4. ROLE OF TRIFLUOROMETHYL GROUP IN POLYIMIDE

Polyimides bearing pendant trifluoromethyl ($-CF_3$) groups are of great interest to the material chemists because the presence of a bulky trifluoromethyl group in the polyimide structure causes a dramatic improvements in their several properties in comparison to their non-fluorinated analogues [47b]. The incorporation of fluorine affects different properties of the polymers in several ways; increase in free volume, which decrease the dielectric constant by reducing the number of polarizable groups in a unit volume. Increase in free volume can be attributed to the bigger size of fluorine atom relative to hydrogen atom that interferes with efficient chain packing. In addition, the significant mutual repulsion of fluorine atoms on different chains may influence the free volume. Furthermore, the electronic polarization is always lowered with fluorine substitution because of the smaller electronic polarizability of the C–F bond relative to C–H. In cases where fluorine is positioned non symmetrically, an increase in the dipole moment would result. This could lead to substantial increases in the dipole orientation polarizability at low frequencies, thereby increasing the low-frequency dielectric constant [52].

The combination of electronic and steric effects reduces the interchain interactions and, particularly, the steric congestion hinders the formation of charge transfer complexes [53]. The C-F bond is a high energy bond, so the polyimides containing fluorine are in general polymers with high T_g and exhibits high thermal stability. Moreover, greater fluorine content improves the flame retardant property and, also increases the hydrophobicity of the polymer. The low water uptake results in decreasing the bulk dielectric constant at higher frequencies. The advantages of fluorinated polyimides in improvement of properties can be summarized as below:

- Increases fractional free volume
- Increases solubility, hence enhances the processability.
- Increases thermal stability
- Increase glass transition temperature
- Decreases dielectric constant.

- Decreases moisture absorption
- Reduce formation of charge-transfer complex
- High optical transparency

Thus all these properties has made the fluorinated polyimides very attractive for various applications in advanced technologies, such as in high performance structural resins, thermally stable coatings and films, polymeric membranes for gas separation, optical waveguides, and other electronic and optoelectronic applications [54-56].

5. FLUORINATED DIAMINE AND DIANHYDRIDE MONOMERS

As mentioned earlier, the significant property improvement on inclusion of $-CF_3$ groups into the polymer find a large corpus of patent and literary data in the last decade by developing newer polyimides in terms of making new fluorinated diamine and dianhydride monomers. Table 4 and Table 5 depict the molecular structures of few fluorinated diamines and dianhydrides, respectively, developed within the last decade.

6. POLYIMIDES FROM FLUORINATED MONOMERS

The successful commercialization of Kapton® by Du Pont was the breakthrough in polyimide world. The PMDA-ODA based polyimide exhibited very high glass transition temperature ($T_g \sim 390$ °C), high mechanical strength (tensile strength, 158 Mpa, and tensile modulus, 3.2 GPA) and accordingly, found wide-spread applications as electrical insulating material e.g., interlayer dielectrics / inter metal dielectrics (ILD / IMD) [91]. However, the processability of Kapton® was the main concern. The major problem of this type of rigid polyimides is insolubility and infusibility in their fully imidizied form, leading to processing difficulties. They are generally processed from their poly (amic acid) precursors and then converted to polyimide via rigorous thermal treatment. This process has several limitations, such as poor shelf life of the poly(amic acid) and emission of volatile byproducts (e.g., water) during imidization.

Accordingly, several approaches have been undertaken by different researchers to increase the polyimide processability. Incorporation of flexible linkages, bulky side groups or bulky units in the polymer backbone, or noncoplanar or alicyclic monomers in the main chain has been investigated in an attempt in this regard. Polyimides containing flexible spacers have gained a significant importance in polyimide processing. Some polyimides containing oligoethylene glycol moiety have been presented as processable polyimides with potential application as thermally stable adhesives and thermoplastics [92,93]. Polyimides containing oligosiloxane segments are also included within this class of polyimides. Other examples of important bridging linkages are -O-, C=O, -S-, -SO$_2$-, -C(CH$_3$)$_2$-, -CH$_2$-, -CHOH-, and -C(CF$_3$)$_2$ [94-99]. The presence of fluorine and/ or trifluoromethyl groups in polyimde considerably improves solubility. Hougham has reported a number of fluiorinated diamines and dianhydrides in his review in 1999 [47b]. Table 4 and Table 5 represent a list of several fluorinatd diamines and dianhydrides developed within the last decade.

Table 4. Molecular structures of fluorinated diamines

Serial No.	Structure	Reference
1		[57]
2		[58]
3		[58]
4		[23]
5		[23]
6		[11]
7		[59]
8		[59]
9		[59]
10		[60]
11		[61]
12		[62]

Serial No.	Structure	Reference
13		[63]
14		[64]
15		[65]
16		[66]
17		[67]
18		[68]
19		[69]
20		[70]
21		[71]
22		[72]
23		[72]
24		[72]
25		[73]

Serial No.	Structure	Reference
26		[74]
27		[75]
28		[76]
29		[77]
30		[78]
31		[78]
32		[79]
33		[80]
34		[81]
35		[82]
36		[83]
37		[84]

Table 5. Molecular structures of fluorinated dianhydrides

Serial No.	Molecular structure	Reference
1		[85]
2		[85]
3		[86]
4		[87]
5		[88]
6		[89]
7		[90]

Generally, the presence of these 'kink' linkages between aromatic rings or between phthalic anhydride units disrupts the planarity of the polymer repeat unit and increases the torsional mobility. Additionally, these additional bonds leads to an increment of the repeating unit chain length, which, in turn, separates the imide rings leading to a relatively lower density of the imide moiety, which may be responsible for the polymer tractability. The suppression of the coplanarity of the structure is maximal when large groups are introduced in the main chain, for instance sulfonyl or hexafluoroisopropylidene groups, or when the monomers are enlarged by more than one flexible linkage. Hexafluoropropylidene, carbonyl and sulfonyl groups are the most advantageously incorporated concerning processability. The poly(ether imide), marketed under the trade name Ultem[®] by General Electric Company is one of the example of engineering thermoplastic. The polymer is soluble in many of the organic solvents and showed T_g values around 217 °C. This polymer posses excellent flow characteristics and melt stability, because of flexible ether linkages in the main chain and serve as true high performance engineering thermoplastic. In the 1970s, Korshak and coworkers first introduced to the incorporation of bulky cyclic side groups (termed as "cardo groups") in the polymer backbone to make polyimides soluble without compromising the higher glass transitions temperatures (T_g) and thermomechanical resistance [18]. Later, Rusanov et al. prepared several organo-soluble aromatic polyimides using side phthalimide groups or by introducing pendent imide groups [100-102]. The phenyl group containing 'cardo' groups are the most promising among various other group. It does not decrease the thermal stability, and provides a measure of molecular irregularity and separation of chains very beneficial in terms of free volume increase and lowering of the cohesive energy density [103-108]. On the other hand, the presence of the bulky side substituents in polyimides or in any other linear polymer causes a lowering of the chain's torsional mobility and generally an increment of the glass transition temperature [109,110]. The discussion on inclusion of trifluoromethyl (-CF$_3$) groups into the polymer needs a separate attention due to its large beneficiary polymer property improvement. Section 4 (Role trifluoromethyl group in polyimide) has been devoted in the context of several diversified property enhancement after inclusion of -CF$_3$ group.

All these approaches were made to reduce several types of polymer interchain interactions, chain packing and the charge transfer electronic polarization interactions or charge transfer complex (CTC) formation. Incorporation of hexafluoroisopropylidine or pendent trifluoromethyl groups is also of great interest which increases the free volume of the polyimides, thereby improving various properties like solubility, electrical insulating properties, and gas permeabilities without forfeiture of thermal stability. These groups decrease crystallinity, color, water absorption as well as increase optical transparency, environmental stability, and flame resistance.

An illustration of sequential development of several fluorinated polyimides in the last decade can be presented from our own research group. Initially, we designed several poly(ether imide)s based on novel diamines 1 – 5, Table 4 (Figure 2) [11, 23, 57, 58].

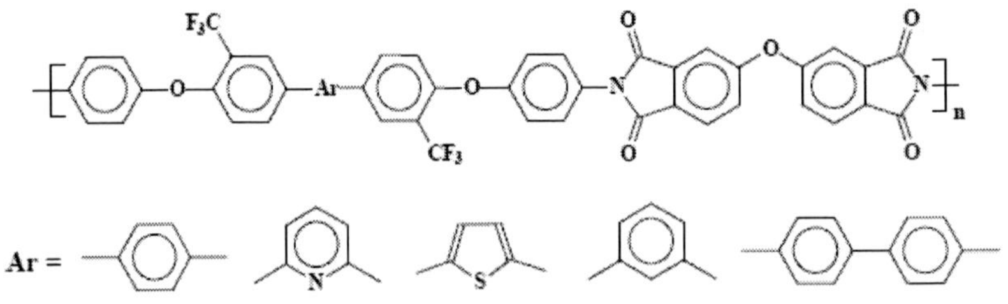

Figure 2. Molecular structures of poly(ether imide)s.

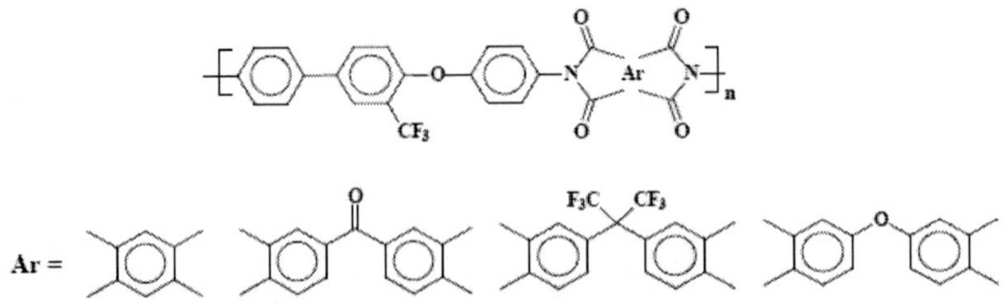

Figure 3. Poly(ether imide)s with unsymmetrical structures.

The polyimides prepared from these monomers (1 – 5) exhibited several interesting properties, e.g. high glass transition temperature (up to 316 °C), high thermal stability in air and reasonably low dielectric constant. However, one of our major interests to prepare organo-soluble polyimide was not successful except some polyimides made from 6FDA as dyanhydride. Kim et al. [111] prepared some processable polyimides based on unsymmetrical diamine that allow us to design the new unsymmetrical monomer 6, Table 4 (Figure 3).

However, the solubility of the poly(ether imide)s derived from this monomer (6) [11] were not up to the satisfaction. Accordingly, further manipulation of the chemical structures of the diamine monomer was done to prepare new polymer structures with better processaibility. Three new diamine monomers (7-9) [59] were prepared by adopting the concept that having both flexible and rigid units in the same molecule with an aim that the flexible ether linkages will allow the free rotation and will lead to better solubility of the prepared polymer in organic solvents and the rigid biphenyl unit linked with the dianhydride moiety through imide linkages will help in getting higher glass transition temperature.

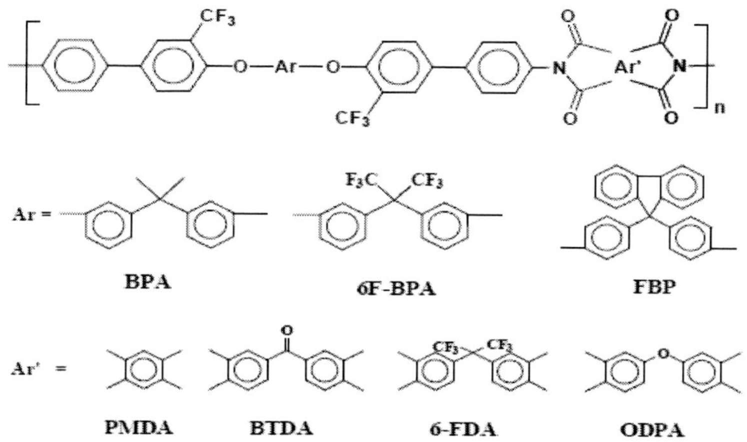

Figure 4. Poly(ether imide)s with both rigid and flexible units.

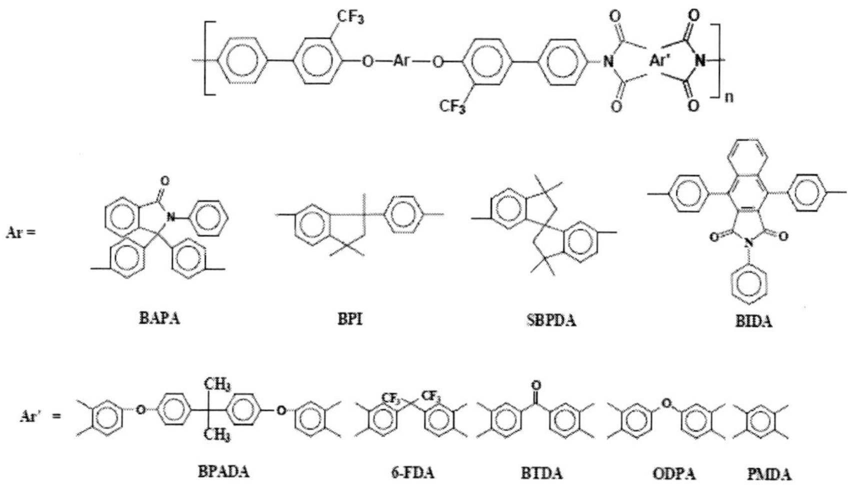

Figure 5. Poly(ether imide)s containing cardo groups.

The poly(ether imide)s (Figure 4) thus prepared showed very good solubility (10 % W/V) in many organic solvents and better than the analogous polymers derived previously from the diamine monomers 1 to 6. The polymers, which are derived from 6FDA and ODA, are soluble even in chloroform and toluene at room temperature. It can be argued that flexible units are more important to get polymer solubility than the unsymmetry of the structure. [59]

In continuation to the development of novel polyimides with better properties several new fluorinated diamine monomers have been developed (10 to 14, Table 4, Figure 5) recently. [60-64]

The last decade is a spectator of a worldwide participation in the research and development of polyimides of better solubility and processability both from the academia and industries. A rigorous discussion on all the development is beyond the scope of this chapter. However, a focus has been intended to highlight few important observations by several reserachers. Y. Li et al. nicely demonstrated the enhancement of polyimide solubility or processability on introduction of trifluoromethyl groups by preparing 4-phenyl-2,6-bis[3-(4'-amino-2'-trifluoromethyl-phenoxy) phenyl] pyridine (m-PAFP) (diamine monomer 16, Table 4) [66] and comparing with its analogue which does not contain any trifluoromethyl group made by Wang et al. [93]. He also synthesized two unsymmetrical diamines with ether–ketone group, 3-amino-4'-(4-amino-2-trifluoromethylphenoxy)-benzophenone, 4-amino-4'-(4-amino-2-trifluoromethylphenoxy)-benzophenone (diamine monomer 20, Table 4) [70] and unsymmetrical ether diamine with a trifluoromethyl pendent group, 1,4-(2'-trifluoromethyl-4',4"-diaminodiphenoxy)benzene (diamine monomer 27, Table 4) [75] which led to polyimides of better processability. H. Han was curious to find the difference of the trifluoromethyl (-CF$_3$) group and ether group affecting the optical property of fluorinated polyimides. Accordingly, he prepared 4,4'-bis(4-amino-2-trifluoromethylphenoxy)diphenyl ether with three ether groups (diamine monomer 22, Table 4) and 2,2-bis[4-(4-amino-2-trifluoromethylphenoxy)phenyl]hexafluoropropane diamine monomer 23, Table 4) with four -CF$_3$ groups with 2-chloro-5-nitrobenzotrifluoride and 4,4'-dihydroxydiphenyl ether or 2,2-bis(4-hydroxyphenol)hexafluoropropane. Based on the comparison of the polyimide series based on 22, 23, and 4,4'-bis(4-amino-2-trifluoromethylphenoxy)biphenyl (24), they claimed that the -CF$_3$ group and ether group on the diamine had almost same effect in lowering the color, but the ether group had better thermal stability [72]. S. Y. Yang et al. prepared two series of polyimides from 4,4'-bis(3-amino-5-trifluoromethylphenoxy) biphenyl (m-6FBAB) (diamine monomer 26, Table 4) and 4,4'-bis(4-amino-5-trifluoromethylphenoxy) biphenyl (p-6FBAB) (diamine monomer 24, Table 4) with various aromatic dianhydrides. It was found out that the polyimides derived from m-6FBAB showed better melt processability and solubility than the p-6FBAB based polymers. [74]. In addition to all these polyimides several new fluorinated polyimides have been developed to ease the processability by developing new fluorinated diamine monomers as depicted in Table 4.

Fluorinated polyimides are also considered to be one of the potential candidates for optical application as they have very good optical transparency. Optical transparency of polyimide films is of significant importance in some application such as in liquid crystal display devices and optical half-waveplates for planar light wave circuits. However, most polyimides have strong absorption between UV and the visible range, rendering their color close to yellow or brown. The transformation of polyimide films from the brownish or yellowish color to light color (more transparent) after the incorporation of the -CF$_3$ groups could be explained by the decreased intermolecular interactions and reduced packing density.

The bulky and electron-withdrawing -CF$_3$ group in the diamine monomer is effective in decreasing CTC (charge transfer complex) formation between polymer chains through steric hindrance and the inductive effect (by decreasing the electron-donating property of diamine moieties). It reduces the intermolecular CTC between alternating electron donor (diamine) and electron acceptor (dianhydride) moieties. C. P. Yang prepared a library of organo-soluble and lightly-colored fluorinated polyimides. It was observed that fluorinated polyimide exhibited lighter color and greater transparency than the analogous nonfluorinated one. [90]

6.1. Properties of Fluorinated Polyimides

All the newly developed fluorinated diamines and dianhydrides were targeted to make processable aromatic polyimides possessing many useful properties, such as high glass transition temperature, excellent dimensional stability, low dielectric constants and outstanding thermal and thermoxidative stabilities. The combination of all these properties leads the polyimides to be widely used as an interlayer dielectrics, flexible circuitry substrates, stress buffers, and passivation layers. In the last decade there have been several research publications on low dielectric constant materials [113]. Currently there is a huge demand for new intermetal dielectric layers (IMD) with lower dielectric constants in order to reduce the signal delays, drive voltages, and power consumption for the high-powered integrated circuits (ICs) used nowadays. Today, silicon dioxide (SiO$_2$) having a dielectric constant (κ) of ~4.0 is mostly used as IMD material. Current specifications for the new generation insulating polymeric film materials is with dielectric constants between 3 and 3.3, and within the next decade, it is expected that devices may require materials with dielectric constants approaching 2.0 or below. While there are a number of possible candidate materials for current uses, the list of viable materials with dielectric constants <3.0 is more limited. The greatest limitations in materials qualification are the stringent IC processing conditions (thermal stability, resistance to chemical/mechanical treatments, mechanical stability, ease of use, etc.) for the ultimate construction of novel molecular devices possessing extraordinary low dielectric properties. A variety of high performance polymers, such as polyimides, aramids, poly(arylene ether)s, and polyesters have been exploited in various electronic devices and components. Among all these class of polymers, polyimide is the most superior as interlayer dielectric materials in electronic devices such as integrated circuits owing to their many favorable properties such as high thermal stability, ease of processing, low water absorption, low stress/coefficient of thermal expansion (CTE), and very good electrical properties (lower dielectric constant, high resistivity and high breakdown voltage) [114, 115]. In electronics packaging, low dielectric materials minimize crosstalk and maximize signal propagation speed in devices. Thermally stable, durable, insulative polyimides are in great demand for the fabrication of microelectronic devices. One of the most commonly employed polyimides, Kapton made from pyromellitic dianhydride/oxydianiline (PMDA/ODA), has a dielectric constant of 3.2 (measured at 1 MHz). Hence the development of polyimides with increasingly lower dielectric constants has been the focus of several recent investigations [116, 117]. Accordingly, a number of fluorinated polyimides have been synthesised to facilitate the study of structure/property relationships of fluorine containing polyimides and to report some tentative generalizations derived from the results.

A variety of strategies have been undertaken to synthesize polyimides with lower dielectric constants . The most common approach has been undertaken to develop materials with high organofluorine content (trifluoromethyl), often by the incorporation of pendant perfluoroalkyl groups into diamine and dianhydride reactants which minimize polarizability and impart a high degree of free volume [47]. As it is discussed earlier that fluorine atom has unique characteristics, such as high electronegativity and low electric polarity. These properties give fluorinated polymers [e.g., poly(tetrafluoroethylene)] attractive features, such as low water uptake, water and oil repellence, low permittivity, low refractive indices, resistance to wear and abrasion, and thermal and chemical stability. Fluorination is also known to enhance the solubility and optical transparency and to lower moisture absorption of polyimides. Therefore, it is expected that fluorinated polyimides will be widely applied in electro-optical and semiconductor industries. The polymer series studied was essentially limited to mainly the 6F dianhydride because it proved to be the only dianhydride with which many of the fluorinated diamines would form polymer films suitable for physical characterization. Table 6 shows some fluorinated polyimides and their repeat unit structures with the respective dielectric constants measured at 1 MHz.

Recent studies [56, 57] demonstrated that polyimides derived from ether-bridged aromatic diamines with trifluoromethyl groups, are soluble high temperature polymer materials with low moisture uptake, low dielectric constant, high optical transparency, and low birefringence. The effect of fluorine substitution on the dielectric constant of the polyimides has been demonstrated for a broad spectrum of fluorine-containing diamine structures (Table 4). The decreased dielectric constants might be partly attributed to the presence of bulky $-CF_3$ groups, which result in a less efficient chain packing and an increase in the free volume. In addition, the strong electronegativity of fluorine results in very low polarizability of the C–F bonds, thus decreasing the dielectric constant. The introduction of – CF_3 group lead to a decrease in dielectric constant but the strong electron withdrawing ability of $-CF_3$ could reduce the reactivity of the monomer leading to low molar mass polymer. $-CF_3$ group located at the ortho position of the amine group is further reduced due to steric hindrance. Minimizing polarizability, maximizing free volume and fluorination all lowers dielectric constants in the polyimides. Polarizability is the primary variable influencing dielectric constants whereas free volume and fluorine content are secondary variables which can alter a polymer's polarizability. Enhanced free volume lowers polarization by decreasing the number of polarizable groups per unit volume. Fluorination increases free volume, lowers electronic polarization and can either increase or have no effect on dipole polarization depending on whether the fluorination is asymmetric or symmetric. Diamines with fluorinated groups in the position ortho to the amino groups were successfully polymerized to high molecular weight. It was found that fluorine substitution generally reduced the dielectric constant substantially in cases with symmetric substitution. It was estimated that around 50% of the observed decreasing trend in the dielectric constant versus percent fluorine can be attributed to a reduction in the absorbed moisture due to an increase in hydrophobicity. The remaining 50% can be apportioned, on average, between increased free volume (~25%) and reduced total polarizability (~25%). Little effect of fluorine substitution on dynamic thermal or thermo oxidative stability was observed for a series of fluorinated polyimides. [46]

Low dielectric constant polyimides and low water absorption polyimides were reported by several researchers [96-99]. A correlation of high free volume and low dielectric constant has been previously reported for polyimides. [124, 125]

Table 6. Properties of few fluorinated polyimides

Serial No.	Polyimide	T_g (°C)	T_{d5} (°C)	TB (MPa)	k	Ref. No.
1		254	531	147	2.82	[118]
2		242	519	129	2.88	[118]
3		211	517	118	2.92	[118]
4		241	514	131	3.24	[118]
5		234	486	116	3.19	[118]
6		283	516	98	3.12	[11]
7		258	514	135	2.91	[11]
8		280	501	84	2.82	[11]
9		252	532	125	2.81	[11]
10		294	532	112	2.94	[57]
11		272	525	100	2.98	[57]
12		278	519	96	2.72	[57]
13		269	513	98	2.93	[23]
14		234	491	87	2.97	[23]

Table 6. (Continued).

Serial No.	Polyimide	T_g (°C)	T_{d5} (°C)	TB (MPa)	k	Ref. No.
15		251	516	97	2.76	[23]
16		316	519	125	2.91	[23]
17		262	531	148	2.98	[23]
18		273	526	109	2.74	[23]
19		295	547 (10%)	118	2.85	[24]
20		292	554	75	2.68	[84]
21		217	545 (10%)	84	2.78	[119]
22		187	532	97	2.84	[66]
23		221	547	108	2.55	[120]
24		264	574	107	2.47	[121]
25		271	565 (10%)	117	2.48	[122]
26		260	519	-	2.68	[105]

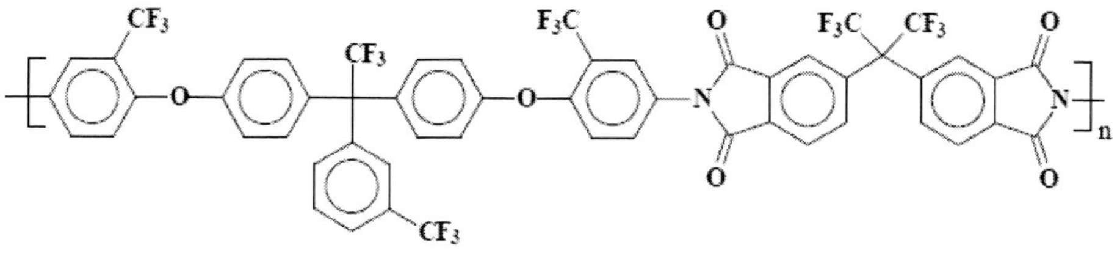

k = 2.71

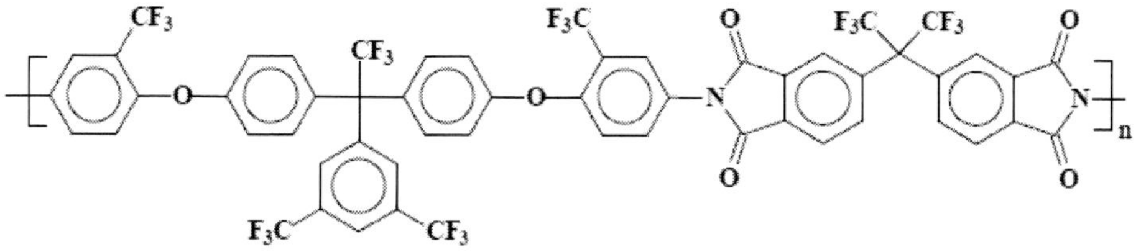

k = 2.49

Figure 6. Effect of symmetric and unsymmetric substitution of -CF$_3$ group.

Positron lifetime spectroscopy and group additivity methods were used to quantify free volume fractions. The introduction of free volume in a polymer decreases the number of polarizable groups per unit volume resulting in lower values for ε_{atomic} and $\varepsilon_{dipolar}$. The addition of pendant groups, flexible bridging units, and bulky groups which limit chain packing density have all been used to enhance free volume in polyimides and are used in this study to examine their effect on dielectric constant. Although incorporation of fluorine into polyimides has been shown to lower dielectric constants, indiscriminate fluorine substitution may actually yield an undesired effect. Symmetric substitution of fluorine does not increase the net dipole moment of the polymer and hence, does not increase the dielectric constant. In fact, the dielectric constant decreases with symmetric fluorine substitution by a combination of lower electronic polarizability and larger free volume (Figure 6). In this study, symmetric and non-symmetric fluorinated groups are used to elucidate the influence of fluorine content on dielectric constant. [108, 109]

We prepared several fluorinated polyimides which showed excellent thermal stability in air as expected in case aromatic polyimides [11, 23, 57, 119]. (1 to 18, Table 6). The higher thermal stability of fluorene containing polyimides is due to their rigid structure and may be due to their higher degree of aromaticity. These poly(ether imide)s showed very high tensile strength and elongation at break up to 72 %. The water absorption value for these polyimides lies between 0.16 to 0.32 %. The 6FDA containing polyimides showed the lowest water absorption that is due to the presence of more number of hydrophobic –CF$_3$ group; in other sense more amount of fluorine content in the polymer structure. A trend (with some deviation) in water absorption vs. polymer structure is found that depends on the % fluorine content in the polymer. In general the water absorption values of these polymers were lower than that of non-fluorinated polyether imides, such as Ultem 1000 (1.52 %) and Kapton H (3%). [117]

A list of several fluorinated polyimides developed by different researchers is depicted in Table 6. The methodology for developing new, highly fluorinated polyimides can be limited, to a certain extent, by synthetic difficulties associated with the incorporation of greater amounts of pendant perfluoroalkyl groups. An alternative approach toward lowering a polymer's bulk dielectric constant is to introduce nanoscopic porosity into the polymer film. The effect of porosity on the dielectric constant of a material can be quite dramatic, with observed lowering of dielectric constant greater than predicted by a linear rule-of-mixtures model. Foamed films with pores in the 10 nm range have been successfully demonstrated to exhibit quite low dielectric constants. The dielectric constant of a fluorinated polyimide with 18% porosity was reduced to 2.35 from 2.85 compared to the parent polyimide [128]. Further increase of the porosity will the morphology and isolated, non-interconnected pores can no longer be expected, as the porosity approaches or exceeds 30%. To be used for interconnect fabrication for circuits with minimum feature sizes; the pores should not be much larger than 10 nm in diameter. In addition, they should be regular with a narrow size distribution, and they must not be interconnected. Further discussions on pores are beyond the scope of this chapter. A porous, low dielectric constant polyimide films have been made by a "nanofoam" approach in a two-step process by K. R. Carter et al [129, 131]. The nanoporous foams were generated by preparing triblock copolymers with the major phase comprising polyimide and the minor phase consisting of a thermally labile block followed by selective removal of labile blocks via thermal treatments, leaving pores the size and shape of the original copolymer morphology. The films were then finally heated incrementally to 310 °C in an inert atmosphere (argon) or under high vacuum to effect solvent removal and densification of the copolymer. The second step involves the thermolysis and removal of the labile block, poly(propylene oxide), by heating the copolymer in the presence of oxygen at 250 °C for 10-12 h. The pore sizes generated in the polymer films are in the tens of nanometers range. A similar type approach was undertaken by several researchers for preparation of this type of nanoporous polymer. [130, 132, 133]

Polymer optical waveguides are attractive as more economical and practical optoelectronic devices and as interconnections in optical communication systems. The polymers have been used in fabrication of waveguides for lightwave circuits, including directional couplers and thermo-optic switches. Several polymers, such as deuterated and fluorinated poly(methyl methacrylate), polystyrene, poly(organosiloxane), polyimide, an acrylate photopolymerizable monomer system, epoxy resin, cross-linked acrylate polymer, and benzocyclobutene have been used as waveguide materials. Conventional silica-based inorganic waveguides have low optical loss, but compared with these waveguides, organic polymer waveguides can be fabricated on larger substrates, lower cost to fabricate, and are more flexible [134]. One of the important factors for such waveguides is that the material should be of high thermal stability for compatibility with high performance electronic device fabrication processes and fluorinated polyimides are a preferred candidate with high thermal stability. Optical loss in polymer at near-infrared (IR) wavelengths is occurred due to the adsorption from the vibrational overtones of C-H bond, which can be overcome by the substitution of hydrogen atoms with chlorine or fluorine. Rigid structure of the polyimide films make them frequently highly birefringent (the difference in refractive index between in plane and out of plane). Different polarized lights would travel at different speeds through these highly birefringent materials. So, the signals become complicated and cause polarization dependent waveguide loss. Accordingly, for the waveguide application low birefringence materials are essential. Refractive indices or birefringence in polymer are affected by several factors, which includes chain rigidity, linearity, geometry of the repeat unit, polarizability, orientation of the bonds etc.

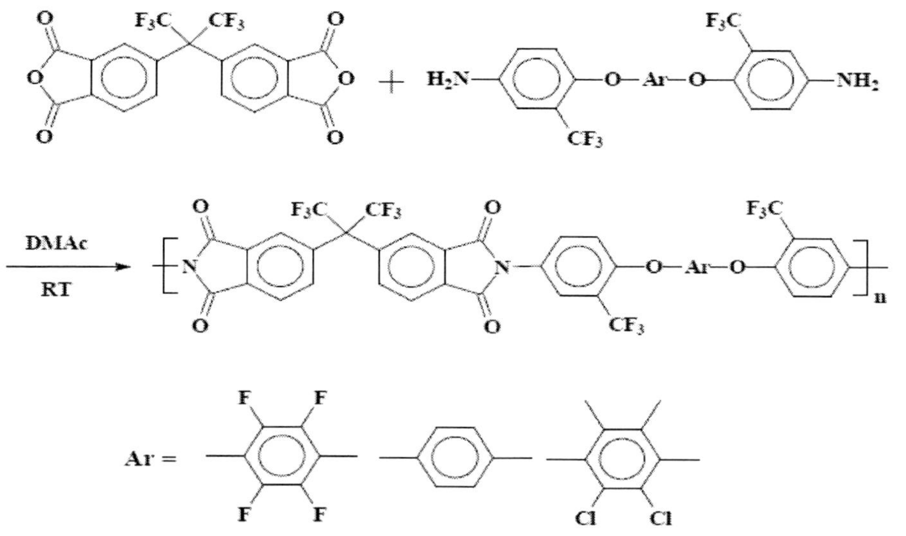

Scheme 6. Fluorinated polyimides for optical waveguide application .

Electronic transition causes absorption in the ultra-violet and visible region. On the other hand, stretching vibration of the chemical bonds causes the absorption in the near-IR region. C-H and O-H bond are strongly affected by the absorption near-IR region. Polymer used in optical devices should have low absorption in optical communication wavelengths (1.3 and 1.55 μm). The absorption due to C-H bond can be diminished by replacing hydrogen to heavier atoms such as fluorine [135]. So, higher the number of C-F bond, lower the absorption near-IR region. Hydrophobic nature of fluorine helps to decrease water absorption in fluorinated polymer. Low water absorption is essential for the low loss optical waveguide. Absorbed water can cause absorption loss at optical communication wavelengths due to O-H absorption and also can change the refractive index. Kwansoo Han [136] reported some fluorinated polyimides for optical waveguide application. These polyimides were prepared by the reaction of 2,2-bis(3,4-dicarboxyphenyl) hexafluoropropane dianhydride (6FDA) with 1,4-bis-(4-amino-2-trifluoromethyl-phenoxy)tetrafluorobenzene (ATPT), 1,4-bis-(4-amino-2-trifluoromethyl-phenoxy)benzene (ATPB), and 1,3–bis-(4-amino-2-trifluoromethyl-phenoxy) 4,6-dicholobenzene (ATPD), respectively (Scheme 6). These polyimides showed glass transition temperature in the range 260-280 °C, low water absorption (< 0.4%), low optical loss (> 0.2 dB/cm) at 1.3 and 1.55 μm and birefringency values in the range 0.0041-0.0066.

6.2. Fluorinated Polyimides for Membrane Based Separation

The increased world-wide competitiveness in production has forced industry to improve current process designs. Consequently, the development of new process designs, and the reorganization of present process designs (with the possible integration of new technologies into them, hybrid systems) is of growing importance to industry [137]. Membrane technologies have recently emerged as an important category of separation processes to the well-established mass transfer processes. Membrane separation technologies offer advantages over existing mass transfer processes [138-140]. Such advantages can comprise

- high selectivity;
- low energy consumption;
- moderate cost to performance ratio;
- compact and modular design.

6.2.1. Polyimides in Gas Separation Applications

The phenomenon of gas permeation through polymer was first observed by Thomas Graham in 1829 when he found an inflation of a wet pig bladder with CO_2. In 1866, Graham first postulated the "solution diffusion process", where he hypothesized that in the permeation process at first the dissolution of the penetrant occurs, and then the dissolved species transmits through the membrane [141, 142]. Currently, membrane based gas separation technology is considered as one of the most innovative and rapidly growing fields across science and engineering. It is an important unit operation technique which can be executed as process intensification strategy to decrease the production costs as well as to reduce equipment size, energy utilization, and waste generation [137]. Membrane based gas separation has been successfully contending over conventional separation techniques like cryogenic distillation and pressure swing adsorption (PSA) processes. Eventually it finds a variety of more economical industrial applications, including hydrogen recovery from reactor purge gas, nitrogen, and oxygen enrichment, and stripping of carbon dioxide from natural gas, etc. [143, 144].

The membrane materials used for the gas separation are glassy polymers and the basis of ability to separate the gas mixtures is the differences in the penetrant size [145]. Membrane based gas separation began in a commercial scale only in late 1970's after the synthesis of asymmetric cellulose acetate membranes by Sourirajan [146]. Membrane based gas separation is now successfully competing over highly sophisticated techniques like cryogenics, adsorption and absorption. The applications, referred to earlier in the introduction, are divided into three main categories as mentioned below [146, 147]:

1. O_2 or N_2 enrichment of air.
2. Acid gas and water vapor separation from natural gas, and
3. H_2 separation from variety of gases, such as CO, CH_4, N_2.

Separation of air into its constituents O_2 and N_2 is an important application of membrane base separation. Other than their individual importance as feedstock chemicals, pure N_2 gas used for making an inert atmosphere for foods and fuels is an important industrial application whereas, oxygen enriched air finds usage in home medical applications (respirators), for improved gas combustion [138, 140]. Isolation of CO_2 from other light gases (usually methane) has been successfully applied in oil recovery from production wells [138, 148, 149]. Reoval of CO_2 increases in mole fraction of condensable components leading to an increase in the dew point of the mixture. Membrane based separation process is successfully used to separate H_2. This process is highly feasible due to the high permeation rate of hydrogen with respect to other supercritical gases like nitrogen (in ammonia purge stream), carbon monoxide (CO/H_2 ratio adjustments), and methane (in refineries and petrochemical process). [150]

The potential application of a polymer as a separation membrane depends upon the possible throughput and the purity of the product. This means that both the permeability coefficient for the gas that is transported more rapidly and the selectivity should be as large as possible. Therefore, in the last three decades, the study of gas permeability and permselectivity of polymer membranes has become a subject of strong research with worldwide participation in both industrial and academic laboratories. Although, the main motivation for these works was primarily empirical search for improved materials for gas separation membranes, another aim, on the other hand, which is often present but not always manifested, is the development of rules for prediction of the permeation parameters of polymers that still have not been investigated or even prepared [151-153]. Such rules, when available, would simplify or significantly facilitate the task of directed synthesis of novel membrane materials with improved properties.

However, to achieve such rules, it is necessary to have a good understanding of the relationship between the properties of the polymers and their gas transport behavior.

Generally it is accepted that bulky groups in the polymer main chain tend to increase the fractional free volume resulting in high permeability coefficients for gases whereas high chain stiffness, as indicated by high glass transition temperature, is expected to result in relatively high selectivities. However, in spite of an exhaustive collection of gas permeability coefficients/selectivities data for nearly all film forming polymer, structure/property correlationship in context to gas separation has not been sufficiently understood. [7, 154–157] Few increment methods are available [151, 152, 158] by which one can calculate the permeability coefficients and selectivities of polymers based on the respective libraries of structural elements. It is well accepted that the gas diffusion in a dense glassy polymer film is controlled by the diffusion jump rate of the gas molecules from one intersegmental channel to other, i.e. from one void of free volume to another. These jumps are controlled by thermally activated motions of certain segments, which open up a sufficiently large "channel" to a neighboring gap. The gas molecules then can diffuse through this channel, and hence, the selectivity for certain gases results from the size of the voids and the width of these channels [155]. In this model, improved permeability and selectivity can be expected only if there is a control over the polymer membrane material and the size distribution of voids of free volume. In our view, to further elucidate the exact structure/property relationships for a series of polymers the chain segments which control the permeation are to be identified [155]. In this regard the gas permeabilities and selectivities of a series of two different classes of polymers; polyimides and poly(ether ketone)s, and the detailed structure/property relationships have already been reported [159, 160]. Investigations till now revealed that gas permeability through polymer membrane is affected by a variety of chemical and physical factors, such as chemical structure of main or side chain, length and size of side chain, stereo configuration, conformation, polarity, cohesive energy density, intra- and inter chain interaction, crystallinity, packing density, glass transition and secondary transitions of polymer chains. Permeability of the gases through the membrane also depends on many other factors such as temperature, pressure, penetrant mobility or solubility effects, composition of the gas mixture, polymer-penetrant interactions etc.

Kim et al. demonstrated how fluorinated alkyl side groups into polyimide membranes affect their physical and gas permeation properties (Table 7) [77]. They demonstrated this with eight polyimides with or without fluorinated side groups by polycondensation of 2-(perfluorohexyl)ethyl-3, 5-diamino benzoate (PFDAB) and *m*-PDA with four aromatic dianhydrides (6FDA, ODPA, BTDA, and PMDA), respectively.

PFDAB   m-PDA

The investigation revealed that the incorporation of fluorinated side groups into the polyimide membranes decreased their surface free energies (T_gs'), increased solubility parameters, and fractional free volume (FFV)s and therefore, substantially enhanced the gas permeabilities for CO_2, O_2, N_2, and CH_4 gases with reduced selectivities for CO_2/CH_4, O_2/N_2, CO_2/N_2 gas pairs depending upon the structure of dianhydride monomers.

Some aromatic polyimides containing $-C(CF_3)_2-$ groups in their dianhydride moieties (e.g., the 6FDA-based polyimides) have been found to be considerably more gas selective, than other glassy polymers with comparable permeabilities. Substitutions of $-C(CF_3)_2-$ groups in the dianhydride moieties appear to be somewhat more effective in enhancing selectivity than substitutions in the diamine moieties of polyimides. It has also been seen that aromatic polyimides that contain $-C(CF_3)_2-$ groups in their both dianhydride moieties and diamine moieties exhibit a high selectivity as well as high permeability [20, 22]. The substitution of $-C(CF_3)_2-$ groups and also of other similarly bulky functional groups in aromatic polyimides help to increase the chain stiffness because these bulky groups inhibit intrasegmental (rotational) mobility. The greater chain stiffness enables the polyimide matrix to better discriminate between permeating molecules of different sizes and shapes and hence selectivity is enhanced. That means it enhances the "molecular sieving" ability of the polymers. Permeability of the $-C(CF_3)_2-$ containing polyimide also increases as the $-C(CF_3)_2-$ groups due to their size inhibit close packing and serve as "molecular spacers" in the polyimide matrix. In polyimide chain packing is strongly affected by interchain interactions, e.g., one factor that can significantly contribute to such interactions in some polyimides is the formation of charge transfer complexes (CTC). St. Clair et al. have shown that the introduction of bulky $-CF_3$ groups into linear polyimide backbones largely eliminate the CTC, resulting in the removal of all color from membranes [161]. The elimination or even weakening of the CTC reduces the chain packing and thereby increases the permeability.

Table 7. A comparison of gas transport properties through fluorinated and non-fluorinated polyimides (Data taken from reference 77)

Polymer	P (CO_2)	P (O_2)	P (N_2)	P (CH_4)	α (CO_2/CH_4)	A (O_2/N_2)	α (CO_2/N_2)
6FDA- PFDAB	17.77	4.74	0.74	0.44	40.4	6.4	24.0
6FDA-*m*- PDA	9.73	2.55	0.38	0.21	46.3	6.7	25.6
ODPA-PFDAB	11.03	2.61	0.56	0.36	30.6	4.9	19.7
ODPA-*m*-PDA	0.301	0.081	0.012	0.0064	47.3	6.8	25.1
BTDA-PFDAB	10.10	2.20	0.48	0.29	34.8	4.6	21.0
BTDA-*m*-PDA	0.428	0.112	0.016	0.0086	49.8	7.0	26.8

Within the last decade our research group has been involved with the development of newer polyimide membrane materials toward selective separation gas mixtures, particularly O_2/N_2 and CO_2/CH_4 gas pair (Table 8 and Table 9) [62, 162-170]. All the synthesized polyimides investigated as gas permeation membranes have been presented in two following schemes (Figure 7 and Figure 8). A systematic investigation of gas transport properties through these polyimides towards four different gases (CH_4, N_2, O_2 and CO_2) was undertaken at three different temperatures (35, 45 and 55 °C) under an applied upstream pressure of 3.5 bar.

Figure 7 represents a series of four fluorinated poly(ether imide)s prepared from four different fluorinated diamines namely, 4,4-bis[3'-trifluromethyl-4'(4"-aminobenzoxy)bezyl] biphenyl (BAQP), 1,4-bis[3'-trifluromethyl-4'(4"-aminobenzoxy)bezyl]benzene (BATP), 2,6-bis[3'-trifluromethyl-4'(4"-aminobenzoxy)bezyl]pyridine (BAPy) and 2,5-bis[3'-trifluromethyl-4'(4"-aminobenzoxy)bezyl] thiophene (BATh) using 4,4'-(hexafluoro-isopropylidene) diphthalic anhydride (6-FDA) as the aromatic dianhydride.

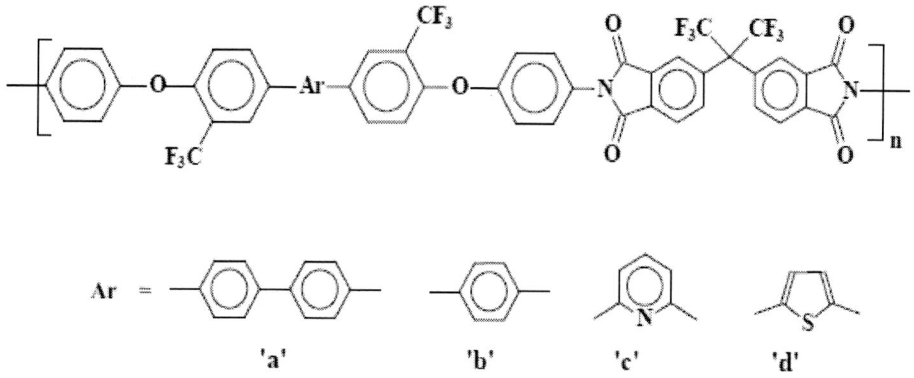

Figure 7. Molecular structure of poly(ether imide) series 'I'.

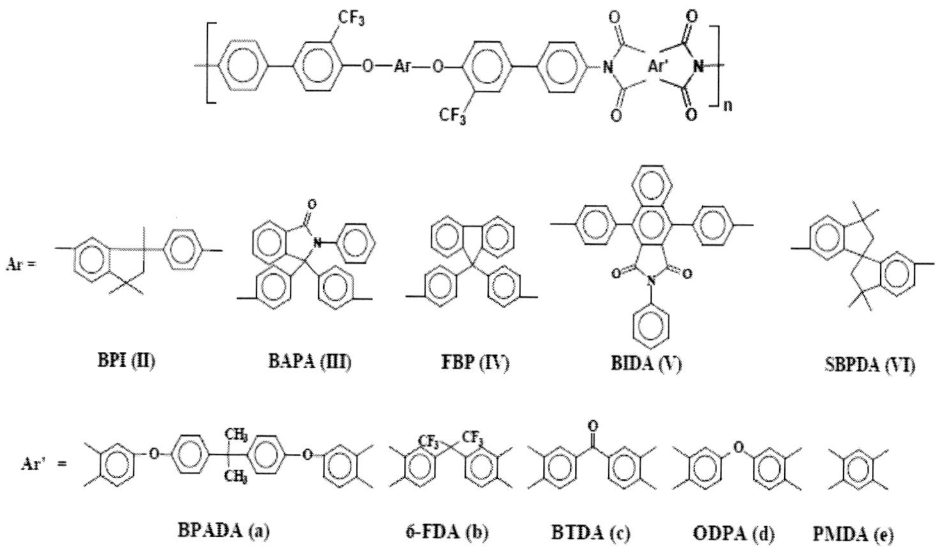

Figure 8. Molecular structure of poly(ether imide) series 'II', 'III', 'IV', 'V' and 'VI'.

Figure 8 represents another five series of polyimides based on five different cardo moiety containing diamines namely, 3-(4'-amino-3-trifluoromethyl-biphenyl-4-yloxy-phenyl)-5-(4'-amino-3-trifluoro methyl-biphenyl-4-yloxy)-1,1,3-trimehylindane (BPI), 3,3-*bis*-[4-{2'-trifluoromethyl 4'-(4"-aminophenyl) phenoxy} phenyl]-2-phenyl-2,3-dihydro-isoindole-1-one (BAPA), 4,4'-bis-((2'-trifluoromethyl-4'-(4"-aminophenyl) phenoxy)-9-fluorenylidene (FBP), 4,9-bis-[4-(4'-amino-3-trifluoromethyl-biphenyl-4-yloxy)-phenyl]-2-phenyl-benzo[f]-isoindole-1,3-dione (BIDA), and 6,6'-bis-[2"-trifluoromethyl 4"-(4'''-aminophenyl)phenoxy]-3,3,3',3'-tetramethyl-1,1'-spirobiindane (SBPDA) and, five aromatic dianhydrides, namely 4,4'-(4,4'-isopropylidenediphenoxy)bis(phthalic anhydride) (BPADA), 4,4'-(hexafluoro-isopropylidene) diphthalic anhydride (6-FDA), 3,3'4,4'-benzophenone tetracarboxylic acid dianhydride (BTDA), 4,4'-oxydiphthalic anhydride (ODPA) and benzene-1,2,4,5-tetracarboxylic dianhydride (PMDA).

A comparison of the results from all these synthesized polyimide membranes with two potentially used polyimide membranes, Matrimid[®] and Ultem[®], is shown in Figure 9 and Figure 10. For a better understanding the comparison of the CO_2/CH_4 and O_2/N_2 separation properties with the commercially used Matrimid[®] and Ultem[®] in respect to the latest upper boundary limit has been presented as a Robeson plot. The gas transport through 'Ia', 'Ib', 'Ic' and 'Id' revealed that incorporation of bond angle into polymer backbone along with an increase in the polarity makes the membrane more permeable to CO_2 (Table 9). As a result we were able to develop membranes with high throughput of CO_2, but at the cost of permselectivity of CO_2 over CH_4. The ideal permselectivity calculated were quiet below to that of Matrimid[®] and Ultem[®]. A new series of polymer, 'II' (bis-phenol indane based) was synthesized by reducing the intersegmental length and incorporating a cardo group in order to get higher throughput. Interestingly, among the five polyimides of this series ('IIa' to 'IIe') BPI-6FDA ('IIb') showed higher flux of CO_2 along with good selectivity which is comparable of Matrimid[®]. A further manipulation of free fractional volume has been done by choosing phenolphthalein anilide as another cardo group ('IIIa' to 'IIIe'). In addition to the reduced intersegmental length, planar and aromatic phenolphthalein anilide helps to optimize the inter chain distance by enhanced charge transfer complex. Accordingly, the permeability of CO_2 as well as the ideal permselectivity of CO_2 over CH_4 were further improved which were more than both Matrimid[®] and Ultem[®]. Later, we prepared two more polyimide series 'IVa' to 'IVe' (FBP based) and 'Va' to 'Ve' (BIDA based) to further manipulate the gas transport properties. But, we could not achieve better CO_2/CH_4 separation as that of polyimide series 'III'. However, we were able to increase the CO_2 permeability upto 71.3 Barrer for 'Vb' (BIDA-6FDA) and O_2/N_2 permselectivity upto 8.4 for 'Va' (BIDA-BPADA). From all these investigation we found that with increasing polarity the CO_2 permeability increases substantially with a loss of selectivity over CH_4 whereas the rigidity controls the slectivity. Therefore, we anticipated that if we can make a polymer with high rigidity along with lower or moderate polarity we can come up with better permselectivity (CO_2/CH_4) along with higher permeability of CO_2. Accordingly, we prepared another series based on rigid bis-spiro indane containing diamine ('VI'). We were surprised to find out that this series of polyimide showed excellent permselectivity for both CO_2/CH_4 (68.2, VIa) and O_2/N_2 (11.7, VIc) gas pairs and of course, with little lose of CO_2 permeability. However, we were able to achieve higher O_2 permeability (36.1 Barrer) for 'VIb' (SBPDA-6FDA).

Table 8. Ideal gas separation properties for O_2/N_2 gas pair

Polymer	P (O_2)	P (N_2)	α_P (O_2/N_2)	D (O_2)	D (N_2)	α_D (O_2/N_2)	Ref.
BAQP-6FDA (Ia)	17.08	3.11	5.5	6.02	1.87	3.2	[164]
BATP-6FDA (Ib)	15.17	2.87	5.3	6.19	2.12	2.9	[164]
BAPy-6FDA (Ic)	12.15	1.91	6.4	4.85	1.35	3.6	[164]
BATh-6FDA (Id)	11.65	1.78	6.6	5.57	1.43	3.9	[164]
BPI-BPADA (IIa)	10.95	2.08	5.26	7.94	5.86	1.35	[162]
BPI-6FDA (IIb)	14.98	2.44	6.14	11.97	7.54	1.59	[162]
BPI-BTDA (IIc)	7.98	1.32	6.05	5.23	3.67	1.43	[162]
BPI-ODPA (IId)	8.95	1.42	6.30	7.08	4.45	1.60	[162]
BPI-PMDA (IIe)	12.22	2.36	5.18	9.34	6.77	1.38	[162]
BAPA-BPADA (IIIa)	4.25	0.95	4.47	4.17	1.07	3.90	[163]
BAPA-6FDA (IIIb)	10.23	1.71	5.98	7.47	1.38	5.41	[163]
BAPA-BTDA (IIIc)	4.33	0.83	5.22	4.81	1.28	3.76	[163]
BAPA-ODPA (IIId)	4.22	0.96	4.40	5.28	1.22	4.33	[163]
BAPA-PMDA (IIIe)	7.62	1.20	6.35	6.20	1.88	3.30	[163]
FBP-BPADA (IVa)	7.01	1.19	5.89	6.99	3.34	2.09	[170]
FBP-6FDA (IVb)	13.46	2.06	6.53	13.11	4.98	2.63	[170]
FBP-BTDA (IVc)	6.24	0.95	6.57	8.22	4.11	2.00	[170]
FBP-ODPA (IVd)	7.86	1.22	6.44	8.08	3.61	2.24	[170]
BIDA-BPADA (Va)	10.32	1.23	8.39	11.44	1.95	5.87	[63]
BIDA-6FDA (Vb)	25.37	4.22	6.01	14.00	3.32	4.22	[63]
BIDA-BTDA (Vc)	6.99	0.98	7.13	6.30	1.59	3.96	[63]
BIDA-ODPA (Vd)	7.74	1.45	5.34	5.87	1.93	3.04	[63]
SBPDA-BPADA (VIa)	9.92	1.02	9.73	9.01	1.96	4.60	[62]
SBPDA-6FDA (VIb)	36.08	3.35	10.77	22.72	4.90	4.64	[62]
SBPDA-ODPA (VIc)	13.07	1.12	11.67	11.41	2.90	3.93	[62]

P = gas permeability coefficients measured at 35 ^{o}C and 3.5 atm.; D = Gas Diffusivity Constants, α_P = Ideal gas permselectivity values; α_D = Ideal gas diffusion selectivity values; 1 barrer = 10^{-10} cm^3 (STP) cm/cm^2 s cm Hg.

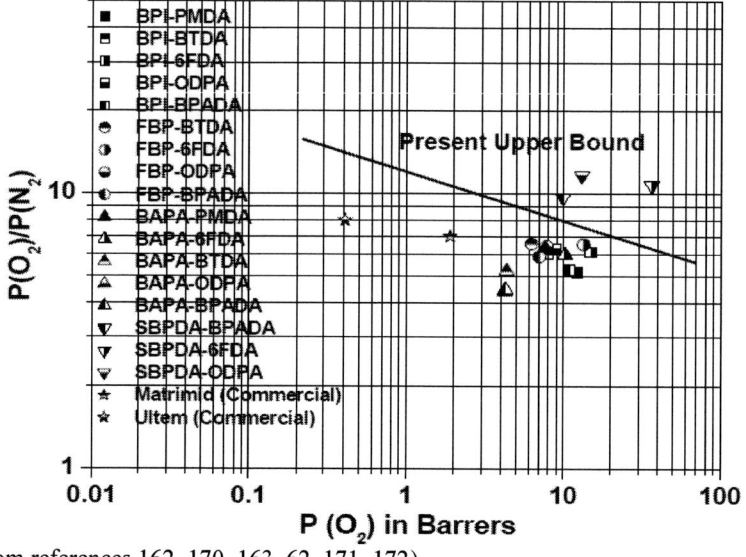

(Data taken from references 162, 170, 163, 62, 171, 172).

Figure 9. Robeson plot for a comparison of the O2/N2 separation properties of the investigated polymers with Matrimid® and Ultem® .

Table 9. Ideal gas separation properties for CO_2/CH_4 gas pair

Polymer	P (CO_2)	P (CH_4)	α_P (CO_2/CH_4)	D (CO_2)	D (CH_4)	α_D (CO_2/CH_4)	Ref.
BAQP-6FDA (Ia)	36.61	1.51	24.3	4.68	0.51	9.2	[164]
BATP-6FDA (Ib)	33.12	1.19	28.2	3.47	0.64	5.4	[164]
BAPy-6FDA (Ic)	51.92	1.95	26.6	2.18	0.31	7.0	[164]
BATh-6FDA (Id)	45.31	1.68	27.0	2.31	0.26	8.9	[164]
BPI-BPADA (IIa)	39.45	1.38	28.59	4.56	0.58	7.86	[162]
BPI-6FDA (IIb)	57.45	1.62	35.46	5.70	0.68	8.38	[162]
BPI-BTDA (IIc)	34.20	0.88	38.86	3.39	0.40	8.48	[162]
BPI-ODPA (IId)	35.78	0.98	36.51	4.24	0.49	8.65	[162]
BPI-PMDA (IIe)	44.68	1.52	29.39	4.83	0.64	7.55	[162]
BAPA-BPADA (IIIa)	16.61	0.84	19.77	1.96	0.55	3.56	[163]
BAPA-6FDA (IIIb)	53.85	1.01	53.32	3.79	0.92	4.12	[163]
BAPA-BTDA (IIIc)	17.09	0.62	27.57	1.46	1.24	1.18	[163]
BAPA-ODPA (IIId)	14.59	0.84	17.37	1.86	1.04	1.79	[163]
BAPA-PMDA (IIIe)	39.57	0.79	50.09	1.74	1.65	1.06	[163]
FBP-BPADA (IVa)	22.52	1.01	22.30	3.79	0.95	3.99	[170]
FBP-6FDA (IVb)	53.09	1.34	39.62	4.83	1.24	3.90	[170]
FBP-BTDA (IVc)	36.07	0.94	38.37	4.02	1.04	3.87	[170]
FBP-ODPA (IVd)	25.91	1.04	24.91	4.01	1.02	3.93	[170]
BIDA-BPADA (Va)	25.65	0.69	37.17	2.54	1.64	1.55	[63]
BIDA-6FDA (Vb)	71.32	1.99	35.84	3.28	1.48	2.22	[63]
BIDA-BTDA (Vc)	16.06	0.55	29.20	1.26	1.54	0.82	[63]
BIDA-ODPA (Vd)	16.99	0.67	25.36	1.85	1.63	1.14	[63]
SBPDA-BPADA (VIa)	23.87	0.35	68.2	3.01	1.83	1.65	[62]
SBPDA-6FDA (VIb)	52.98	1.21	43.79	6.25	3.79	1.65	[62]
SBPDA-ODPA (VIc)	22.64	0.41	55.22	3.85	2.28	1.69	[62]

P = gas permeability coefficients measured at 35 °C and 3.5 atm.; α = gas selectivity values; 1 barrer = 10^{-10} cm^3 (STP) cm/cm^2 s cm Hg.

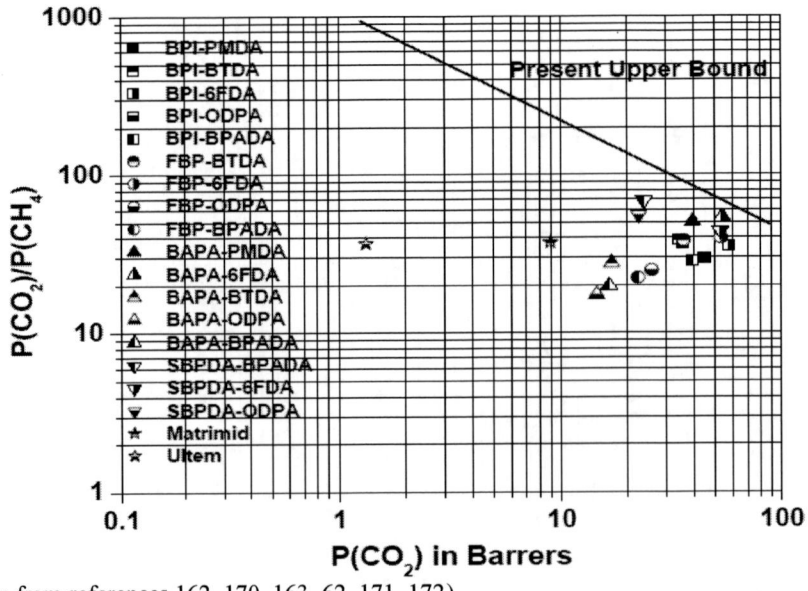

(Data taken from references 162, 170, 163, 62, 171, 172).

Figure 10. Robeson plot for a comparison of the CO_2/CH_4 separation properties of the investigated polymers with Matrimid® and Ultem® ® .

6.2.2. Polyimides in Pervaporation Applications

Recently, pervaporation (PV) has emerged itself as one of the most promising membrane technologies. It finds a wide range of potential applications starting from the well-established dehydration of organic compounds to the recovery of organic compounds from water and the separation of organic mixtures. PV is a single process which has often successfully competed with conventional processes like distillation, adsorption, and stripping, which are mostly used as reliable processes in industry [173]. In a pervaporation process a membrane is kept in contact with a liquid mixture, one component is preferentially removed from the mixture due to its higher affinity and/or quicker diffusivity through the membrane. Consequently, both the more permeable species in the permeate and the less permeable species in the feed can be concentrated. In order to ensure the continuous mass transport through the membrane, a very low absolute pressures (e.g., 133.3–400.0 Pa (1–3 mmHg)) is maintained to the downstream side of the membranes, removing all the molecules migrating to the face, and thus rendering a concentration difference across the membrane. Alternatively, sweeping gas in the downstream side of the membrane also serves the purpose in stead of the generally used vacuum operation. Sometime pervaporation is combined with distillation or with a chemical reactor known as hybrid systems for the supply of products suitable for further processing or waste disposal in accordance with environmental standards.

Separation and purification of mixtures are major problems in the chemical industry. The separation of aromatics/aliphatics is receiving more and more attention as, e.g. the benzene content in gasoline legally will be limited to less than 1% in the benzene content in Europe. Pervaporation appears to be very suitable for this separation problem, because a total removal of benzene is not necessary. Furthermore membrane separation processes require less energy compared with the conventional separation which is currently carried out by extractive distillation. Various types of polymers have been developed for the pervaporation of benzene/cyclohexane separation. Polyimides are known for their excellent thermal, chemical and physical properties and have been extensively investigated for the application in various separation processes. It is well known that the chain stiffness and packing density of polyimides influence selectivity and permeability [174, 175]. It has been found that using dianhydrides with -CF_3 groups like 6FDA the chain mobility and chain packing can be restricted and, therefore, significantly improved selectivity is reached. At the same time the permeability is improved due to the high free volume produced by the bulky -CF_3 groups. However, if high concentrations of aromatics are present in the feed mixture most of the 6FDA based polyimide membranes significantly swell. As a result, with increasing aromatic content in the feed the flux increases while the selectivity strongly decreases. Incorporation effects of fluorinated alkyl side groups into polyimide membranes on their pervaporation properties were investigated by S. Y. Kim et al. for several dilute aqueous organic solutions [176]. They assured that polyimide membranes containing fluorinated side groups showed high pervaporation selectivity and high permeating flux towards hydrophobic solvents. This can be explained in terms of their enhanced sorption/sorption selectivity and concentration-average diffusion coefficient/diffusivity selectivity towards organic solvent over water due to enhanced hydrophobicity and increased fractional free volume (FFV) obtained by the introduction of fluorinated side groups.

Recently, C. Staudt et al. studied fluorinated non-cross-linked and crosslinked 6FDA–4MPD/DABA (m:n) copolyimides prepared from 6FDA (4,4'-hexafluoroisopropylidene diphthalic anhydride), DABA (3,5-diamino benzoic acid) and 4-MPD (2,3,5,6-tetramethyl-

1,4-phenylene diamine) for pervaporation experiments (Figure 11) [177]. These 6FDA–4MPD/DABA (m:n) copolyimides consist of different ratios of the two diamines 4MPD and DABA leading to different numbers of cross-linking sites. They found 40% higher flux of benzene through the pre-treated 6FDA–4MPD/DABA 4:1 membrane crosslinked with ethylene glycol than the flux achieved with an untreated membrane.

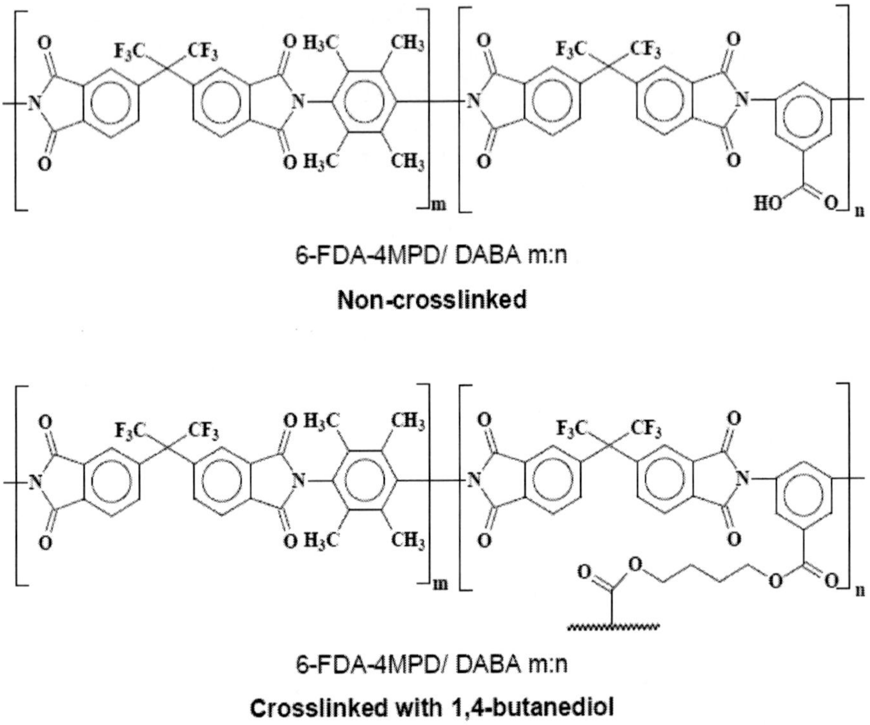

6-FDA-4MPD/ DABA m:n

Non-crosslinked

6-FDA-4MPD/ DABA m:n

Crosslinked with 1,4-butanediol

Figure 11. Molecular structures of crosslinked and non-crosslinked fluorinated polyimide.

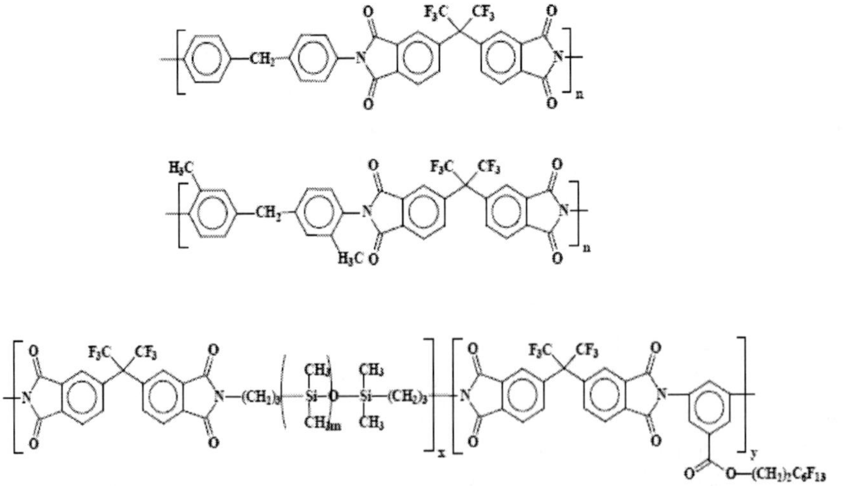

Figure 12. Molecular structures of few fluorinated polyimide membrane used for pervaporation.

Some more molecular structure of fluorinated polyimides used for pervaporation application within the last decade has been presented below (Figure 12) [178, 179].

6.3. Polyimides in Fuel Cell Applications

The technology of fuel cells was known to mankind as one of the oldest electrical energy conversion. Although the principle for fuel cell as energy conversion was discovered by Christian Friedrich Schönbein, the first fuel cell system was invented a quite back in the middle of the 19th century by Sir William Grove [180]. The development of fuel cells was lacking during their first century because of abundant primary energy sources. In the 20th century, a concern about the environmental consequences of fossil fuel use in production of electricity and for the propulsion of vehicles drives the development of polymer electrolyte membrane fuel cells (PEMFC). PEMFC have been recognized as a promising power-source for various applications due to their low emission and high conversion efficiencies [181-183]. Currently, various class of fuel cells are available based on the operating temperature (low-temperature and high-temperature fuel cells) and the electrolyte employed in the cell. Alkaline Fuel Cell (AFC), the Polymer Electrolyte Fuel Cell (PEMFC), the Direct Methanol Fuel Cell (DMFC) and the Phosphoric Acid Fuel Cell (PAFC) fall under the class of low-temperature fuel cells. Molten Carbonate Fuel Cell (MCFC) and the Solid Oxide Fuel Cell (SOFC) are high-temperature fuel cells operating at temperatures approx. 600±1000 °C.

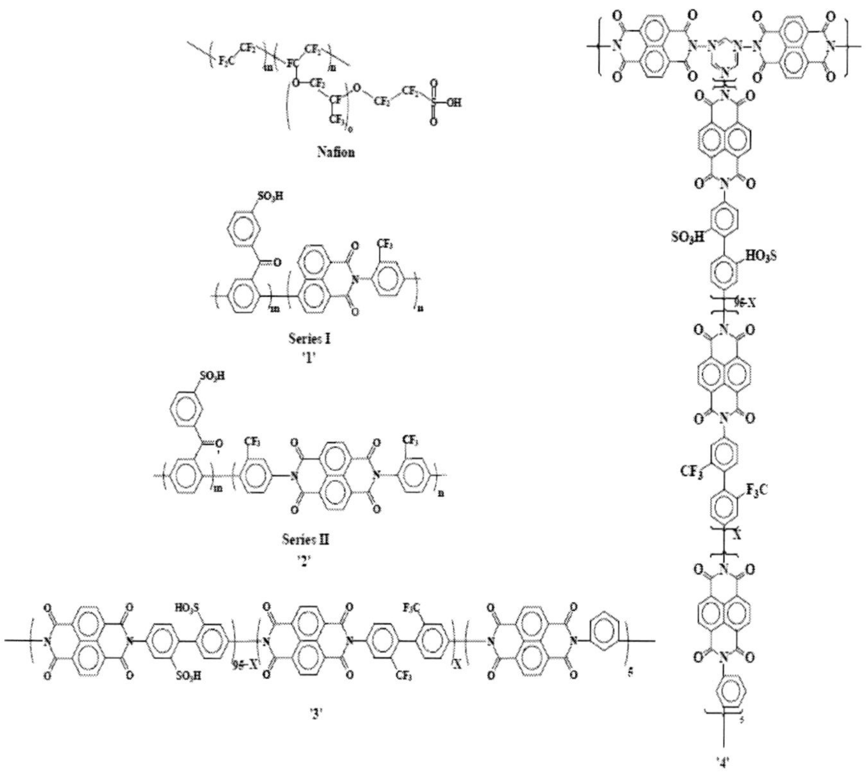

Figure 13. Fluorinated polyimides as PEMFC.

Fluorinated polyimides have attracted considerable interest due to their potential applicability as proton conductive polymers in polymer electrolyte membrane fuel cells (PEMFC). Nafion® (Dupont), a perfluorosulfonic acid polymer possessing good chemical resistance, oxidative stability, and good proton conductivity, is commercially used as a PEM in commercial systems. Nonetheless, it has some specific limitations, e.g., very high cost, high gas permeability, and property deterioration at high temperatures. All these discrepancies prompted an intense research in the development of new PEM materials with low cost and improved properties. A significant amount of development on sulfonated aromatic polyimides and poly(arylene ether)s have been done on potential PEM materials. However, most of them failed to meet the stringent requirements of high conductivity, with good mechanical properties and oxidative stability. Sulfonated aromatic polyimides have attracted significant attention in this regard, due to their superior solvent-resistance, excellent thermal stability and mechanical strength as well as good film forming ability. Few fluorinated polyimides with potential performance as PEMFC are depicted in Figure 13.

Zhang et al. synthesized novel sulfonated poly(arylene-co-naphthalimide)s containing -CF_3 substituents on the naphthalimide unit (Figure 13) by the copolymerization of sulfonated aromatic dichloride monomer, sodium 3-(2,5-dichlorobenzoyl) benzenesulfonate with N-(4-chloro-2-trifluoromethylphenyl)-5-chloro-1,8-naphthalimide ('1') and with bis(N-(4-chloro-2-trifluoromethyl phenyl)-1,4,5,8-naphthalimide ('2') [184]. Two series of sulfonated copolymers of high molecular weight and different degree of sulfonation were synthesized by Ni(0)-catalysed coupling of the monomers with different feed ratios. Introduction of the hydrophobic -CF_3 groups on the ortho-position of imido groups resulted in low water uptake and excellent oxidative stabilities of the copolymer membranes. Water uptake helps to dissociate the acid functionalities, thus improving the conductivity due to increased proton transport. However, too high water uptake leads to deterioration of the mechanical properties of the membranes. The $-CF_3$ groups functions in a beneficial way by protecting the polymer main chains from attack by water molecules containing highly oxidizing radical species. Among the two series of polymers, series I with 1,8-napthalimide units, exhibited higher oxidative and hydrolytic stabilities, as well as higher proton conductivity than series II, containing 1,4,5,8-napthalimide units. This was attributed to the difference in the chemical structure of the polymer chains with respect to the naphthalimide units and the number of carbonyl groups, which in turn determined the flexibility of the polymer chains. A copolymer of series I with 50% incorporation of sulfonated monomer exhibited the best combination of properties, with proton conductivity (2.6×10^{-1} S/cm at 80 °C), even higher than Nafion® 117 (1.5×10^{-1} S/cm at 80 °C). Such materials could be identified as potential alternative to Nafion® membranes.

Watanabe and coworkers synthesized a series of sulfonated polyimide copolymers containing bis(trifluoromethyl)biphenylene groups with the aim of improving the hydrolytic and oxidative of the polyimides while retaining their superior proton-conductive properties [185]. Copolymerization of 4,4'-diamino-2,2'-biphenyldisulfonic acid and 3,3'-bis(trifluoromethyl) benzidine was carried out using 1,4,5,8-naphthalenetetracarboxylic dianhydride along with a small amount of m-phenylene diamine ('3'). The synthesis of a similar polymer ('4') with a branched structure was also attempted using a small amount of a trifunctional monomer, melamine. The copolyimides (Figure 13) were soluble in organic solvents, prepared therefrom were used to study water-uptake, hydrolytic stability, mechanical properties and proton conductivity of the polymers. A typical trend was observed in the water-uptake behaviour of these polyimides as a function of the ion-exchange capacity and bis(trifluoromethyl) biphenylene content. Initially the water-uptake value decreased from 102% to 68% on increasing the bis(trifluoromethyl) biphenylene content (x), increasing to 73% with further increase in x, before decreasing to 13% at x = 60. The molecular bulkiness of the hydrophobic component led to confinement of water in the interchain spaces, and the observed effect. Introduction of $-CF_3$ groups was highly effective in improving the oxidative stability of the sulfonated polyimides due to the repulsion of water by fluorine. Branching also lowered the water uptake from 53 wt% to 31 wt% due to suppressed molecular motion, thus improving oxidative stability as well. Proton conductivities increased linearly with sulfonic acid content up to 80 °C, whereas above 100 °C, the conductivities started falling rapidly possibly due to evaporation of water from the membrane. The presence of fluorine-containing groups, along with branching was seen to improve the tensile properties under heated and humid conditions. The copolymer with 30% of the bis-(trifluoromethyl) biphenylene was found to be the optimum composition for achieving a good combination of oxidative stability, proton conductivity, and mechanical properties.

Recently, Yamazaki and Kawakami demonstrated that the graft copolymer membrane could realize a high proton conductivity and low gas permeability of the fuel. They synthesized a high proton conductive and low gas permeable sulfonated graft copolyimides (Figure 14) [186]. The grafted copolyimide membrane exhibited significantly higher proton conductivity in comparison to that of the Nafion®. The proton conductivity measured at 90 °C and at 98% relative humidity was 1.2 S/cm, which is more than 7 times to that of Nafion®. Furthermore, the polymeric membrane made from the polymer possessed an excellent gas barrier property. The oxygen permeability through the membrane was observed to be approximately one eighth to that of the Nafion®. The selectivity ratio calculated from the proton and oxygen transports of the graft copolyimide membrane was sixty six (66) times to that determined in Nafion®.

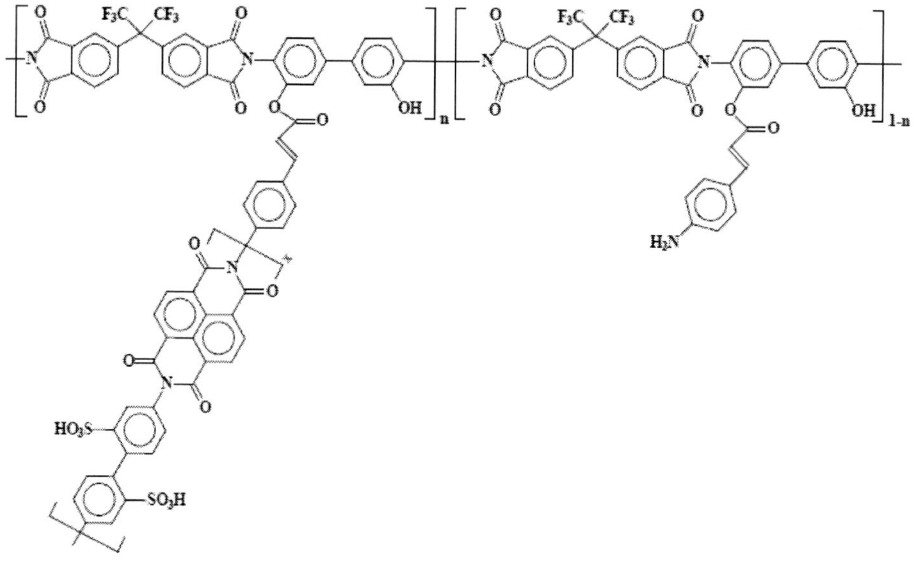

Figure 14. High proton conductive and low gas permeable sulfonated graft copolyimides.

SUMMARY

In the last decade several fluorinated polyimides have been prepared. The polymers showed improved solubility, exhibited outstanding optical transparency as well as excellent thermal, mechanical, and electric properties in comparisons to their non fluorinated analogues. The resulting polyimides have been investigated for low dielectric constant application and for membrane based applications. It was observed that in general the fluorinated polyimides showed low dielectric constant and reduction in dielectric constant depends on the fluorine content. However the use of fluorinated polymers as insulating material is not known due to the evolution of HF at high temperatures that considered a serious problem. Fluorinated polyimides are also considered to be one of the potential candidates for optical application as they have very good optical transparency in the visible and near infrared region. Many optical components and optical devices have been already developed using fluorinated polyimides and their optical applications are coming to extensive use in near future.

Fluorinated polymers are interesting membrane forming material and are potential candidate for membrane based separations. Fluorinated polyimides bearing –CF$_3$ groups showed very high gas permeability and their permselectivity values for a pair of gases also manipulated by designing new polyimides. They also used for pervaporation application for separation of close boiling solvent mixture and for dehydration. In spite of their various advantages for gas separation and pervaporation, polyimides are expensive materials with high production costs. Therefore, the use of fluorinated polyimides must identify suitable or value-added separations for specific applications to compensate or justify their use. Also, the use of molecular simulation along with experiments to create a comprehensive material-performance data base is needed to better understand the structure-property correlation.

Sulfonated polyimides are good candidate for fabrication of proton exchange membrane and their application in fuel cell, due to their superior solvent-resistance, excellent thermal stability and mechanical strength as well as good film forming ability. It is observed that sulfonated polyimides with pendent –CF$_3$ groups enhance the performance of membrane due their decreased water uptake and oxidative stability in contrast to their non fluorinated analogues. At the same time the high gas permeability and property deterioration at high temperatures of commercial Nafion® can be circumvented by proper selection of trifluoromethyl containing sulfonated polyimides.

This article covered a majority of the work in the recent decade involving the synthesis and property evaluation of novel –CF$_3$ based polyimides. The information is intended to assist polymer/material scientists to give a direction for further research with high-performance polyimides.

REFERENCES

[1]   Bogert, T. M.; Renshaw, R. R. *J. Am. Chem. Soc.* 1908, *30*, 1135-1144.

[2]   Edwards, W. M.; Robinson, I. M. *U. S. Patent.* 1955, *No. 2 710 853*.

[3]   Ghosh, M. K.; Mittal, K. L. In *Polyimides: Fundamentals and Applications;* Plastics engineering and 36; Marcel Dekker: New York, U.S., 1996.

[4]   Abadie, M. J. M.; Sillion, B. In *Polyimides and other high temperature polymers;* Elsevier: Amsterdam, 1991.

[5]   Feger, C.; Khojasteh, M. M.; McGrath, J. E. In *Polyimides: Chemistry and Characterization;* Elsevier: Amsterdam, 1994.

[6]   Feger, C.; Khojasteh, M. M.; Htoo, M. S. In *Advances in Polyimide Science and Technology;* Technomic: Lancaster, PA, 1993.

[7]   Stern, S. A. *J. Membr. Sci.* 1994, *94,* 1-65.

[8]   Hsiao, S. H.; Chung, C. L.; Lee, M. L. *J. Polym. Sci. Part A: Polym. Chem.* 2004, *42,* 1008-1017.

[9]   Vora, R. H.; Sawant, P. D.; Goh, S. H.; Vora, M. In *Polyimides and other high temperature polymers*; Mittal, K. L.; Ed.; VSP Publishers: Leiden, The Netherland, 2005; Vol. 3, pp 199-265.

[10]   Liaw, D. J.; Chang, F. C. *J. Polym. Sci. Part A: Polym. Chem.* 2004, *42,* 5766-5774.

[11]   Kute, V.; Banerjee, S. *Macromol. Chem. Phys.* 2003, *204,* 2105-2112.

[12]   White, D. M.; Takehoshi, T.; Williams, F. J.; Relles, H. M.; Donahue, P. F.; Klopfer, H. J.; Loucks, G. R.; Manello, J. S.; Mathews, R. O.; Schluens, R. W. *J. Polym. Sci. Polym. Chem. Ed.* 1981, *19,* 1635-1658.

[13]   Johnson, R. O.; Burlhis, H. S. *J. Polym. Sci. Polym. Symp.* 1983, *70,* 129-143.

[14]   Eastmond, G. C.; Paprotny, J.; Irwin, R. S. *Macromolecules,* 1996, *29,* 1382-1388.

[15]   Hasio, S. H.; Yang, C. P.; Yang, C. Y. *J. Polym. Sci. Part A: Polym. Chem.* 1997, *35,* 1487-1497.

[16]   Zeng, H. B.; Wang, Z. Y. *Macromolecules,* 2000, *33,* 4310-4312.

[17]   Hasio, S. H.; Li, C. T. *Macromolecules,* 1998, *31,* 7213-7217.

[18]   Korshak, V. V.; Vinogradova, S. V.; Vygodski, Y. S. *J. Macromol. Sci. Rev: Macromol. Chem.* 1974, *C11,* 45-142.

[19]   Coleman, M. R.; Koros, W. J. *J. Membr. Sci.* 1990, *50,* 285-297.

[20]   Tanaka, K.; Kita, H.; Okano, M.; Okamoto, K. I. *Polymer* 1992, *33,* 585-592.

[21]   Burma, M.; Fitch, J. W.; Cassidy, P. E. *J. Macromol. Sci., Rev. Macromol. Chem. Phys.* 1996, *C36,* 119-159.

[22]   Tanaka, K.; Kita, H.; Okamoto, K. I. *J. Polym. Sci., Part B: Polym. Phys.* 1993, *31,* 1127-1133.

[23]   Banerjee, S.; Madhra, M. K.; Salunke, A. K.; Jaiswal, D. K. *Polymer* 2003, *44,* 613-622.

[24]   Yang, C. P.; Chen, R. S.; Chen, K. H. *Colloid Polym. Sci.* 2003, *281,* 505-515.

[25]   Yang, S. Y.; Ge, Z. Y.;Yin, D, X.; Liu, J. G.; Li, Y. F.; Fan, L. *J. Polym. Sci. Part A: Polym. Chem.* 2004, *42,* 4143-4152.

[26]   Ando, S.; Sawada, T.; Inoue, Y. *Electron Lett.* 1993, *29,* 2143-2144.

[27]   Sonnett, J. M.; McCulloung, R. L.; Beeler, A. J.; Gannett, T. P. In *Advances in Polyimide Science and Technology,* Feger, C.; Khojasteh, M. M.; Htoos, M. S.; Eds.; Technomic Publishing Co.: Lancaster, 1993.

[28]   Takekoshi, T. In *Encyclopedia of Chemical Technology;* Kirk, Othmer; Ed.; John Wiley and Sons: New York, 1996; Vol. 19, pp. 813–837.

[29]   Kricheldorf, H. R. In *Progress in Polyimide Chemistry II*; Advances in Polymer Science and 141; Springer- Verlag: Berlin, New York,1999.

[30] Frost, L. W.; Kesse, J. *J. Appl. Polym. Sci.* 1964, *8*, 1039.

[31] Feger, C.; Khojasteh, M. M.; McGrath, J. E. In *Polyimides: Chemistry, materials, and characterization*, Elsevier: Amsterdam, 1989.

[32] Hsu, L. –C. (1991), Ph. D. thesis, University of Akron.

[33] Dhara, M. G.; Banerjee, S. *Prog. Polym. Sci.* 2010, *35*, 1022-1077.

[34] Synder, R. W.; Thomson, B.; Bartges, B.; Czeriwski, D.; Painter, P. C. *Macromolecules* 1989, *22*, 4166-4172.

[35] Scroog, C. E. In *Encyclopedia of Polymer Science and Technolog;*, Bikales, N. M.; Ed.; Interscience: New York, 1969; Vol. 11, p. 247.

[36] Cassidy, P. C.; Fawcett, N. C. In *Encyclopedia of Chemical Technology;* Wiley, New York, 1982, p. 704.

[37] Angelo, R. J.; Golike, R. C.; Tatum W. E.; Kreuz, J. A. In *Advances in Polyimide Science and Technology;* Weber, W. D.; Gupta, M. R.; Eds.; Plas. Eng., Brookfield: CT, 1985; p. 67.

[38] Wilson, D.; Stenzenberger, H. D.; Hergenrother, P. M. In *Polyimides*, Blackie and Son Ltd.: Glasgow and London, 1990.

[39] Johnston, J. C.; Meador, M. A. B.; Alston, W. B. (1987), *J. Polym. Sci. Part A: Polym. Chem.* 1987, *25*, 2175-2183.

[40] Sroog, C. E. *J. Polym. Sci. Macromol. Rev.* 1976, *11*, 161.

[41] Kailani, M. H.; Sung, C. S. P.; Haung, S. *Macromolecules* 1992, *25*, 3751-3757.

[42] Reynolds, D. W.; Cassidy, P. E.; Johnson, C. G.; Cqmerson, M. L. *J. Org. Chem.* 1990, *55*, 4448-4454.

[43] Seebach, D. *Angew chem.* 1990, *102*, 1363; Angew chem. Int. Ed. 1990, *29*, 1320-1367.

[44] Huheey, J. E. *J. phys. Chem.* 1965, *69*, 3284-3291.

[45] Yokozeki, A.; Baurer, S. H. *Top. Curr. Chem.* 1975, *53*, 71-119.

[46] Hougham, G.; Tesoroj, J. G.; Shawt, J. *Macromolecules*, 1994, *27*, 3642-3649.

[47] a) Ando, S.; Matsuura, T.; Sasaki, S. In *Fluoropolymers 2. Properties;* Chapter 14, Hougham, G.; Cassidy, P. E.; Johns, K.; Davidson, T.; Eds.; Plenum Publishers, New York, 1999. b) Hougham, G.; In *Fluoropolymers 2. Properties;* Chapter 13, Hougham, G.; Cassidy, P. E.; Johns, K.; Davidson, T.; Eds.; Plenum Publishers, New York, 1999.

[48] Fujita, T. *Prog. Phys. Org. Chem.* 1983, *14*, 75-113.

[49] Simons, J. H.; Mc Arthur, R.E. *Ind. Eng. Chem*, 1947, *39*, 364-367.

[50] Kitazume, T.; Ohnogi, T. *Synthesis*, 1988, 614-615.

[51] Kimoto, H.; Cohen, L. A. *J. Org. Chem.* 1979, *44*, 2902-2906.

[52] Verbicky, J. W. In *Encyclopedia of Polymer Science and Engineering;* Mark, H. F.; Bikales, N. M.; Overberger, C. G.; Menges, G.; Eds.; Wiley: New York, 1988; Vol. 12, p 364.

[53] Cassidy, P. E.; Aminabhavi, T. M.; Farley, J. M. *J. Macromol. Sci. Rev.,* 1989, *C29*, 365-372.

[54] Xiao, Y. C.; Low, B. T.; Hosseini, S. S.; Chung, T. S.; Paul, D. R. *Prog. Polym. Sci.,* 2009, *34*, 561-580.

[55] Wang, Y.; Goh, S. H.; Chung, T. S.; Peng, N. *J. Membr. Sci.* 2009, *326*, 222-233.

[56] Low, B. T.; Xiao, Y. C.; Chung, T. S. *Polymer* 2009, *50*, 3250-3258.

[57] Banerjee, S.; Madhra, M. K.; Salunke, A. K.; Maier, G. *J. Polym. Sci. Part A: Polym. Chem.* 2002, *40*, 1016-1027.

[58] Madhra, M. K.; Salunke, A. K.; Banerjee, S.; Prabha, S. *Macromol. Chem. Phys.* 2002, *203*, 1238-1248.

[59] Kute, V.; Banerjee, S. *J. Appl. Polym. Sci.* 2007, *103*, 3025-3044.

[60] Maji, S.; Sen, S. K.; Dasgupta, B.; Chatterjee, S.; Banerjee, S. *Polym. Adv. Technol.* 2009, *20*, 384-392.

[61] Dasgupta, B.; Sen, S. K.; Maji, S.; Chatterjee, S.; Banerjee, S. *J. Appl. Polym. Sci.* 2009, *112*, 3640-3651.

[62] Sen, S. K.; Banerjee, S. *J. Membr. Sci.* 2010, *365*, 329-340.

[63] Sen, S. K.; Banerjee, S. (Unpublished result).

[64] Ghosh, A.; Banerjee, S. *High Perform. Polym.* 2009, *21*, 173-186.

[65] Hsiao, S. -H.; Guo, W.; Chung, C. -L.; Chen, W. -T. *Eur. Polym. J.* 2010, *46*, 1878-1890.

[66] Ma, T.; Zhang, S.; Li, Y.; Yang, F.; Gong, C.; Zhao, J. *J. Fluorine Chem.* 2010, *131*, 724-730.

[67] Zhao, X.; Wang, C.; Chen, L.; Zhu, M. *Colloid Polym. Sci.* 2009, *287*, 1331-1337.

[68] [Wang, C. -Y.; Li, G.; Jiang, J. -M. *Polymer* 2009, *50*, 1709-1716.

[69] Liu, Y.; Zhang, Y.; Guan, S.; Li, L.; Jiang, Z. *Polymer* 2008, *49*, 5439-5445.

[70] Yang, F.; Li, Y.; Ma, T.; Bu, Q.; Zhang, S. *J. Fluorine Chem.* 2010, *131*, 767-775.

[71] Velez-Herrera, P.; Ishida, H. *J. Fluorine Chem.* 2009, *130*, 573-580.

[72] Jang, W.; Shin, D.; Choi, S.; Park, S.; Han, H. *Polymer* 2007, *48*, 2130-2143.

[73] Oishi, Y.; Kikuchi, N.; Mori, K.; Ando, S.; Maeda, K. *J. Photopolym. Sci. Tec.* 2002, *15*, 213-214.

[74] Wang, K.; Fan, L.; Liu, J. -G.; Zhan, M. -S.; Yang, S. -Y. *J. Appl. Polym. Sci.* 2008, *107*, 2126-2135.

[75] Shao, Y.; Li, Y. -F.; Zhao, X.; Wang, X. -L.; Ma, T.; Yang, F. -C. *J. Polym. Sci. Part A: Polym. Chem.* 2006, *44*, 6836-6846.

[76] Li, Z. X.; Fan, L.; Ge, Z. Y.; Wu, J. T.; Yang, S. Y. *J. Polym. Sci. Part A: Polym. Chem.* 2003, *41*, 1831-1840.

[77] Kim, J. -H.; Lee, S. -B.; Kim, S. Y. *J. Appl. Polym. Sci.* 2000, *77*, 2756-2767.

[78] Yang, C. -P.; Hsiao, S. -H.; Tsai, C. -Y.; Liou, G. -S. *J. Polym. Sci. Part A: Polym. Chem.* 2004, *42*, 2416-2431.

[79] Xie, K.; Zhang, S. Y.; Liu, J. G.; He, M. H.; Yang, S. Y. *J. Polym. Sci., Part A: Polym. Chem.* 2001, 39, 2581-2590.

[80] Yang, C. -P.; Chen, R. -S.; Chiang, H. -C. *Polym. J.* 2003, *35*, 662-670.

[81] Yang, C. -P.; Hsiao, S. -H.; Chen, K. -H. *Polymer* 2002, *43*, 5095-5104.

[82] Yang, C. -P., Chen, R. -S. and Chen, K. -H. (2005), J. Appl. Polym. Sci., 95, 922-935.

[83] Yang, C. -P.; Chen, R. -S.; Chen, K. -H. *J. Polym. Sci., Part A: Polym. Chem.* 2003, *41*, 922-938.

[84] Yang, C. -P.; Chiang, H. -C. *Colloid Polym. Sci.* 2004, *282*, 1347-1358.

[85] Liu, Y.; Xing, Y.; Zhang, Y.; Guan, S.; Zhang, H.; Wang, Y.; Wang, Y.; Jiang, Z. *J. Polym. Sci. Part A: Polym. Chem.* 2010, *48*, 3281-3289.

[86] Li, H.; Liu, J.; Wang, K.; Fan, L.; Yang, S. *Polymer* 2006, *47*, 1443-1450.

[87]   Yang, S. Y.; Ge, Z. Y.; Yin, D. X.; Liu, J. G.; Li, Y. F.; Fan, L. *J. Polym. Sci. Part A: Polym. Chem.* 2004, *42*, 4143-4152.

[88]   Myung, B. Y.; Ahn, C. J.; Yoon, T. H. *Polymer* 2004, *45*, 3185-3193.

[89]   Myung, B. Y.; Kim, J. J.; Yoon, T. H. *J. Polym. Sci. Part A: Polym. Chem.* 2002, *40*, 4217-4227.

[90]   Yang, C. -P.; Su, Y. -Y.; Hsu, M. -Y. *Polym. J.* 2006, *38*, 132-144.

[91]   Takekoshi, T. In *Encyclopedia of Chemical Technology*, 4th Ed.; Kirk – Othmer; Ed.; John Wiley and Sons; NY, 1996; Vol. 19, pp 813-837.

[92]   De Visser A. C.; Gregonis, D. E.; Driessen, A. A. *Makromol. Chem.* 1978, *179*, 1855-1859.

[93]   Feld, W. A.; Ramalingam, B.; Harris F. W. *J. Polym. Sci. Part A: Polym. Chem.* 1983, 21, 319-328.

[94]   Adrova, N.; Bessonov, M.; Laius, L. A.; Rudakov, A. P. In *Polyimides: A new class of heat resistant polymers,* IPST Press: Jerusalem, 1969.

[95]   Dine-Hart, R. A.; Wright, W. W. *Makromol. Chem.* 1972, *153*, 237-254.

[96]   Huang, W.; Tong, Y.; Xu, J.; Ding, M. *J. Polym. Sci., Part A: Polym. Chem.* 1997, *35*, 143-151.

[97]   Bell, V. L.; Stump, B. L.; Gager, H. *J. Polym Sci. Part A: Polym. Chem.* 1976, *14*, 2275-2291.

[98]   Lin, T.; Stickney, K. W.; Rogers, M.; Riffle, J. S.; McGrath, J. E.; Marand, H.; Yu, T. H.; Davis, R. M. *Polymer* 1993, *34*, 772-777.

[99]   Cassidy, P. E.; Aminabhavi, T. M.; Farley, J. M. *J. Macromol. Sci. Rev.* 1989, *C, 29*, 365-429.

[100] Korshak, V. V.; Rusanov, A. L. *Bull. Acad. Sci. USSR Div. Chem. Sci.* 1968, *17*, 2298.

[101] Rusanov, A. L.; Komarova, L. G.; Sheveleva, T. S.; Prigozhina, M. P.; Shevelev, S. A.; Dutov, M. D.; Vatsadze, L. A.; Serushkina, O. V. *React. Funct. Polym.* 1996, *30*, 279-292.

[102] Rusanov, A. L.; Shifrina, Z. B. *High Perform. Polym.* 1993, *5*, 107-121.

[103] Harris, F. W.; Norris, S. O. *J. Polym. Sci. Polym. Chem. Ed.* 1973, *11*, 2143-2151.

[104] Harris, F. W.; Hsu, S. L. -C. *High Perform. Polym.*, 1989, *1*, 3-16.

[105] Giesa, R.; Keller, U.; Eiselt, P.; Schmidt, H. -W. *J. Polym. Sci. Part A: Polym. Chem.* 1993, *31*, 141-151.

[106] Imai, Y.; Maldar, N. N.; Kakimoto, M. -A. *J. Polym. Sci. Part A: Polym. Chem.* 1984, *22*, 2189-2196.

[107] Akutsu, F.; Kataoka, T.; Shimizu, H.; Naruchi, K.; Miura, M. *Macromol. Chem. Rapid Commun.* (1994), *15*, 411-415.

[108] Spiliopoulos, I. K.; Mikroyannidis, J. A. *Macromolecules* 1996, *29*, 5313-5319.

[109] Lozano, A. E.; de la Campa, J. G.; de Abajo, J.; Preston, J. *Polymer* 1994, *35*, 872-877.

[110] Kricheldorf, H. R.; Schwarz, G.; Volker, E. Makromol. *Chem. Rapid. Commun.* 1989, *10*, 243-248.

[111] Chung, I. S.; Kim, S. Y. *Macromolecules.* 2000, 33, 3190-3193.

[112] Wang, X.; Li, Y.; Gong, C.; Zhang, S.; Ma, T. *J. Appl. Polym. Sci.* 2007, *104*, 212-219.

[113] Maier, G.; *Prog. Polym. Sci.* 2001, 26, 3-65.

[114] Licari, J.; Hughes, L. A. In *Handbook of Polymer Coating for Electronics;* Chemistry, Technology, and Applications, 2nd ed.; Noyes Data: Park Ridge, New Jersey, 1990; p 1.

[115] Gain, O.; Boiteux, G.; Seytre, G.; Garapon, J.; Sillion, B. *Sixth International Conference on Dielectric Materials, Measurements and Applications.* 1992, 401.

[116] Ukishima, S.; Iijima, M.; Sato, M.; Takahashi, Y.; Fukada, E. *Thin Solid Films* 1997, *308*, 475-479.

[117] Zhang, X.; You, L.; Dabral, S.; Chiang, C.; Yaney, D. S.; Joshi, R. V.; Yang, G. R.; Lu, T. M.; McDonald, J. F. *Planarizing Techniques for Parylene as an Interlayer Dielectric, Proc. VMIV.* 1993, 168.

[118] Banerjee, S.; Madhra, M. K.; Kute, V. *J. Appl. Polym.* Sci. 2004, *93*, 821-832.

[119] Yang, C. P.; Su, Y. Y.; Chen, Y. C. *Eur. Polym. J.* 2006, *42*, 721-732.

[120] Zhou, H.; Liu, J.; Qian, Z.; Zhang, S.; Yang, S. *J. Polym. Sci. Part A: Polym. Chem.* 2001, *39*, 2404-2413.

[121] Hsiao, S. H.; Yang, C. P.; Chung, C. L. *J. Polym. Sci. Part A: Polym. Chem.* 2003, *41*, 2001-2018.

[122] Yang, C. P.; Hsiao, S. H; Chung, C. L. *Polym. Int.* 2005, *54*, 716-724.

[123] Chung, C. L.; Hsiao, S. H. *Polymer* 2008, *49*, 2476-2485.

[124] Eftekhari, A.; St.Clair, A. K.; Stoakley, D. M.; Kuppa, S.; Singh, J. J. *J. Polym. Mater. Sci. Eng.* 1992, *66*, 279-280.

[125] Hougham, G.; Tesoro, G.; Viehbeck, A. *Macromolecules* 1996, *29*, 3453-3456.

[126] Tao, L.; Yang, H.; Liu, J.; Fan, L.; Yang, S. *Polymer.* 2009, *50*, 6009-6018.

[127] Ge, Z.; Fan, L.; Yang, S. *Eur. Polym. J.* 2008, *44*, 1252-1260.

[128] Hedrick, J. L.; Carter, K. R.; Labadie, J. W.; Miller, R. D.; Volksen, W; Hawker, C. J.; Yoon, D. Y.; Russell, T. P.; McGrath, J. E.; Briber, R. M. *Adv. Polym. Sci.* 1998, *141*, 1-43.

[129] Carter, K. R.; DiPietro, R. A.; Sanchez, M. I.; Swanson, S. A. *Chem. Mater.* 2001, *13*, 213-221.

[130] Chen, Y. W.; Wang, W. C.; Yu, W. H.; Kang, E. T.; Neoh, K. G.; Vora, R. H.; Ong, C. K.; Chen, L. F. *J. Mater. Chem.* 2004, *14*, 1406-1412.

[131] Hedrick, J. L.; Carter, K. R.; Richter, R.; Miller, R. D.; Russell, T. P. *Chem. Mater.* 1998, *10*, 39-49.

[132] Fu, G. D.; Wang, W. C.; Li, S.; Kang, E. T.; Neoh, K. G.; Tseng, W. T.; Liaw, D. J. *J. Mater. Chem.* 2003, *13*, 2150-2156.

[133] Wang, W. C.; Vora, R. H.; Kang, E. T.; Neoh, K. G.; Ong, C. K.; Chen, L. F. *Adv. Mater.* 2004, *16*, 54-57.

[134] Ando, S. *J. Photopolym. Sci. Technol.* 2004, *17*, 219-232.

[135] Han, K.; Lee, H. J.; Rhee, T. H., *J. Appl. Polym. Sci.* 1999, *74*, 107-112.

[136] Han, K. *Korean polym. J.* 2000, *8*, 165-171.

[137] Dautzenberg, F. M.; Mukherjee, M. *Chem. Eng. Sci.* 2001, *56*, 251-267.

[138] Baker, R. W. *Ind. Eng. Chem. Res.* 2002, *41*, 1393-1411.

[139] Baker, R. W. *Gas Separation;* In *Membrane Technology and Applications*; 2nd ed.; John Wiley: Chichester, UK, 2004; pp 287-335.

[140] Strathmann, H. *AIChE J.* 2001, *47*, 1077-1087.

[141] Stannett, V. *J. Membr. Sci.* 1978, *3*, 97-115.

[142] Boddeker, K. W. *J. Membr. Sci.* 1995, *100*, 1-3.

[143] Spillman, R. W. *Chem. Eng. Prog.* 1989, *85,* 41-62.

[144] Mazur, W. H.; Chan, M. C.; *Chem. Eng. Prog.* 1982, *78,* 38-43.

[145] Pinnau, I; Freeman, B. D. *ACS Symposium Series* 1999, *744,* 1-22.

[146] Koros, W. J.; Chern, R. T. In *Handbook of Separation Process Technology;* Rousseau, R. W.; Ed.; Wiley-Interscience: New York, 1987; pp. 862-953.

[147] Ghosal, K.; Freeman, B. D. *Polym. Adv. Technol.* 1993, *5,* 673-697.

[148] Schell, W. J. *J. Membr. Sci.* 1985, *22,* 217-224.

[149] Semenova, S. I. *J. Membr. Sci.* 2004, *231,* 189-207.

[150] Ockwig, N. W.; Nenoff, T. M. *Chem. Rev.* 2007, *107,* 4078-4110.

[151] Alentiev, A.Yu.; Loza, K. A.; Yampolskii, Yu. P. *J. Membr. Sci.* 2000, *167,* 91-106.

[152] Robeson, L. M.; Smith, C. D.; Langsam, M. *J. Membr. Sci.* 1997, *132,* 33-54.

[153] Park, J. Y.; Paul, D. R. *J. Membr. Sci.* 1997, *125,* 23-39.

[154] Koros, W. J.; Fleming, G. K. *J. Membr. Sci.* 1993, *83,* 1-80.

[155] Maier, G. *Angew. Chem. Int. Ed.* 1998, *37,* 2960-2974.

[156] Paul, D. R.; Yampolskii, Yu. P. *Polymeric Gas Separation Membranes;* CRC Press: Boca Raton, FL, 1994; pp 17.

[157] Kesting, R. E.; Fritzsch, A. K. In *Polymeric Gas Separation Membranes*; John Wiley and Sons: New York, 1993.

[158] Yampolskii, Y. P.; Shishatskii, S.; Alentiev, A. Y.; Loza, K. *J. Membr. Sci.* 1998, *148,* 59-69.

[159] Maier, G.; Wolf, M.; Bleha, M.; Pientka, Z. *J. Membr. Sci.* 1998, *143,*105-113.

[160] Maier, G.; Wolf, M.; Bleha, M.; Pientka, Z. *J. Membr. Sci.* 1998, *143,*115-123.

[161] St. Clair, A. K.; St. Clair T. L.; Slemp, W. W. In *Recent Advances in Polyimide Science and Technology;* Weber, W. D.; Gupta, M. R. Eds.; Society of Plastic Engineers; Poughkeepsie: New York, 1987, pp. 16.

[162] Dasgupta, B.; Sen, S. K.; Banerjee, S. *J. Membr. Sci.* 2009, *345,* 249-256.

[163] Sen, S. K.; Banerjee, S. *J. Membr. Sci.* 2010, *350,* 53-61.

[164] Sen, S. K.; Dasgupta, B.; Banerjee, S. *J. Membr. Sci.* 2009, *343,* 97-103.

[165] Dasgupta, B.; Sen, S. K.; Banerjee, S. *Mat. Sci. Eng. B.* 2010, *168,* 30-35.

[166] Banerjee, S.; Maier, G.; Dannenberg, C.; Spinger, J. *J. Membr. Sci.* 2004, *229,* 63-71.

[167] Xu, Z.; Dannenberg, C.; Springer, J.; Banerjee, S.; Maier, G. *J. Membr. Sci.* 2002, *205,* 23-31.

[168] Xu, Z.; Dannenberg, C.; Springer, J.; Banerjee, S.; Maier, G. *Chem. Mater.* 2002, *14,* 3271-3276.

[169] Ghosh, A.; Sen, S. K.; Dasgupta, B.; Banerjee, S.; Voit, B. *J. Membr. Sci.* 2010, *364,* 211-218.

[170] Dasgupta, B.; Banerjee, S. *J. Membr. Sci.* 2010, *362,* 58-67.

[171] Bos, A.; Pu¨nt, I. G. M.; Wessling, M.; Strathmann, H. *Sep. Purif. Technol.* 1998, *14,* 27-39.

[172] Xia, J.; Liu, S.; Pallathadka, P. K.; Chng, M. L.; Chung, T. S. *Ind. Eng. Chem. Res.* 2010, *49,* 12014-12021.

[173] Lipnizki, F.; Field, R.W.; Ten, P.K. *J. Membr. Sci. 1999, 153,* 183-210.

[174] Stern, S. A.; Mi, Y.; Yamamoto, H.; St. Clair, A. K. *J. Polym. Sci., Part B: Polym. Phys.* 1989, *27,* 1887-1909.

[175] Kim, T. H.; Koros, W. J.; Husk, G. R.; O'Brien, K. C. *J. Membr. Sci.* 1988, *37,* 45-62.

[176] Kim, J. H.; Chang, B. J.; Lee, S. B.; Kim, S. Y. *J. Membr. Sci.* 2000, *169,* 185-191.

[177] Katarzynski, D.; Staudt, C. *J. Membr. Sci.* 2010, *348,* 84-90.

[178] Wang, L.; Zhao, Z.; Li, J.; Chen, C. *Eur. Polym. J.* 2006, *142,* 1266-1272.

[179] Chang, B. J.; Chang, Y. H.; Kim, D. K.; Kim, J. H.; Lee, S. B. *J. Membr. Sci.* 2005, *248,* 99-107.

[180] Bossel, U. (2000), Eur. Fuel Cell Forum: Oberrohrdorf.

[181] Souzy, R.; Ameduri, B. *Prog. Polym. Sci.* 2005, *30,* 644-685.

[182] Carrette, L.; Friedrich, K. A.; Stimming, U. *Fuel Cells.* 2001, *1,* 5-39.

[183] Steele, B. C. H.; Heinzel, A. *Nature* 2001, *414,* 345-352.

[184] Qiu, Z.; Wu, S.; Li, Z.; Zhang, S.; Xing, W.; Liu, C. *Macromolecules.* 2006, *39,* 6425-6432.

[185] Miyatake, K.; Zhou, H.; Matsuo, T.; Uchida, H.; Watanabe, M. *Macromolecules,* 2004, *37,* 4961-4966.

[186] Yamazaki, K.; Kawakami, H. *Macromolecules.* 2010, *43,* 7185-7191.

In: Polymer Synthesis
Editor: E. Kowsari

Chapter 5

SYNTHESIS AND APPLICATIONS OF HYPERBRANCHED POLYMERS

*E. Kowsari**

Department of Chemistry, Amirkabir University of Technology,
Tehran, Iran,

ABSTRACT

With the rapid development of synthetic chemistry, many new molecular structures can be designed to investigate the role of polymer topology on the physical and chemical properties of macromolecules in traditional, block, hyperbranch, and, in particular, dendritic copolymers. Hyperbranched polymers are highly branched macromolecules with three-dimensional dentritic architectures. Due to their unique physical and chemical properties and potential applications in various fields from drug-delivery to coatings, interest in hyperbranched polymers is rapidly growing, as confirmed by the increasing number of publications. This chapter reviews the synthesis, modifications, and applications of hyperbranched and hyper grafted polymers, focusing on the recently developed novel synthetic strategies for hyperbranched polymers.

1. INTRODUCTION

Hyperbranched polymers are highly branched macromolecules with three-dimensional dentritic architectures. Due to their unique physical and chemical properties and potential applications in various, fields from drug-delivery to coatings, interest in hyperbranched polymers is experiencing rapid growth, as confirmed by the increasing number of publications. As the fourth major class of polymer architecture, coming after traditional types which include linear, cross-linked, and branched architectures, dendritic architecture consists of six subclasses: (a) dendrons and dendrimers, (b) linear-dendritic hybrids, (c) dendri grafts

---

* Department of Chemistry, Amirkabir University of Technology, No. 424, Hafez Avenue, 1591634311, Tehran, Iran, E-mail address: kowsarie@aut.ac.ir. Corresponding author: Fax: +98 (21)64542762, Tel.: +98 (21) 64542769.

or dendronized polymers, (d) hyperbranched polymers, (e) multi-arm star polymers, (f) hypergrafts or hypergrafted polymers. The first three subclasses exhibit perfect structures with a degree of branching (DB) of 1.0, while the latter three exhibit a random branched structure. Dendrons and dendrimers with high regularity and controlled molecular weight are prepared step by step via convergent and divergent approaches. A linear polymer linked with side dendrons is known as dendronized polymer. Dendronized polymers can be obtained by direct polymerization of dendritic macromonomers or by attaching dendrons to a linear polymeric core. The development of hyperbranched polymers is a rapidly expanding field in the area of macromolecular science. Hyperbranched polymers can be prepared by means of single monomer methodology (SMM) and double-monomer methodology (DMM). In SMM, the polymerization of an AB_n or latent ABn monomer leads to hyperbranched macromolecules. SMM consists of at least four components: (1) polycondensation of AB_n monomers; (2) self-condensing vinyl polymerization; (3) self-condensing ring-opening polymerization; (4) proton-transfer polymerization. In DMM, direct polymerization of two suitable monomers or a monomer pair gives rise to hyperbranched polymers. A classical example of DMM, the polymerization of A_2 and B_n (n >2) monomers, is well known. Recently, a novel DMM based on the in situ formation of AB_n intermediates from specific monomer pairs has been developed. This form of DMM is designated as 'couple-monomer methodology' (CMM) to clearly represent the method of polymerization. Many commercially available chemicals can be used as the monomers in these systems, which should extend the availability and accessibility of hyperbranched polymers with various new end groups, architectures and properties [1]. This review highlights some of the notable examples in the synthesis of hyperbranched polymers and some of the key advances that have been made in the application of these hyperbranched materials in the areas of material property modifications and in high value technologies.

2. DISSCUSSION

2.1. Synthesis of Hyperbranch Polymers and their Applications in Drug Delivery System and Medicine

The design and synthesis of hyperbranched molecules, which can be imaged in ViVo using ^{19}F MRI in under 10 min, have been demonstrated by K.J. Thurecht and coworkers [2]. The hyperbranched polymers were synthesized using a variant of published RAFT routes (Figure1). Except for the mannose-derivative (P4), the particles were in the range of 10 nm. This suggests they exist as discrete macromolecules in solution while P4 forms agglomerated particles.

Such hyperbranched polymers hold promise as new generation tracking and targeting MRI contrast agents; further tests on the in ViVo efficiency of a variety of specific cell-targeted polymers are currently underway in our laboratory.

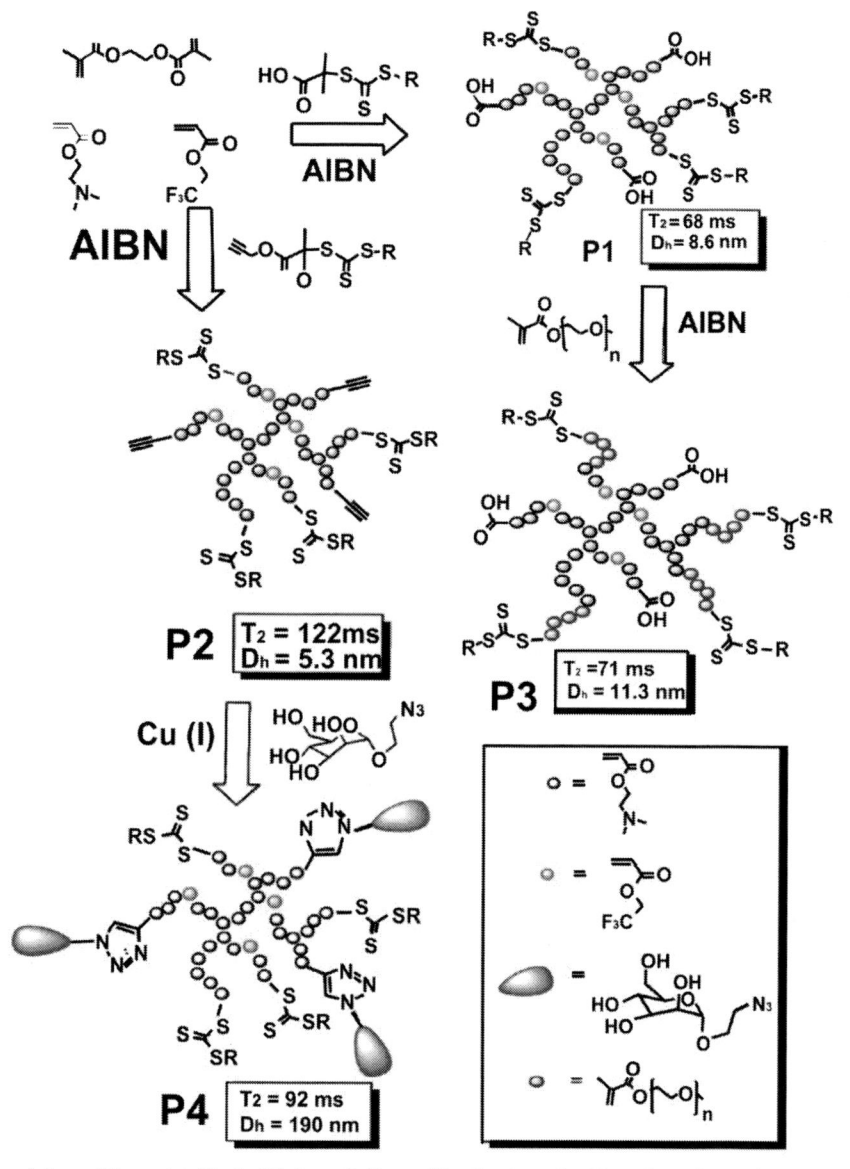

(Reproduced from Thurecht, K. J.; Blakey, I; Peng, H.; Squires,O.; Hsu, S.; Alexander, C.; Whittaker, A. K. *J. Am. Chem. Soc*.2010, *132*, 5336.., Copyright (2010), with permeation from American Chemical Society)).

Figure 1. Synthetic Scheme and Physical Properties of Polymers Having Acid (P1), Alkyne (P2), and Mannose (P4) End Groups.

Figure 2 shows an overlay of the ^{1}H and ^{19}F images 2 h after P1 injection. In the frontal slice, the bladder and intestine are clearly delineated while a slice through the transverse direction shows the spinal column and the bladder. In both cases, the ^{19}F signal was clearly observed, predominantly within the bladder. This suggests that they are able to successfully image these polymers *in ViVo* and that the blood-borne polymer can be excreted by kidneys. This is the first example of functionalizable copolymers successfully engineered for *in ViVo* ^{19}F MRI.

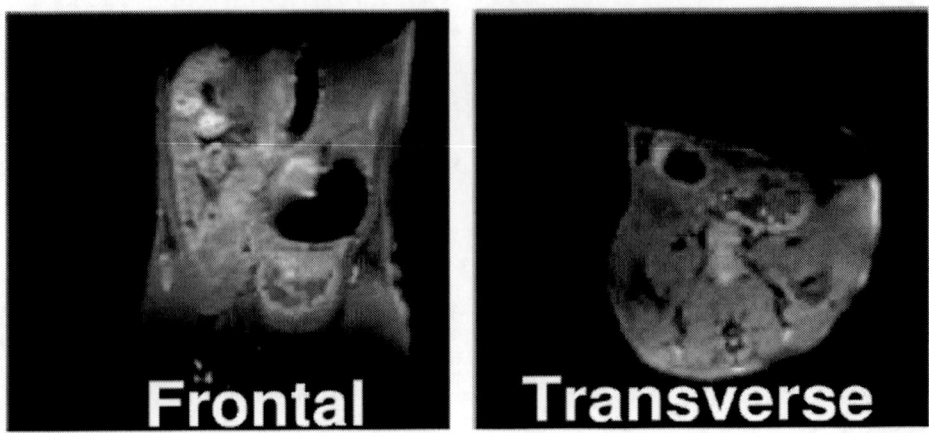

(Reproduced from Thurecht, K. J.;Blakey, I. ; Peng, H.; Squires, O.; Hsu, S. ; Alexander, C.; Whittaker, A. K. *J. Am. Chem. Soc.* 2010, *132*, 5336.., Copyright (2010), with permeation from American Chemical Society)).

Figure 2. MRI images of mouse abdominal region 2 h following injection of P1 into mouse tail vein. 1H image is shown in grayscale while 19F image is overlaid. P1 appears accumulated in the bladder.

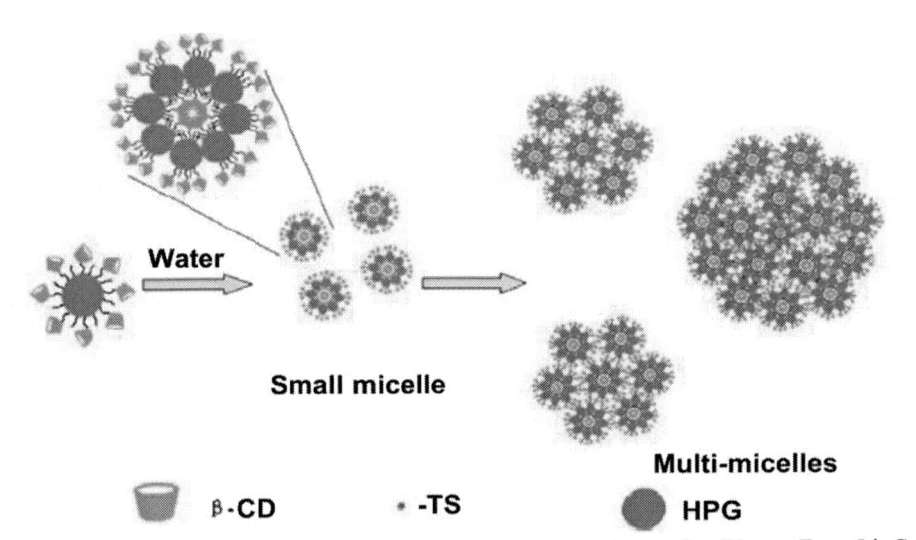

(Reproduced from Zhang, X.; Zhang, X.; Wu, Z.; Gao, X.; Cheng, C.; Wang, Z.; , Li C. *Acta Biomateriallia*, 2011,7,582., Copyright (2011), with permeation from Elsevier)).

Figure 3. A possible self-assembly mechanism of HPG-g-CD.

Hyperbranched poly (NIPAM) polymers functionalized have been developed with the antibiotics Vancomycin and Polymyxin-B that are sensitive to the presence of bacteria in solution by Shepherd and coworkers [3]. Binding of bacteria to the polymers causes a conformational change, resulting in collapse of the polymers and the formation of insoluble polymer/bacteria complexes. They have applied these novel polymers to their tissues engineered human skin model of a burn wound infected with Pseudomonas aeruginosa and Staphylococcus aureus. When the polymers were removed from the infected skin, either in a polymer gel solution or in the form of hydrogel membranes, they removed bound bacteria, thus reducing the bacterial load in the infected skin model. These bacteria-binding polymers have many potential uses, including coatings for wound dressings. The development of successful formulations for poorly water soluble drugs remains a longstanding, critical, and challenging issue in cancer therapy. A b-cyclodextrin (CD) functionalized hyperbranched poly-glycerol (HPG) has been prepared as a potential water insoluble drug carrier by Zhang and coworkers [4]. The HPG-g-CD molecules could self-assemble into multimolecular spherical micelles in water, the size of which ranged from 200 to 300 nm, with good dispersity. A possible self-assembly mechanism of HPG-g-CD shows in Figure 3.

A high loading capacity and high encapsulation efficiency of paclit- axel, as a model, were obtained. The release profiles of different co-polymer compositions showed a burst release followed by continuous extended release. Furthermore, MTT analysis showed that HPG-g-CD had good biocompatibility, indicating that HPG-g-CD may be considered a promising hydrophobic drug delivery system. The detailed synthesis route is shown in Figure 4.

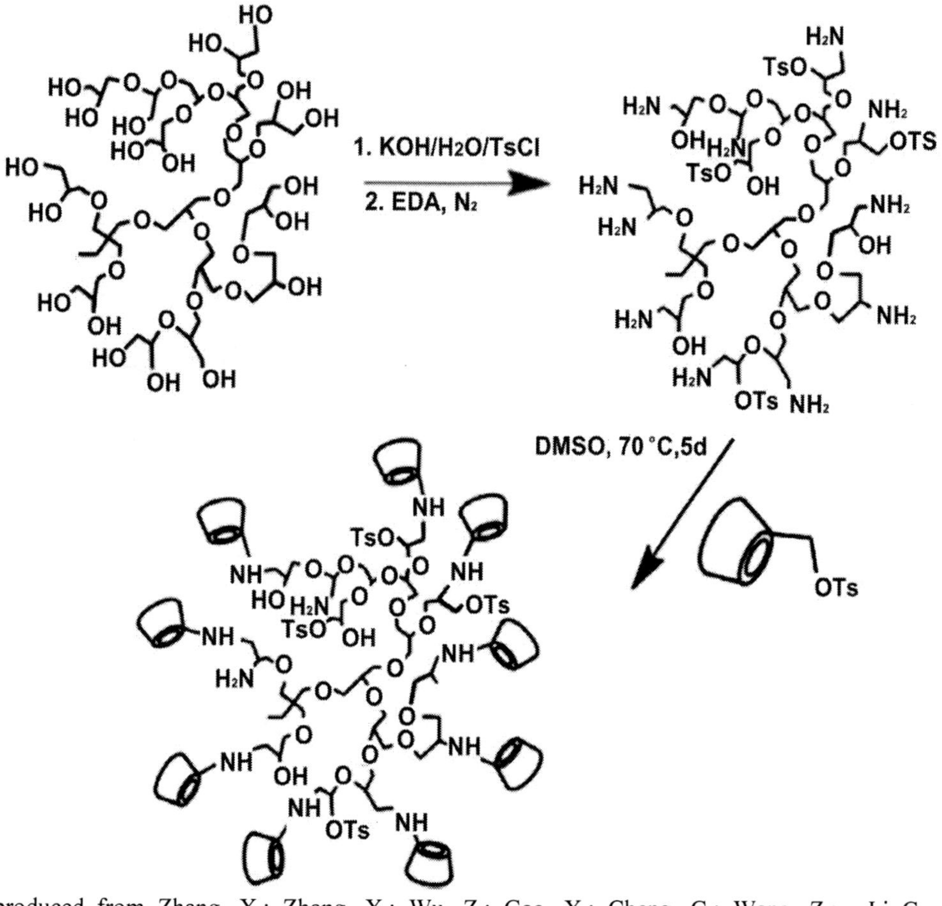

(Reproduced from Zhang, X.; Zhang, X.; Wu, Z.; Gao, X.; Cheng, C.; Wang, Z.; , Li C. *Acta Biomateriallia*, 2011,7,582., Copyright (2011), with permeation from Elsevier)).

Figure 4. Synthetic route of the HPG-g-CD co-polymer.

For the construction of a well-defined antibody surface, protein A was used as a binding material to immobilize antibodies onto gold-derivatized transducers. The traditional method tends to assemble protein A directly onto the gold-derivatized transducers, Shen and coworkers [5] tried to indirectly bind protein A onto sensors through hyperbranched polymer (HBP) which was synthesized from p-phenylenediamine and trimesic acid. The three-dimensional structure of HBP and the characteristics including orientation control and biocompatibility of protein A led to highly efficient immunoreactions and enhanced detection system performance.

Mugabe and coworkers [6] reported the development and in vitro characterization of paclitaxel (PTX) and docetaxel (DTX) loaded into hydrophobically derivatized hyper-branched polyglycerols (HPGs). Several HPGs derivatized with hydrophobic groups ($C_8/_{10}$ alkyl chains) (HPG–C8/10–OH) and/or methoxy polyethylene glycol (MePEG) chains (HPG–$C_8/_{10}$–MePEG) were synthesized. PTX or DTX were loaded into these polymers by a solvent evaporation method and the resulting nanoparticle formulations were characterized in terms of size, drug loading, stability, release profiles, cytotoxicity, and cellular uptake. PTX and DTX were found to be chemically unstable in unpurified HPGs and large fractions (80%) of the drugs were degraded during the preparation of the formulations.

Höbel and coworkers [7] systematically analyzed physicochemical and biological properties of DNA and siRNA complexes prepared from a set of maltose, maltotriose or maltoheptaose-modified hyperbranched PEIs (termed (oligo-)maltose-modified PEIs; OM-PEIs). They showed that pH-dependent charge densities of the OM-PEIs correlate with the structure and degree of grafting, and the length of the oligomaltose. Importantly, upon their systemic application in vivo, OM-PEI/siRNA complexes show marked differences in the siRNA biodistribution profile with, for example, substantially decreased siRNA levels in the liver and increased siRNA levels in the muscle. Taken together, it is demonstrated that OM-PEI complexes show structure-dependent physicochemical and biological properties and may represent promising, tailor-made platforms for the delivery of siRNAs, particularly for in vivo applications.

2.2. Synthesis of Hyperbranched Polymers and Applications in Membranes

The membranes of hyperbranched polyglycidol (HPG) and hyperbranched poly (amine-ester) (HPAE) were prepared by crosslinking terminal hydroxyl groups with 4,4'-oxydiphthalic anhydride (ODPA) and glutaraldehyde (GA), respectively, by Wei and coworkers [8]. The intrinsic hydrophilicity of the membranes makes them have water permeation preference in pervaporation process. On the basis of their swelling behavior, the pervaporation properties of these two crosslinked membranes in separating water from water/isopropanol mixture were discussed and compared.

Scanning electron microscope (SEM) revealed that the crosslinked HPG-ODPA and HPAE-GA membranes had the dense and homogenous matrices. The intrinsic hydrophilicity of the membranes makes them having water permeation preference in pervaporation process. On the basis of their swelling behavior, the pervaporation properties of these two crosslinked membranes in separating water from water/isopropanol mixture were discussed and compared. Synthesis processes of HPG and HPAE are shown in Figure 5.

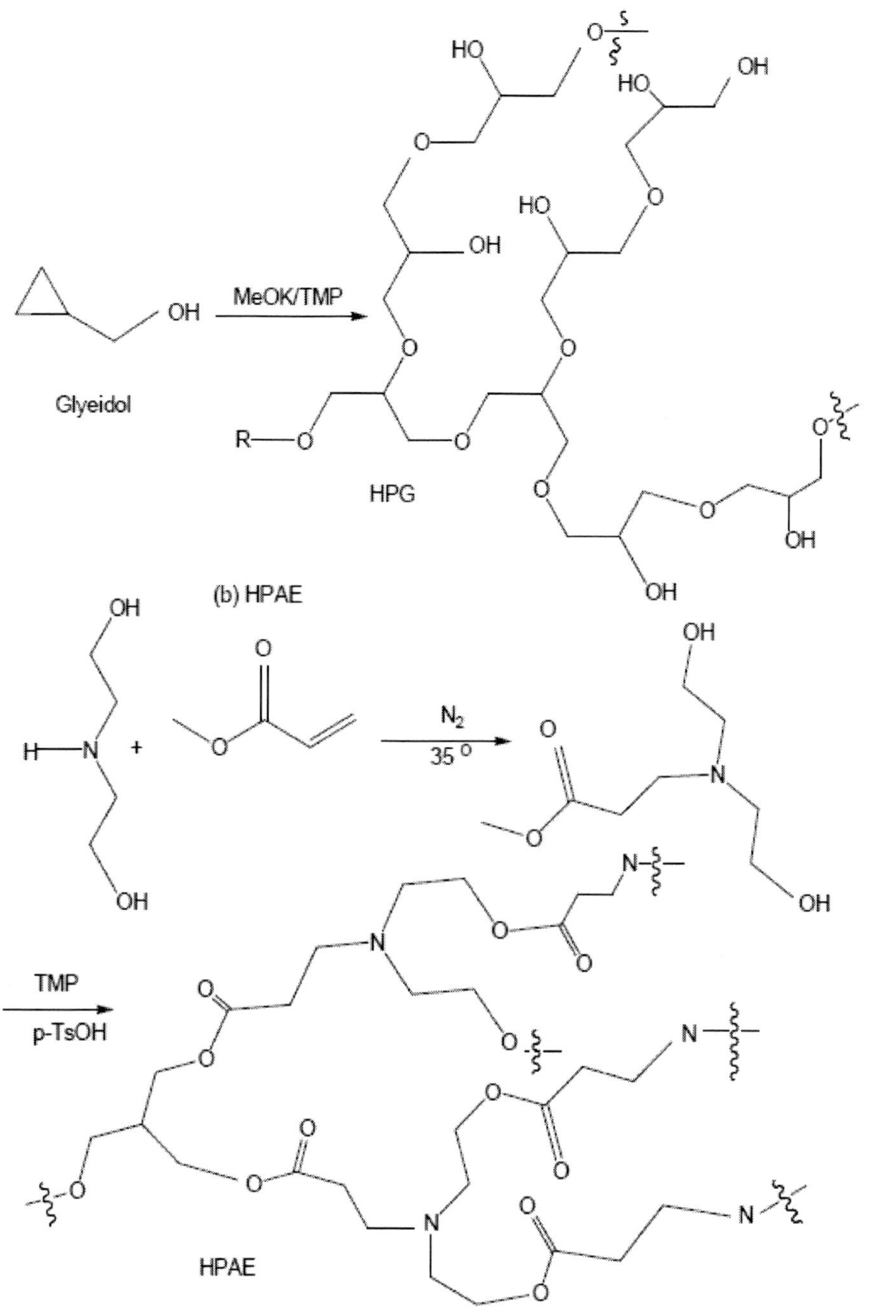

Figure 5. Schematic synthesis processes of HPG and HPAE.

By controlling the chemical structure of hyperbranched polymer, linkage unit and crosslinking degree, crosslinked HBP membranes might be a new kind material for pervaporation dehydration. Schematic for crosslinked membranes of HBP-HPG-ODPA and HBP-HPAE-GA are show in Figure 6.

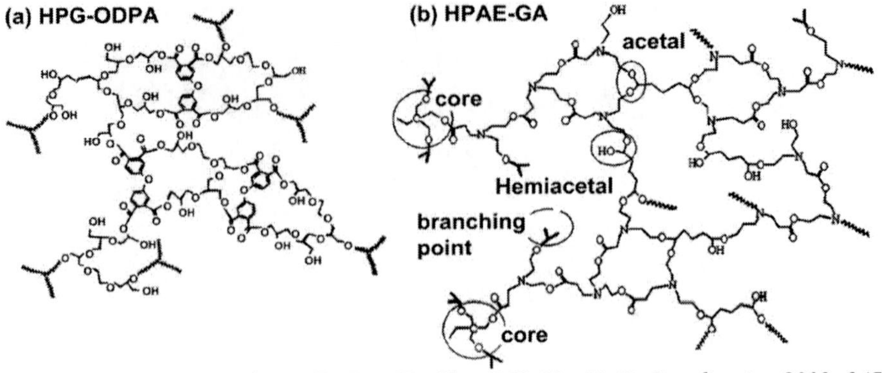

(Reproduced from Wei, X. Z.; Liu, X. F.; Bao-Ku Zhu, B.K; Xu, Y. Y. *Desalination* 2009, *247*, 647., Copyright (2009), with permeation from Elsevier).

Figure 6. Schematic for crosslinked membranes of HBPs: (a) HPG-ODPA and (b) HPAE-GA.

Chu and coworkers [9] showed that physical cross-linking of sulfonated poly(ether ether ketones) sPEEKs with hyper-branched bismaleimide oligomer (modified bismaleimide, mBMI) leads to densely packed polymer. The branched structure and their degree of entanglement with sPEEK polymer matrix increased with longer curing time. The results showed that physical cross-linking with highly branched mBMI was effective in reducing water uptake, lower methanol permabiity with reduced sPEEK membrane swelling. Except for heavily entangled sample (sPEEK/mBMI(30)) annealed for 20 h, all membranes displayed fair proton conductivity above 10^{-2} S/cm at room temperature. The polymerization of BMI monomer is shown in Figure 7.

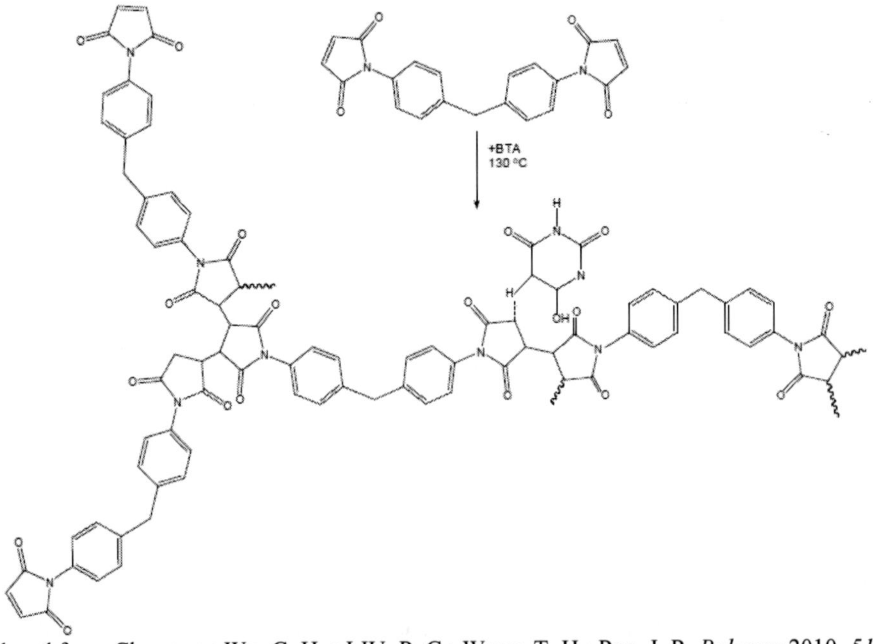

(Reproduced from Chu, p. p.; Wu, C. H. ; LIU, P. C.; Wang, T. H.; Pan, J. P. *Polymer* 2010, *51*,1386., Copyright (2010), with permeation from Elsevier).

Figure 7. The polymerization of BMI monomer.

A new proton conducting membrane where the poly(ether ether ketone ketone) copolymer containing unsaturated norbornene unit and pendant sulfonic acid was synthesized by aromatic substitution polymerization reaction using 1,3-bis(4-fluorobenzoyl)benzene (DFBP), 6,7-dihydroxy- 2-naphthalenesulfonate (DHNS), and 1,4-dihydro-1,4-methanonaphthalene-5,8-diol (NB-ph-diOH) monomers by Chen and coworkers [10]. The increase of proton conductivity, water uptake and ion exchange capacity (IEC) are found to corroborate with increasing the DHNS feed concentration. Furthermore, cross-linking through the double bond in norbornene unit reduces both the methanol permeability and methanol solvent uptake.

The electrochemical behaviors of hyperbranched poly(ferrocenyl-methylsilane) (HPFMS) and linear oligo-(ferrocenyldimethylsilane) (LOFS) films were studied systematically by cyclic voltammetry and chronocoulometry under different polymer coverage and solvents Huo and coworkers. [11] Both poly(ferrocenylsilanes) show stable cyclic voltammographs in $LiClO_4$ solutions. It was also found that a solvent with the appropriate solubility parameter and polarity, lower viscosity, and higher dielectric constant is in favor of charge transport through polymer films, which is consistent with the proposed model of electrode process for HPFMS films. These results imply that hyperbranched ferrocenyl polymers have the potential to be excellent chemical sensor materials with convenient synthesis and high sensitivity.

2.3. Other Synthesis Methods and Applications of Hyperbranch Polymers

Sol–gel derived organo-silicate hybrid coatings prepared using hyperbranched poly(ethylene imine) with or without organic corrosion inhibitors (MBT and MBI) were developed on AA2024-T3 to provide active corrosion protection when the integrity of coating starts to fail [12].

The effect of different neutral polymers on the self-assemblies of hyperbranched poly(ethyleneimine) (PEI) and sodium dodecyl sulfate (SDS) has been investigated at different ionization degrees of the polyelectrolyte molecules by Pojják and Mészáros. [13] Chemical transformations of small molecules have served as a rich source of reactions for the development of new polymerization processes, and "click" reaction has the potential to become a powerful polymerization technique. Qin and coworkers [14] give a brief account of the research efforts devoted to the development of click reaction into a new polymerization process. Remarkable progresses have been made in recent years in the exploration of metal-mediated and metal-free click polymerization systems and in the syntheses of linear and hyperbranched polytriazoles with regioregularmolecular structures and advanced functional properties.

Li and co-workers [15], for example, successfully prepared soluble hyperbranched PTAs from the click polymerization of an AB_2 monomer (Figure 8) by decreasing the amount of water in the water/DMF mixture to 1:20 by volume.

The strategy of using an $A_2 + B_3$ monomer combination can solve the problem of self-oligomerization (Figure 9), although there is a concern that such process may involve cross-linking reaction, leading to the formation of insoluble networks.

The strategy of using an $A_2 + B_3$ monomer combination can solve the problem of self-oligomerization, although there is a concern that such process may involve cross-linking reaction, leading to the formation of insoluble networks.

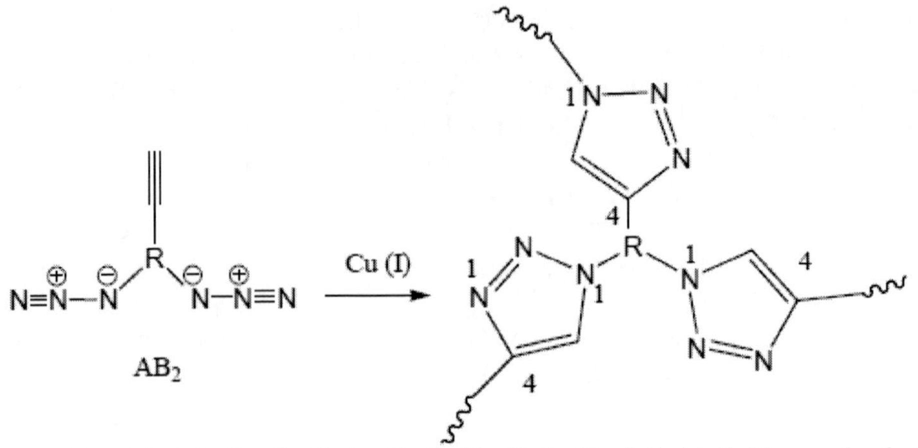

(Reproduced from Li, Z. A.; Yu, G.; Hu, P.; Ye, C.; Liu, Y. Q.; Qin, J. G.; Li, Z. *Macromolecules* 2009, 42, 1589, Copyright (2009), with permeation from American Chemical Society)).

Figure 8. Synthesis of 1,4-regioregular hyperbranched PTA by Copper(I)-mediated click polymerization of ethynylene diazides (AB2) monomer.

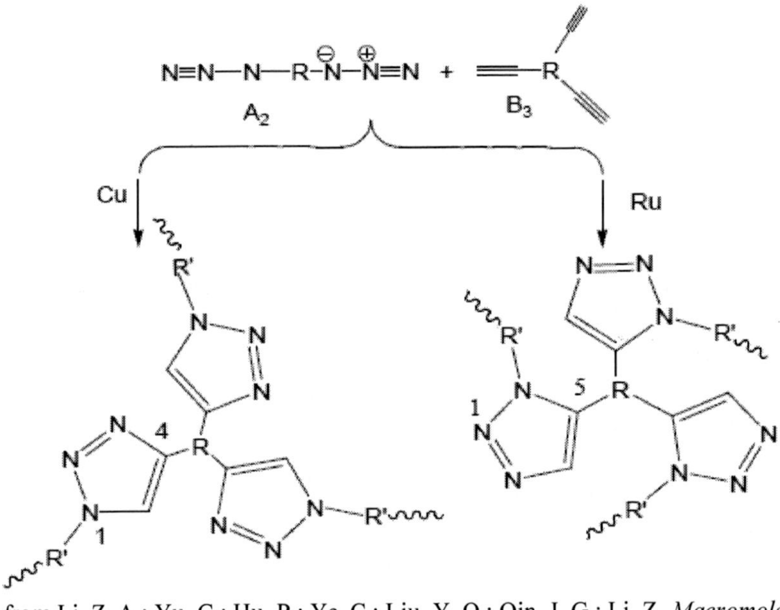

(Reproduced from Li, Z. A.; Yu, G.; Hu, P.; Ye, C.; Liu, Y. Q.; Qin, J. G.; Li, Z. *Macromolecules* 2009, 42, 1589, Copyright (2009), with permeation from American Chemical Society)).

Figure 9. Syntheses of 1,4- and 1,5-Regioregular Hyperbranched PTAs by Cu- and Ru-Catalyzed Click Polymerizations of Diazide (A$_2$) and Triyne (B$_3$) Monomers.

Thermocurable hyperbranched polystyrenes were synthesized using atom transfer radical polymerization and exhibited superior ultrathin film formation capabilities in comparison with the linear analogues, as assessed by the minimal film thickness attainable by spin-coating without dewetting by Yoon and coworkers. (Figure 10) [16]

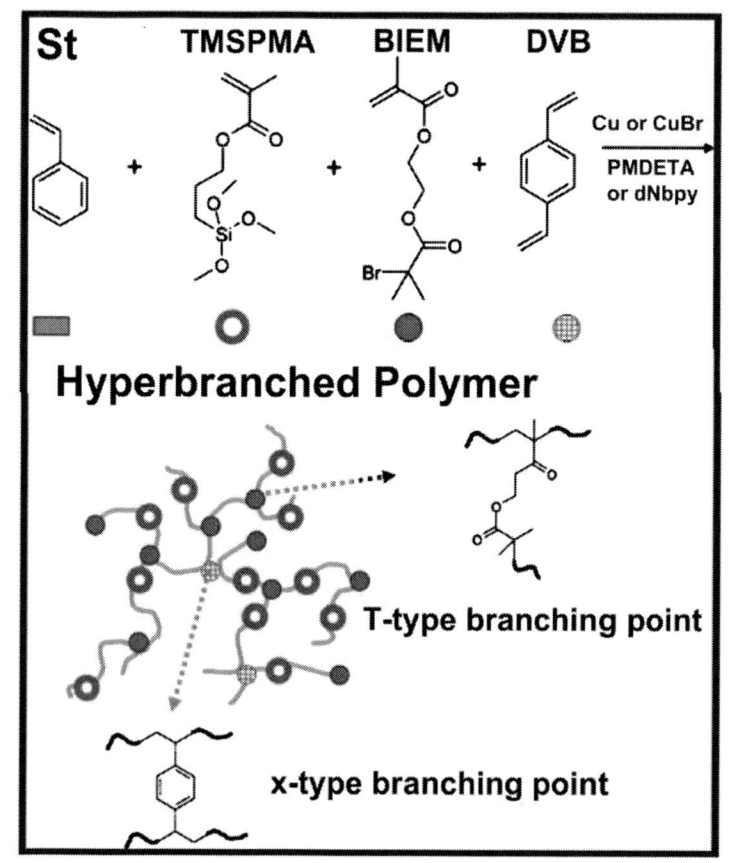

(Reproduced from Yoon, J. A.; Young, T.; Matyjaszewski, K.; Kowalewski T. *ACS applied materials and interfaces*. 2010, *2*, 2475, Copyright (2010), with permeation from American Chemical Society)).

Figure 10. Synthesis of a thermocurable hyperbranched polystyrene-copoly (3-(trimethoxysilyl)propyl methacryl.

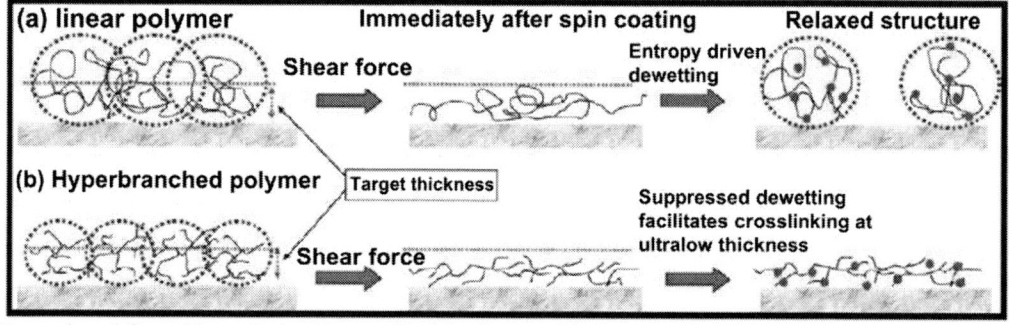

(Reproduced from Yoon, J. A.; Young, T.; Matyjaszewski, K.; Kowalewski T. *ACS applied materials and interfaces*. 2010, *2*, 2475, Copyright (2010), with permeation from American Chemical Society)).

Figure 11. Proposed mechanism of anticipated improvement of ultrathin film forming capabilities of hyperbranched vs linear polymer.

They were suitable as ultrathin film organic dielectrics, with parallel plate specific capacitances as high as ~680 nF/cm^2. Similar to high performance inorganic dielectrics, capacitance measurements pointed to the presence of "dead" interfacial capacitance, which could be accounted for by considering the geometric effect of roughness "incommensurability" between metal electrode and polymer film.

They demonstrate that the undesirable dewetting can be reduced through the use of highly branched cross-linkable polymers, presumably because of their more compact shape, allowing achievement of lower thickness without extensive coil stretching (Figure 11).

CONCLUSION

Hyperbranched polymers are by now an established class of polymeric materials and can be considered as highly functional specialty products. It is verified that they offer the chance for the development of new products but at the same time they present a challenge due to their complex branched structure. Examples out of a broad variety of structures as well as several possible applications are discussed.

ACKNOWLEDGMENTS

The author wishes to express her gratitude to National Elite Foundation for the financial support.

REFERENCES

[1]   Gao, C.; Yan, D. Prog. Polym. Sci. 2004, 29, 183.
[2]   Thurecht, k. j.;Blakey, I; Peng, H.;Squires,O.;Hsu, S.; Alexander, C.;Whittaker, A. K. J. Am. Chem. Soc.2010, 132, 5336.
[3]   Shepherd, J.; Sarker, P.; Rimmer,S.; Swanson, L.; MacNeil, S.; Douglas, I. Biomaterials 2011, 32, 258.
[4]   Zhang, X.; Zhang, X.; Wu, Z.; Gao, X.; Cheng, C.; Wang, Z.; , Li C. Acta Biomateriallia, 2011,7,582.
[5]   Shen,G.; Cai, C.; Wang, K.; Lu, J. Anal Biochem 2011,409, 22.
[6]   Mugabe,C.; Liggins,R. T.; Guan, D.; Manisali, I.; Chafeeva, I.; Brooks, D. E.; Heller, M.; Jackson, J.K.; Burt, H.M. International Journal of Pharmaceutics 2011, 404, 238.
[7]   Höbel, S.; Loos, A.; Appelhans, D.; Schwarz, S.; Seidel, J.; Voit, B.; Aigner A., Journal of Controlled Release 2011, 149, 146.
[8]   Wei, X. Z.; Liu, X. F.; Bao-Ku Zhu, B.K; Xu, Y. Y. Desalination 2009, 247, 647.
[9]   Chu, p. p.;Wu, C. H.; LIU, P. C.;Wang,T. H.;Pan, j. P. Polymer 2010, 51, 1386.
[10]   Chen,L. K.; Wu,C.S.; Chen, M. C.; Hsu,K. L.; Li, H. C.; Chi-Han Hsieh,C. H.; Hsiao,M. H.; Chang,C. L.; Chu, P.P.J. Journal of Membrane Science 2010, 361, 143.
[11]   Huo,J.; Wang ,L.; Yu, H.; Deng, L.; Ding,J.; Tan,Q.; Liu,Q.; Xiao,A.; Ren,G.; J. Phys. Chem. B 2008, 112, 11490.

[12] Roussi, E.; Tsetsekou, A.; Tsiourvas, D.; Karantonis, A. Surface and Coatings Tech. 2011, 205, 3235.

[13] Pojják, K.; Mészáros, R. J. Colloid and Interface Sci 2011, 355, 410.

[14] Qin, A.; Lam, J.W.Y.; Tang, B.Z. Macromolecules 2010, 43, 8693.

[15] Li, Z. A.; Yu, G.; Hu, P.; Ye, C.; Liu, Y. Q.; Qin, J. G.; Li, Z. Macromolecules 2009, 42, 1589

[16] Yoon, J. A.; Young, T.; Matyjaszewski, K.; Kowalewski T. ACS applied materials and interfaces. 2010, 2, 2475.

PART 2. CONDUCTIVE POLYMERS

In: Polymer Synthesis
Editor: E. Kowsari

ISBN 978-1-61324-672-6
© 2012 Nova Science Publishers, Inc.

Chapter 6

SYNTHESIS, CHARACTERIZATION AND APPLICATIONS OF CONDUCTING POLYMER NANOFIBERS

Gustavo Morari do Nascimento[*]
Universidade Federal de Minas Gerais, Instituto de Ciências
Exatas-Departamento de Química, MG-Belo Horizonte, Brazil, SA

ABSTRACT

In recent years the synthesis of conducting polymers with controlled morphology is one of the big deals in the polymer science and technology. By the emergence of nanotechnology, researchers become more interested in studying the unique properties of nanoscale materials. There are different ways to produce polymeric nanofibers, as example the polymerization of aromatic monomers into media having large organic acids. These acids form micelles upon which the monomer is polymerized and doped. Fiber diameters are observed to be as low as 30-60 nm and are highly influenced by reagent ratios. Uniform nanofibers (from 30 to 200 nm) can also be obtained when the polymerization is done at an aqueous-organic interface. It is hypothesized that migration of the product into the aqueous phase can suppress uncontrolled polymer growth by isolating the fibers from the excess of reagents. Thus, the template-free methods, such as interfacial, seeding and micellar can be employed as different "bottom-up" approaches to obtain pure polymeric nanofibers. The possibility to prepare nanostructured conducting polymers by self-assembly with reduced post-synthesis processing warrants further study and application of these materials, especially in the field of electronic nanomaterials. The notable applications include in tissue engineering, biosensors, filtration, wound dressings, drug delivery, and enzyme immobilization. In this chapter this amazing new area of polymeric nanofibers will be reviewed concerning the state-or-art results of synthesis, spectroscopic characterization and applications. The discussion will be centered in the previous and new results obtained by our group, using mainly resonance Raman and X-ray absorption techniques applied to the study of polyaniline nanofibers. Special attention will be given in the role of the synthetic pathways in the control of the electrical, thermal and mechanical properties and also the morphological aspects of the polymeric material.

---

[*] E-mail: morari@yahoo.com.

The main goal of this work is to contribute in the rationalization of some important results obtained in the open area of polymeric nanofibers.

1. Synthesis

1.1. Conducting Polymers

The conducting polymers are one of the most important classes of molecular conductors or synthetic metals. The study of conjugated polymers started at 70s,[1,2] when the Shirakawa and collaborators[1,2] synthesized more stable films of semiconducting poly(acetylene). However, just in 1977,[3-7] it was found that the conductivity of the poly(acetylene) increase by 13 orders of magnitude when its films are doped with acid (or base) of Lewis. The concept of doping is unique and has central importance, because it is what differentiates the conducting polymers from all other types of polymers.[8-10] During the doping process, an insulating or semiconducting organic polymer with low conductivity, typically ranging from 10^{-10} to 10^{-5} Scm^{-1}, is converted into a polymer which shows conductivity in a "metallic" regime (ca. 1-10^4 Scm^{-1}). The addition of non-stoichiometric chemical species in quantities commonly low ($\leq 10\%$), results in dramatic changes in electronic properties, electrical, magnetical, optical and structural of the polymer. Contrarily to the classical doping in inorganic semiconductors, for organic polymers, the dopants chemically react with the chain and causes disturbance in the crystalline structure of the polymer. Doping is a reversible process, and the polymer can return to its original state without major changes in its structure. In the doped state, the presence of counter ions stabilizes the doped state. By adjusting the level of doping, it is possible to obtain different values of conductivity, ranging from the state or non-doped insulating state to the highly doped or metallic. The major classes of conducting polymers can be doped by p (oxidation) or n (reduction) through chemical and/or electrochemical process.[8-10]

1.2. Polyaniline (PANI)

Polyaniline (PANI) is one of the most important conducting polymers owed to its easy preparation and doping process, environmental stability, and potential use as electrochromic device, as sensor and as corrosion protecting paint. These properties turned PANI attractive to use in solar cells, displays, lightweight battery electrodes, electromagnetic shielding devices, anticorrosion coatings and sensors. The recent research efforts are to deal with the control and the enhancement of the bulk properties of PANI, mainly by formation of organized PANI chains in blends, composites and nanofibers.[11-14] The fully reduced leucoemeraldine base form (LB, see Figure 1.1 for y = 1) and the fully oxidized pernigraniline base form (PB in Figure 1.1 for y = 0) are non-conducting forms of PANI. The half-oxidized emeraldine base (EB in Figure 1.1 for y = 0.5) is a semiconductor but after protonation it becomes a conducting emeraldine salt form of PANI (ES, see structure in Figure 1.1).[11,12] EB and ES can also assume two different types of crystalline arrangements depending on the synthetic route used.[15,16]

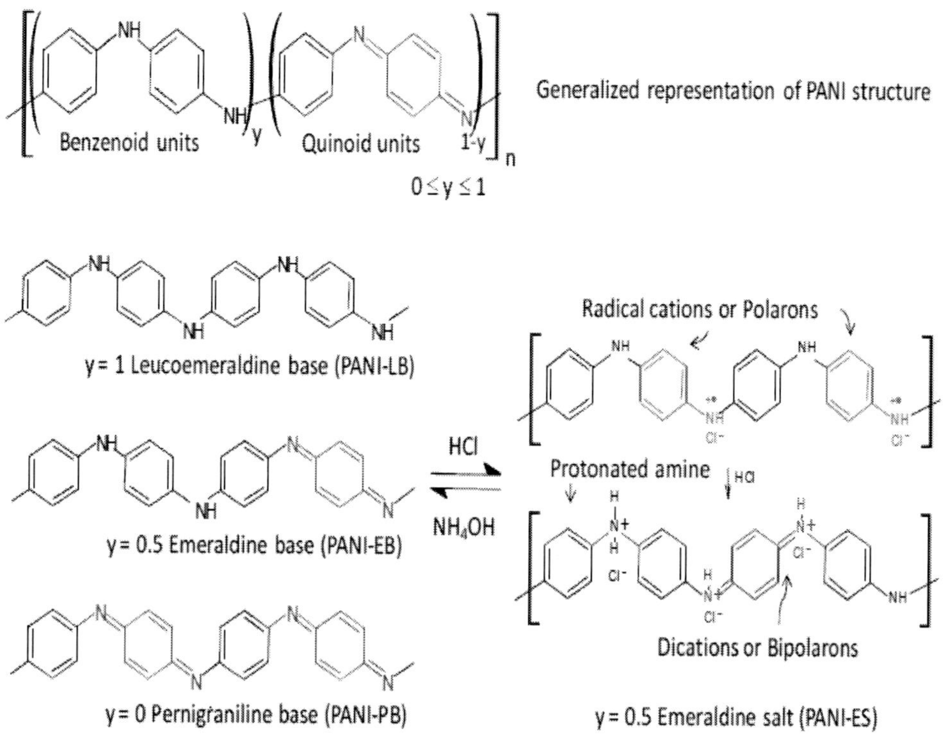

Figure 1.1. Chemical representation of generalized PANI structure and its most common forms.

Polyaniline can be synthesized by two main methods, by chemical or by electrochemical polymerization of aniline in acidic media. The chemical oxidation is commonly performed using ammonium persulfate in aqueous acidic media (hydrochloric acid, sulfuric, nitric or perchloric acid) containing aniline. This is the conventional synthetic route of PANI, but one of disadvantages of this route is the presence of excess of oxidant and salts formed during the synthesis, leading to a polymeric sample that is practically insoluble in majority of solvents, making its processing very difficult.[17] During the oxidative polymerization of aniline, the solution becomes progressively colored resulting in a solid dark green. The color of the solution is owing to the presence of soluble oligomers formed by coupling of radical cations of aniline. The intensity of the color depends on the environment and also the concentration of oxidant.[17] There are many variations of the chemical synthesis of PANI, however there is a certain consensus that there are four main parameters that affect the course of the reaction and the nature of final product, being: (1) nature of the synthetic medium, (2) concentration of the oxidant, (3) duration of the reaction, and (4) temperature of the synthetic medium.[17] The polymerization of aniline was observed as an autocatalytic process.[18,19] Kinetic studies suggest that initial oxidation of aniline leads to the formation of dimeric species, such as p-amino-biphenyl-amine, N,N'-Biphenyl-hidrazine, and Benzidine.[20] The oxidation of aniline and its derivatives in strongly acidic media favor the formation of benzidine,[21,22] while in a slightly acid or neutral prevails p-amino-Biphenyl-amine, but in basic medium the formation of azo bonds, resulting from the head-to-head coupling is favored. These dimeric species have lower oxidation potential than aniline and are oxidized immediately after its formation (the N,N'-biphenyl-hidrazine is converted to benzidine through rearrangement that occurs in acid medium,[23] resulting two types of charged quinoid-di-imine species.

Afterwards, electrophilic attacks in these species, followed by deprotonation, are responsible for the growth of oligomers with subsequent formation of polymer chains of PANI.[24]

1.3. Nanostructured Polyaniline

The synthesis of nanostructured PANI, especially as nanofibers, can improve its electrical, thermal and mechanical stabilities. These materials can have an important impact for application in electronic devices and molecular sensors owing their extremely high surface area, synthetic versatility and low-cost. The conventional synthesis of polyaniline, based on the oxidative polymerization of aniline in the presence of a strong acid dopant, typically results in an irregular granular morphology that is accompanied by a very small percentage of nanoscale fibers.[25,26] However, different approaches have been developed in order to produce PANI and many other polymers with nanostructured morphology. In this chapter will be analysed the synthetic routes that produce nanostructured PANI, mainly as nanofiber or nanotube morphology, without the use of rigid templates.

The nanostructured PANI has been prepared by different synthetic ways. Nevertheless, these approaches can be grouped into two general synthetic routes, as can be seen in the Figure 1.2. Uniform nanofibers of pure metallic PANI (30-120 nm diameter, depending on the dopant) have also been prepared by polymerization at an aqueous-organic interface.[15] The first step (see item a) of the interfacial polymerization), the oxidant and monomers (aniline), dissolved in immiscible solvents, are put together without external agitation. Afterwards, some aniline monomers are oxidized in the interfacial region between the two solutions, being formed some oligomers of the interfacial polymerization). It is hypothesized that migration of the product into the aqueous phase can suppress uncontrolled polymer growth by isolating the fibers from the excess of reagents. Afterwards, the initial chains grow up and more PANI chains are formed. Interfacial polymerization can therefore be regarded as a non-template approach in which high local concentrations of both monomer and dopant anions at the liquid–liquid interface might be expected to promote the formation of monomer-anion (or oligomer-anion) aggregates. These aggregates can act as nucleation sites for poly-merization, resulting in powders with fibrillar morphology. It has recently been demonstrated that the addition of certain surfactants to such an interfacial system grants further control over the diameter of the nanofibers. An important part that is frequently neglected or not deeply explained in details is the isolation of the nanostructured PANI from the solution. But, generally, the nanofibers are isolated by filtration in a nanoporous filters, being the isolated polymer washed with different solutions with the aim to clean it up. The solution can be also dialyzed and the cleaned solution containing the nanofibers is centrifugated in order to separate the nanofibers apart from the solution.

PANI nanofibers or nanotubes HY can be obtained by making use of large organic acids (see Figure 1.2, item b)). These acids form micelles upon which aniline is polymerized and doped. Fiber diameters are observed to be as low as 30-60 nm and are highly influenced by reagent ratios.[27-30] Ionic liquids (ILs) have also been used as synthetic media for the preparation of nanostructured conducting polymers.[31-33] Ionic liquids are organic salts with low lattice energies, which results in low melting points and many ILs are liquids at room temperature.[34] There is a large variety of ionic liquids and the most used ILs are derived from imidazolium ring, pyridinium ring, quaternary ammonium and tertiary phosphonium cations. The usual differentiation between conventional molten salts and ionic liquids is based on the melting point. While most molten salts have melting points higher than 200°C, ionic liquids normally melt below 100°C .[35] The most unusual characteristic of these systems is that, although they are liquids, they present features similar to solids, such as structural organization at intermediate distances[36] and negligible vapor pressure.[37] This structural organization can act as a rigid template, and PANI nanofibers are acquired when the aniline is polymerized in these media.

Thus, the template-free methods can be employed as different "bottom-up" approaches to obtain pure PANI nanofibers. The possibility to prepare nanostructured PANI by self-assembly with reduced post-synthesis processing warrants further study and application of these materials, especially in the field of electronic nanomaterials.

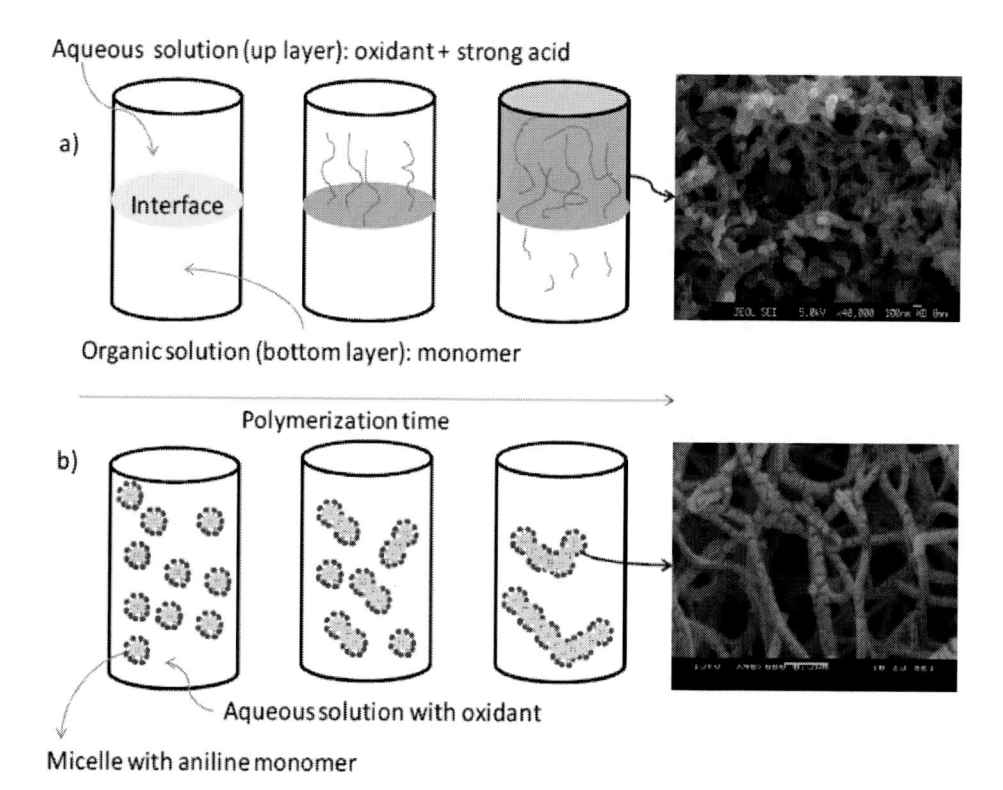

Figure 1.2. Schematic representation of: a) interfacial polymerization and b) micellar polymerization. In the interfacial polymerization the top layer is an aqueous solution containing HCl acid and $(NH_4)_2S_2O_8$ (others acids or oxidants can be used); the bottom layer has aniline dissolved in the chloroform (others solvents immiscible in water can be used). Starting the polymerization and migration of oligomers from organic bottom layer to the aqueous top layer and formation of PANI. The scanning electron microscopic (SEM) image was obtained from PANI nanofibers obtained from interfacial polymerization using HCl, $(NH_4)_2S_2O_8$, and chloroform. The nanofibers have ca. 30 nm of diameter. In the micellar polymerization the solubilization of aniline is in an aqueous solution containing organic acids that act as surfactants. After added the oxidant the polymerization starts and depending on the concentration of aniline in solution, it is possible to form hollow nanofibers (as named nanotubes) or nanofibers. The SEM image obtained from the PANI powder obtained from micellar polymerization using β-naphtalenesulfonic acid (β-NSA), $(NH_4)_2S_2O_8$, and molar ratio of β-NSA:aniline of 1:4. The nanofibers have ca. 93 nm of diameter.

2. CHARACTERIZATION

Different characterization techniques have been employed in determination of the structure and also the behavior of nanostructured conducting polymers. Among these, the spectroscopic techniques have been essential in understanding the molecular structure and interactions resultant of the arrangement at nanometric level. In this section, a survey of the main characteristics of the spectroscopic techniques that have been used by our group in the structural investigation of nanostructured polyanilines will be given. The main goal of this work is to contribute in the rationalization of some important results obtained in the open area of PANI nanofibers.

2.1. Vibrational Spectroscopy: Infrared and Raman

2.1.1. Background

The vibrational spectroscopy (Infrared absorption (IR) and Raman spectroscopy) has been largely used in the characterization of complex materials, such as conducting polymers. Detailed information about the structure of the polymer can be obtained, because each group of atoms show a band position in the spectra that is related to the type of atom and the strenght of the chemical bonds. In addition, both position and intensity of the vibrational bands are influenced by the chemical environment or in other words by the type of interaction that between the molecules in a system. The IR and Raman are vibrational techniques that differ in its physical origin. When a molecule is exposed to the infrared radiation (ν), there is an absorption only if the molecule has available states (E_1, E_2, E_3,... E_N) where the energy between gap between the states has the same value of the incident radiation (ν). So, this statement can be summarized by the expression: $E_N - E_1 = h\upsilon$, being h the Planck's constant. For Raman scattering, the observed bands have a different origin; they are derived from an inelastic scattering process. Typically, a molecule can scatter a monochromatic ν_o frequency from elastic and/or in an inelastic way. The elastic scattering is called Rayleigh scattering and is much more intense than the inelastic process or Raman scattering (approximately 10^{-8} lower than the intensity of the incident radiation). The Raman spectrum appears in a wavelength (λ_o) slightly higher or lower than the incident radiation.[38-42]

When a photon of wavelength (λ_o) interacts with a molecule, there is a perturbation of the molecule which *eigenstate* can be described as the linear combination of the *stationary-state* (*or time-independent*) wave functions of the molecule. Sometimes the states of the pertubated molecule are called "*virtual*" because they are not states from the stationary condition (*eigenstates*). Usually an electronic state that is not an eigenstate of the system (*virtual state*) is generated by the annihilation of photon; therefore it is not a resonance phenomenon, like observed in the IR absorption. The photon created when the molecule returns to ground state has, more often, the same incident photon energy, characterizing then an elastic collision between the photon and the molecule (Rayleigh scattering). In some cases, the photon emerges with an energy different from initial (Raman scattering), the result of an inelastic interaction, and this can be exemplified in the energy diagram for a diatomic molecule, which is represented in Figure 2.1. The photon energy $(v_o - v_m)$ appears in the region of the spectrum called Stokes region; in addition the photons with energy $(v_o + v_m)$ appears in the anti-Stokes region. The bands observed in anti-Stokes region are less intense than those in the Stokes region since the Stokes population of molecules that have the energy corresponding to the vibrational level $v = 0$ (v is the vibrational quantum number) is much than the molecules that occupy the level $v = 1$ (according to the *Maxwell-Boltzmann* distribution), which makes the Stokes transitions occur with greater probability, making the bands corresponding to the Stokes region more intense than the anti-Stokes Raman region.[43-45]

A vibration is said to be active in the infrared when the dipolar momentum of the molecule changes (μ) with vibration coordinate (ϱ), summarized as: $\left(\frac{\partial \mu}{\partial \varrho}\right) \neq \emptyset$. In Raman, the vibration is active when causes a change in the polarizability (α) (or induced dipolar momentum) of the molecule; being described as: $\left(\frac{\partial \alpha}{\partial \varrho}\right) \neq \emptyset$. Hence, as the nature of the Raman effect is physically different from the infrared, the selection rules for these two forms of vibrational spectroscopy are different, which leads to different spectra. The two techniques are complementary, being used combined in the elucidation of the vibronic structure of compounds.

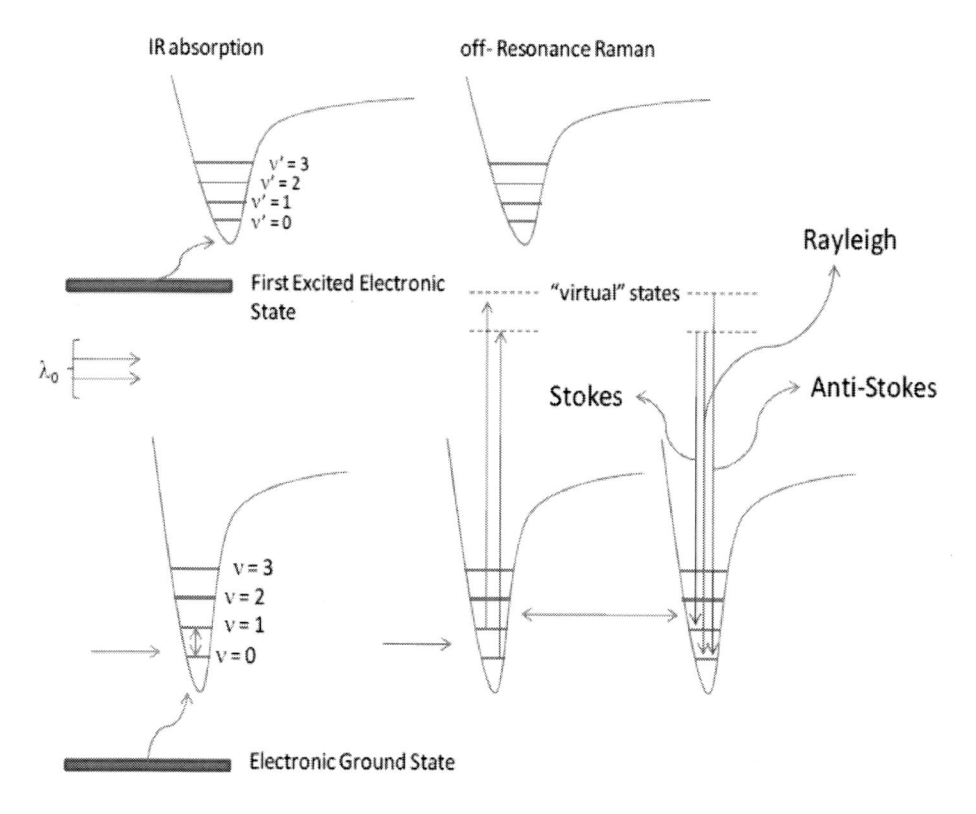

Figure 2.1. Schematic representation of the diagram of energy for a diatomic molecule. The figure represents the IR absorption and the Raman resonance using a radiation with energy that is not enough to excite the molecule to other electronic states.

Nowadays, the Raman spectra are collected in instruments having microscope. The principle is illustrated in Figure 2.2, the laser excitation source come to the sample on the microscope stage through optical elements. The Raman scattered radiation is collected in a scattering angle of 180° by the same microscope objective and captured by an opening of the spectrometer using a beam splitter. It is necessary that the instrument has a high lighting efficiency as well the collection of the scattered radiation must be precisely done, owing to the very small Raman cross section (typically a factor of 10^{-6} to 10^{-12} of the incident radiation) and the small volume of the sample. Raman microscopy is a non destructive technique and usually has no requirement for sample preparation. The main advantage however is the ability to focus the laser on a very small part of the sample (1 µm approximately or smaller). The high lateral resolution and depth of field (the order of a few micrometers) are very useful for the study of multilayered polymeric thin films.[43-45]

Many other factors distinguish the IR absorption spectroscopy to the Raman scattering; one of the most important is that the water has a very low Raman cross section, being possible to study aqueous solutions or *in situ* reactions, as the formation of polymeric films over the electrode in an aqueous solution. On contrary, in the IR spectroscopy the water is a serious interfering (due to strong absorption of the IR radiation by the water molecules). From the perspective of the instrumentation and sampling, the IR spectrometer is limited by the absorption of its optical components, which are commonly Potassium Bromide (KBr), which is transparent only above 350 cm^{-1}. In the Raman spectrometer the spectral limit is defined by the notch filters used for the screening of the Rayleigh scattering, thus the spectra can be obtained without major problems from 50 cm^{-1}.

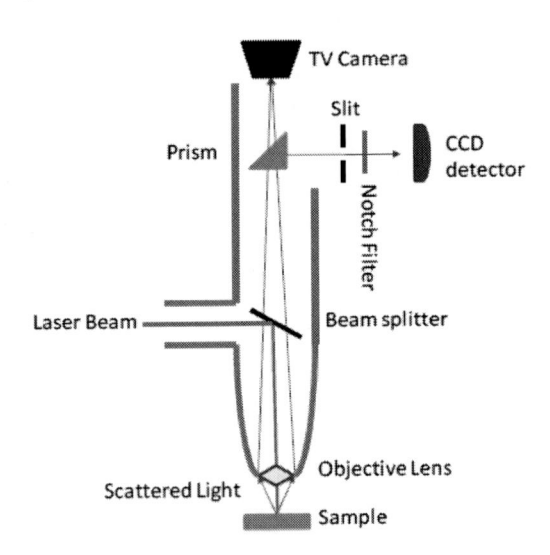

Figure 2.2. Conventional Raman microscope.

In addition, generally, for the acquisition of the Raman spectra are not necessary the treatment or handling of the sample, whereas for measuring the typical IR spectrum is necessary to disperse the sample in a solid matrix, such as KBr or by using dispersions in mineral oil. However, there is the possibility to measure the IR spectra by Attenuated Total Reflectance (ATR). In the ATR method it is possible to measure the IR spectra direct from the sample, only a small manipulation is necessary in the sample, such as spraying or compression over the ATR support.

The ATR technique can be used to obtain the IR spectra of solids, liquids or thin films, the main part of the ATR accessory is a transparent crystal having high reflection index. The materials frequently are: zinc selenide (ZnSe), KRS-5 (TlI$_4$/TlBr$_4$) and germanium (Ge). Mirrors inside the ATR support conduct the IR beam to a focus on crystal phase. After passing through the crystal, owing to the higher refractive index of the crystal, the IR radiation experiences the phenomenon of total internal reflectance. The IR spectrum of the sample is acquired through the existence of an evanescent wave, as shown in Figure 2.3. The evanescent wave is attenuated by the absorption of the sample; hence the name attenuated total reflectance (ATR). A good contact between the sample and the crystal is essential to ensure the penetration of the evanescent wave into the sample. This is the fundamental reason why the crystal is kept clean and free of risks. A certain pressure is sometimes applied in order to permit a good evanescent wave coupling with the sample.

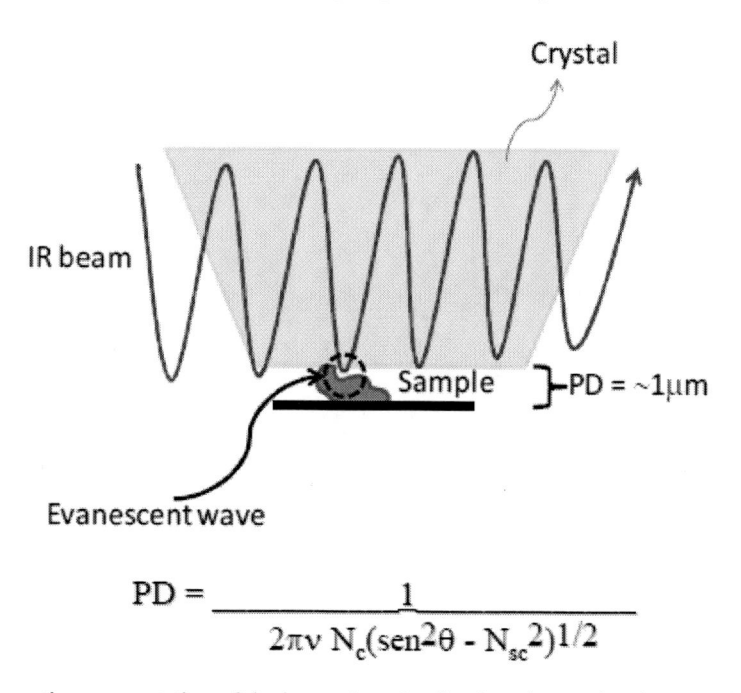

$$PD = \frac{1}{2\pi v \, N_c (sen^2\theta - N_{sc}^2)^{1/2}}$$

Figure 2.3. Schematic representation of the internal total reflection observed at the crystal interface. The DP of the evanescent wave is 1μm.[41] The penetration depth of IR radiation is given by the formula above, being PD the penetration depth, θ = angle of incidence, v = wave number, $N_{sc}= N_s / N_c$; being N_s = refractive index of the sample and N_c = refractive index of crystal. Note that the PD (penetration depth) is dependent on the wave number of the radiation, hence lower wave number penetrate beyond the higher wave numbers, as result, the intensity of the ATR peaks are higher to lower wave numbers.

2.1.2. Resonance Raman

In the off-resonance Raman spectroscopy (sometimes called normal Raman spectroscopy) the intensities of the Raman bands are linearly proportional to the intensity of the incident light (I_o, see Figure 2.4), proportional to the fourth power of the wavelength of the scattered light (λ_s^4 or v_s in wavenumber units, see Figure 2.4), and proportional to the square of the polarizability tensor ($[\alpha]^2$).[43-46] The situation changes dramatically, when the laser line falls within a permitted electronic transition. The Raman intensities associated with vibrational modes which are tightly coupled or associated with the excited electronic state can suffer a tremendous increase of about 10^5 powers; this is what characterizes the resonance Raman effect. (see Figure 2.4).

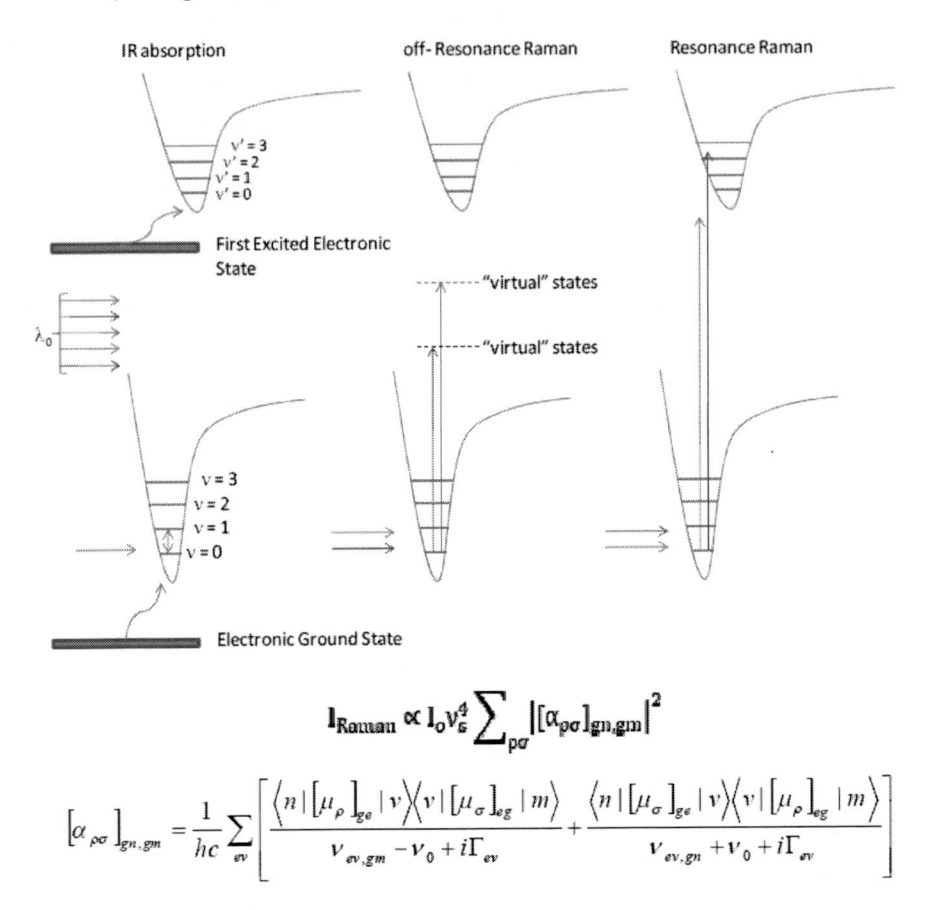

$$I_{Raman} \propto I_0 v_s^4 \sum_{\rho\sigma} \left| [\alpha_{\rho\sigma}]_{gn,gm} \right|^2$$

$$[\alpha_{\rho\sigma}]_{gn,gm} = \frac{1}{hc} \sum_{ev} \left[\frac{\langle n | [\mu_\rho]_{ge} | v \rangle \langle v | [\mu_\sigma]_{eg} | m \rangle}{v_{ev,gm} - v_0 + i\Gamma_{ev}} + \frac{\langle n | [\mu_\sigma]_{ge} | v \rangle \langle v | [\mu_\rho]_{eg} | m \rangle}{v_{ev,gn} + v_0 + i\Gamma_{ev}} \right]$$

Figure 2.4. Schematic representation of two electronic states (ground and excited) and their respective vibrational levels. The arrows indicated the types of transitions that can be occurred among the different levels. It is important to say that in the case of Raman scattering, if the used laser line (λ_0, or as wave number, represent by v_0) has energy similar to one electronic transition of the molecule, the signal can be intensified, known as resonance Raman Effect. In the Figure v_0 and v_s (the scattered frequency is composed by: $v_{ev,gm}$ and $v_{ev,gn}$, the stokes and anti-stokes components, respectively) are the laser line and the scattered frequencies. It was given the equations that describe the Raman Intensity and also the tensor of polarizability. The equation is formed in the numerator part by transition dipole moment integrals between the electronic ground state (g, for the vibrational m or n states) and an excited electronic state (e, for any vibrational v states). The sum is done over all possible (e,v) states. In the denominator part is the difference or sum of the scattered and incident light, added by the dumping factor ($i\Gamma_{ev}$) that contents information about the lifetime of the transition states.

The mathematical and theoretical backgrounds used to the interpretation of the resonance Raman behavior can be found extensively in the literature.[43-46] Generally, the tensor of polarizability is described as shown in the Figure 2.4. The equation is formed in the numerator part by transition dipole moment integrals between the electronic ground state (g, for the vibrational m or n states) and an excited electronic state (e, for any vibrational v states). The sum is done over all possible (e,v) states. In the denominator part is the difference or sum of the scattered and incident light, added by the dumping factor ($i\Gamma_{ev}$) that contents information about the lifetime of the transition states. The theoretical formalism developed by Albrecht et al. is commonly employed to describe the resonance Raman process.[43-46] This enormous intensification makes, in principle, the Raman spectrum easy to be acquired. But, in a state of resonance, a lot of radiation is absorbed, leading to a local heating and frequently can be observed a decomposition of the conducting polymer. Despite of this problem, the RR spectroscopy has been largely used in the study of the different chromophoric units present in polyaniline and others conducting polymers, just by tuning an appropriate laser radiation to an electronic transition of the polymer.

PANI shows a characteristic Raman spectrum for each oxidized or protonated form (see Figure 2.5).[47,48] The Raman spectrum of fully reduced PANI (applied potential of -100 mV) was identified as being formed by benzenoid rings. In contrast, the intensity of the Raman spectra obtained for PANI at 632.8 nm (E_{laser}= 1.97 eV) increased when PANI was oxidized. At applied potential of +600 mV three Raman bands (1160, 1490 and 1595 cm^{-1}) were identified as characteristics of the quinoid structure of PANI. Figure 2.5 presents the segments of PANI and its characteristic Raman bands at their corresponding exciting radiation.[47-50] The PANI-LB is characterized by the vibrational modes of the benzene ring in 1618 and 1181 cm^{-1}, attributed to the νCC and βCH, respectively. The amine group is characterized by νCN stretch at 1220 cm^{-1}. For PANI-PB the βCH band value is at 1157 cm^{-1}, and another characteristic band of PANI-PB is the stretch of C=N bond at 1480 cm^{-1}. Another way to determine the degree of oxidation of PANI,[49] consists in determination of the intensities of the bands at about 1500 cm^{-1} for PANI-LB (νCC) and the band around 1600 cm^{-1} for PANI-PB (νC=C) observed in the IR spectra. The intensity ratio between these two bands ($I(1600)/I(1500)$) is a way to determine qualitatively the degree of oxidation in the chain of PANI.

The Raman studies of PANI-ES suggest the existence of bipolaronic segments (dications or protonated imines).[51] The presence of these segments was also indicated by UV-VIS-NIR data [52] and by EPR[53] The origin of doublet nature of the CN stretch (ca. 1320-1350 cm^{-1}) remains unclear. But, some authors suggested,[48] that the doublet may be associated with the existence of two different conformations of PANI. The Raman study of PANI doped with camphorsulfonic acid (CSA) and dissolved in m-cresol[54,55] revealed a conversion of dications to radical cations. This behavior is associated with changes in the electronic structure, leading to the appearance of new Raman bands and the modification of others, due to, the high charge delocalization on the polymeric backbone.[56,57]

The Raman studies of PANI using near-infrared (NIR) laser line is also found.[58-61] The most peculiar feature observed at 1064.0 nm is the presence of a sharp band around 1375 cm^{-1} in PANI-EB spectrum, which was correlated to polaronic segments localized at two benzene rings. On the other hand,[58] it was proposed that this band was not correlated with protonated segments but with over-oxidized segments such as those present in PANI-PB. Some controversial aspects about the Raman spectra of PANI at NIR excitation were recently re-examined.[61] The bands from 1324 to 1375 cm^{-1} were associated to νC–N of polarons with different conjugation lengths and with the presence of charged phenazine-like and/or oxazine-like rings in PANI-ES as chemically prepared. The formation of cross-linking structures is associated with the ES form of PANI. The bands from 1450 to 1500 cm^{-1} in the

PANI-EB and PANI-PB spectra were associated with the νC=N mode of the quinoid units having different conjugation lengths.

The thermal behavior of PANI revealed that there is the appearance of intense bands at 574, 1393 and 1643 cm^{-1} in the Raman spectra at 632.8 nm during heating.[54,55,62] The same behavior is observed in the poly(diphenylamine) doped with HCSA (PDFA-CSA) during heating. By comparing the results obtained from the thermal monitoring of PANI-CSA and PDFA-CSA, it was possible to assign these bands to the reaction of the polymer with oxygen, with formation of chromophores with oxazine-like rings. It was also demonstrated that the increase of laser power at 1064.0 nm causes deprotonation of PANI-ES and formation of cross-linking segments having phenazine and/or oxazine-like rings. The formation of cross-linking structures is associated with the ES form of PANI.

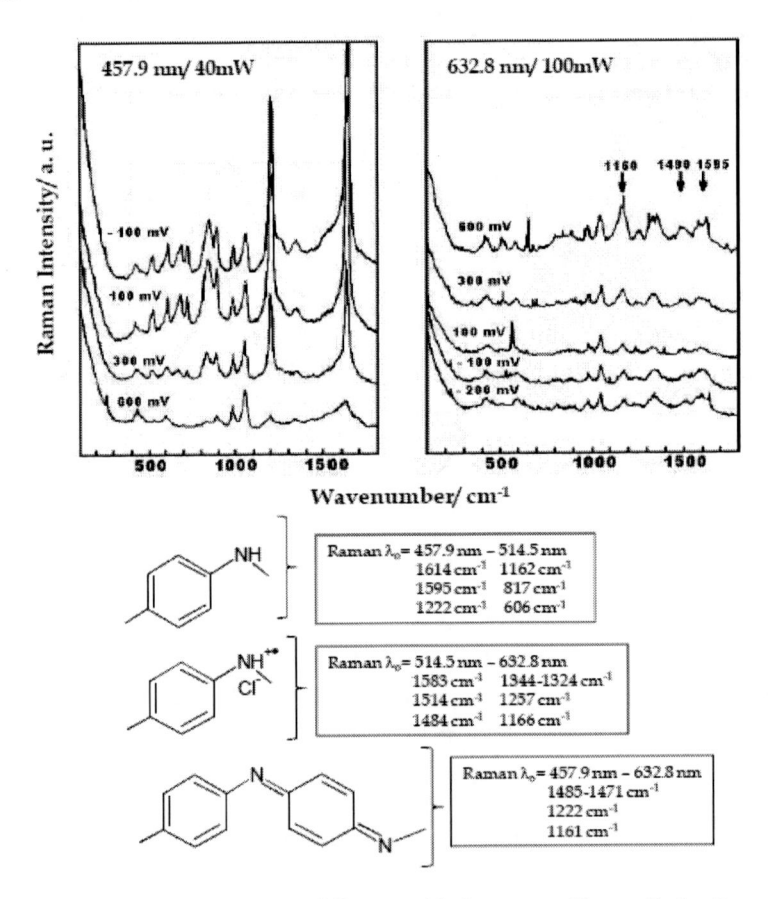

Figure 2.5. Top: Raman spectra of PANI in different oxidation stages (the applied voltage is indicated in the figure) at indicated laser line (457.9 nm and 632.8 nm). Reproduction authorized by Elsevier.[47] Bellow: schematic representation of segments of PANI and its characteristic Raman bands at indicated laser lines.[48].

A broad variety of organic acids have been employed in order to modulate the diameter of PANI nanofibers.[27-30] The FTIR spectra of PANI doped with various organic acids, containing SO$_3$-H groups, show broad bands at about 3430 cm^{-1}, 1560 cm^{-1}, 1480 cm^{-1}, 1130 cm^{-1}, and 800 cm^{-1}, which are related to emeraldine PANI salt.[63] The UV-vis spectra of all doped PANI samples show two polaronic absorptions around 400 and 800 nm. The position of polaronic bands shifts to a long wavelength when the size of organic dopant increases. For instance, the polaron absorption for PANI doped with smaller dopant (α-NSA, α-naphtalenesulfonic acid) is located at 800-900 nm. On the other hand, the polaron absorption for the doped PANI with larger dopant (β-NSA) is shifted to 1060-1118 nm.

The resonance Raman spectra for PANI-β-NSA nanofibers having different diameters show the same profile, it indicates that the morphological differences in PANI-NSA nanofibers have small influence in the Raman spectra from 1000 to 1800 cm^{-1}. Comparing the RR spectra of PANI-NSA fibers to PANI-ES spectrum, bands at 1163 and 1330 cm^{-1} in PANI-NSA spectra can be associated with those at 1165 and 1317-1337 cm^{-1} in PANI-ES spectrum. These bands have been assigned to βC-H and νC-N of polaronic segments, respectively.[30] Their relative intensities in PANI-NSA spectra increase as the molar ratio of β-NSA:aniline increases. Hence, the RR data of the PANI-NSA nanofibers show that the spectral changes observed among the as-prepared PANI-NSA samples are owing to differences in the protonation degrees. The same behavior was observed for PANI nanofibers prepared with stearic acid.[64]

The Raman spectra of PANI nanofibers prepared in micellar media also show the presence of bands at ca. 578, 1400, and 1632 cm^{-1}. These bands were strictly correlated with the formation of cross-linking structures in PANI chains after heating in the presence of air.[62] Different studies show that the bands at ca. 578, 1400, and 1632 cm^{-1} are similar to those observed for dyes with phenoxazine ring. The presence of phenoxazine rings in PANI backbone was also observed in the study of formation of polyaniline nanotubes under different acidic media.[65,66] The authors concluded that the presence of phenoxazine units is crucial for stacking and stabilization of the nanotube wall of PANI.[67]

On the comparison of the spectral behavior of PANI nanofibers/nanotubes prepared with NSA (β-naphtalenesulfonic acid) or with DBSA (dodecybenzenesulfonic acid) indicates that polymeric chains have a certain degree of extended conformation due to the presence of free-carrier absorption in the UV-VIS-NIR spectra. Hence, the presence of 609 cm^{-1} band in the PANI-NSA and PANI-DBSA Raman spectra indicates that these samples have a certain degree of extended conformation. The band at 609 cm^{-1} can be assigned to a vibrational mode related to benzene deformations or torsions. Probably, this mode is sensible to changes of the dihedral angle between neighbors benzene rings, or in other words, sensible to the conformation of the PANI chains.[67]

Electron microscopic images reveal the loss of the fibrous morphology of PANI after treatment of PANI-NSA samples with HCl solution in order to acquire higher doping state.[30] However, further studies reveal that submitting the PANI-NSA to heating treatment at 200°C, occurs the formation of a high degree of cross-linking structures, verified by the appearance of characteristic RR bands at 578, 1398 and 1644 cm^{-1}, hence the fibrous morphology is retained after the doping process.[30,68] PANI nanofibers synthesized in ionic liquids have been studied by Raman spectroscopy. PANI nanofibers were obtained by electropolymerization of aniline in BMIPF$_6$ (1-butyl-3-methyl-imidazolium hexafluoro-phosphate).[69] The Raman spectra show that the PANI is similar to the emeraldine salt form. However, the intensity of the quinoid ring stretching at 1578 cm^{-1} is higher than that of the benzenoid band at 1469 cm^{-1}, indicating the existence of a higher amount of quinoid structures. The authors suggest that the PANI film synthesized in this ionic liquid media is formed by small amount of non-conducting forms such as PANI-EB and PANI-PB.[69]

PANI nanofibers prepared from interfacial polymerization were also characterized by Raman spectroscopy. It was observed that the bands at 200 and 296 cm^{-1}, related to C_{ring}-N-C_{ring} deformation and lattice modes of polaron segments of PANI with type-I crystalline arrangement,[16] practically disappears in the Raman spectra of PANI nanofibers. This effect is very pronounced for the nanofiber sample prepared using 5.0 mol.L^{-1} HCl aqueous solution. The bands at about 400 cm^{-1} indicates the increase of the torsion angles of the C_{ring}-N-C_{ring} segments. The FTIR spectra for PANI nanofibers display higher changes in the region from 2000 to 4000 cm^{-1}.[70] Mainly the bands related to NH_2^+ modes at 2480, 2830, and 2920 cm^{-1} increase in their intensities for PANI samples prepared with higher HCl concentration (higher than 1.0 mol.L^{-1}), consequence of the increase of protonated imine and amine nitrogens in the structure of PANI. The band at 3200 and 3450 cm^{-1}, also change their relative intensities, can be assigned to bonded N-H and free N-H stretching modes.[71,72] The changes in the IR bands associated with an increase in the torsion angles of C_{ring}-N-C_{ring} segments is owing to the formation of bipolarons (protonated, spinless units) in the PANI backbone higher than the PANI samples prepared by the conventional route. The nanostructured surface of PANI permits major diffusion of the ions inside the polymeric matrix leading to a more effective protonation of the polymeric chain than the PANI prepared in the conventional way, leading to the reduction of crystallinity of PANI, and the decrease in the amount of nanofibers.[70]

The screening of the electronic and vibrational structure of the polyaniline nanofibers has been decisive in the studies related to the formation, interactions between the chains, properties and stabilities of the nanostructured polyaniline. Nowadays, two great approaches are used to acquire the PANI with nanostructured morphology without the use of rigid hosts: (i) polymerization of aniline in a micellar media and (ii) polymerization of aniline on the interface between two solvents. However, the morphology of PANI obtained without rigid hosts is more susceptible to the synthetic conditions (such as pH) and also post-synthesis procedures. Mainly, it is observed shifts in the vibrational frequencies of polyaniline and also variations in their intensities. The presence of bands owed to phenoxazine rings is observed in PANI backbone formed in micellar media. The presence of phenoxazine units is crucial for stacking and stabilization of the nanotube wall of PANI. Probably, The π-π stacking formed by phenoxazine rings in the PANI backbone prepared in micellar media is one of the driving forces for the formation of PANI chains with extended conformation and PANI particles with one-dimensional (needles and/or nanofibers) morphology. The changes in the intensities of the vibrational spectra at low energies are associated with an increase in the torsion angles of C_{ring}-N-C_{ring} segments due to the formation of bipolarons (protonated, spinless units) in the PANI backbone higher than the PANI samples prepared by the conventional route. The nanostructured surface of PANI permits major diffusion of the ions inside the polymeric matrix leading to a more effective protonation of the polymeric chain than the PANI prepared in the conventional way, leading to the reduction of crystallinity of PANI, and the decrease in the amount of nanofibers.

2.2. X-Ray Absorption Spectroscopy

2.2.1. Background

The use of electromagnetic radiation as a probe to study the structure of matter has been proved to be of great importance in many areas of chemistry. Since a long time ago the infrared radiation has been used to investigate the vibrations of molecules as well the UV and visible light have been used in the study of the electronic structure of molecules and atoms. Nowadays, the use of electromagnetic radiation with energy ranging from vacuum UV (~ 10-40 eV, 125-31 nm), including the soft X-ray (40-1500 eV, from 31 to 0.8 nm), and reaching to hard X-rays (1500-105 eV, from 0.8 to 0.01 nm),[73] are extremely important for elucidating the electronic and magnetic structure of complex materials.

One of the largest difficulties for the development of spectroscopic techniques that use this range of electromagnetic radiation is to obtain a source capable of generating this type of radiation. Only in the last thirty years with the construction and expansion of the called synchrotron light ring was possible to generate electromagnetic radiation having high intensity between the vacuum UV to hard X-rays.[73-75] This technological advancement has consolidated X-ray absorption techniques as one of the most powerful tools in investigation of the structure of complex materials.

In these accelerators the electrons circulate in relativistic velocities and when they pass through magnetic devices placed in the electron trajectory, it occur the alteration of its orbit and the emission of electromagnetic radiation is observed, this emission has been called as synchrotron light. This light has peculiar characteristics such as bright emission, high brightness and polarization from UV to hard X-rays. The synchrotron accelerators are multi techniques laboratories where different experiments can be performed at same time, such as measures of X-ray absorption and fluorescence of molecules, X-ray crystallography of proteins, diffraction of thin films and X-ray scattering at small angles (SAXS).

The National Synchrotron Light Laboratory (LNLS), located in Campinas, Brazil,[76] has been used for our group in the experiments of X-ray absorption at Nitrogen K edge (NK edge XANES) for different conjugated molecules, such as conducting polymers and organic dyes with the main objective to characterize the oxidation states of nitrogen atoms present in their structures. The measurements have been performed at the SGM beam line (Spherical Grating Monochromator). The importance in determining the oxidation states of nitrogen atoms in PANI is owing to the fact that the electronic structure is intimately associated with the nitrogen oxidation states present in the structure of PANI.[11-14] In the characterization of the oxidation states of nitrogen atoms present in the PANI and its derivatives was necessary to built a spectral database with different molecules containing nitrogen atoms in a very large variety of chemical bonds. This approach made possible to investigate modifications in the PANI structure and formation of new segments in nanostructured PANI.

2.2.2. X-Ray Absorption Process

It is well known that an electromagnetic wave can be scattered and/or absorbed by interacting with the surface of a material. In the case of absorption, the X-ray intensity (I) is attenuated when penetrate into the solid material. This decrease is analogous to the Beer-Lambert Law,[77] i.e. : $I_{(x)} = I_o e^{(-\alpha x)}$ showing that the light intensity decreases due to the penetration into the material (x), since the argument (-αx) is a negative function. The decrease is higher when the magnitude of the absorption coefficient (α) is higher. The value of α is a function of the material structure and also the wavelength of electromagnetic field. Photons in the UV region are strongly absorbed by all solid materials, unlike hard X-rays (λ $\ll 1$ nm) whose absorption coefficient increases with increased density of solid material with atomic number of elements.[78] Therefore materials made of heavy atoms, such as lead, absorb more than the materials made of light atoms like beryllium.[78]

The X-ray absorption occurs if the incident photon energy is transferred to an electron strongly bound to the atom. In Figure 2.6. is schematically represented the absorption of a K shell electrons (1s level) of an atom bonded in a solid material. The absorption coefficient decreases with increasing incident photon energy, but there are sudden changes. These variations correspond to different absorption edges present in the material. Considering photons with energy lower than the ionization threshold (hv_1), they are poorly absorbed by the material since there are no unoccupied states below this energy. But, when the energy of photon reaches to the hv_2 value, there is a sharp increase in the absorption, corresponding to the K edge absorption, this energy is named ionization threshold for the 1s electron.[78-80] If the photon energy continue to rise (hv_3) the electron can escape from the atom, leaving it ionized. This explains why the absorption coefficient is very similar to the cross section of the photoelectric effect, i.e., the probability of ejection of electrons from the atom due to the absorption of photons.[78-80] If the value of the photon energy continues to increase, absorption begins to degrade, but there may be new sudden jumps, since there are other edges in the absorption material.

Similar to the X-ray emission, the X-ray absorption spectra can also be used just for detection of a specific chemical element, because the energy values of the edges are characteristic of each chemical element. Compare, for example, the elements beryllium and lead. The beryllium atom has a small atomic mass, having only four electrons occupying the 1s and 2s levels. These levels have little ionization energy, the threshold level for the 1s electron occurs near to 111.5 eV,[81] compared to the same levels in the atom of lead. For higher energies, the beryllium is transparent for the radiation, because there are no levels that can absorb the energy significantly. Hence, beryllium is an excellent material for making windows in the beam lines, because the X-rays can pass through them. Moreover, lead has 82 electrons, which electrons occupy the K, L, M, N and O shells. Some of these shells have high ionization energy and therefore absorb photons with small wavelengths (around 0.014 nm, this corresponds to an energy of 88 keV). This explains why the lead is a good material for blocking X-rays.

The detection of a chemical element in a material is only the simplest information that is available from the X-ray absorption spectra. In fact, the phenomenon of X-ray absorption is much more complex, and therefore carries much more information. Generally, the absorption spectra are complex, possessing a set of variations that extend over a wide range of energy (tens of units of eV). Figure 2.7. represents a typical X-ray absorption spectrum, have a large absorption near the edge and a series of oscillations that will lose intensity as they move away from the absorption edge. This set of oscillations is divided into two regions, one region near the absorption edge, and one that goes beyond the edge. The structure close to the so-called edge is named NEXAFS (Near-Edge X-ray Absorption Fine Structure") or better known as XANES (X-ray Absorption Near-Edge Structure"), the second region comprises the so-called EXAFS (Extended X-ray Absorption Fine Structure"). The XANES region includes a range of energy before the absorption edge up to the beginning of the EXAFS region. The definition of the boundary between these two regions is arbitrary, but there is some consensus that XANES region extends to 50 eV after the absorption edge. The EXAFS region can be defined as the point where the wavelength of ejected electrons is equal to the distance between the absorber atom and its neighbor atoms, this region can extend up to 1000 eV after the edge.[82]

The EXAFS spectrum originates from effects of interference between the excited atoms.[82-84] The wave function of the excited electron propagates beyond the atom and is partially reflected by neighboring atoms. The interference between the wave that spreads and that is reflected by neighboring atoms causes ripples in the absorption spectrum.[83,84] The interference can be constructive or destructive depending on the wavelength associated with the electron and with the interatomic distance. Thus, through the treatment of EXAFS data is possible to determine the interatomic distances between the atom that has suffered excitation and its neighbor atoms.[85,86]

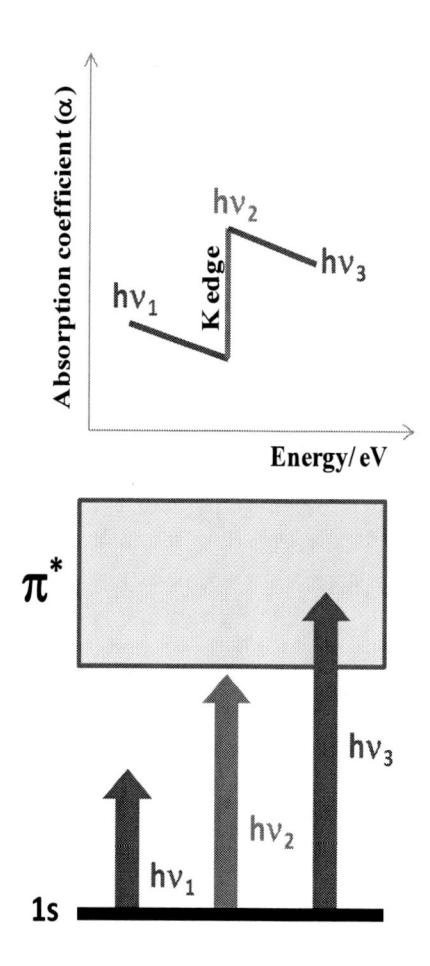

Figure 2.6. Schematic representation of photons with different energies compared in relation to the ionization threshold for a given material. The $h\nu_1$ photon has low energy to produce ionization; the photon $h\nu_2$ has the exact energy for the ionization, so there is a sudden jump in the absorption coefficient that is the experimental characterization of the absorption edge, and finally the photon $h\nu_3$ has much more energy than the edge.

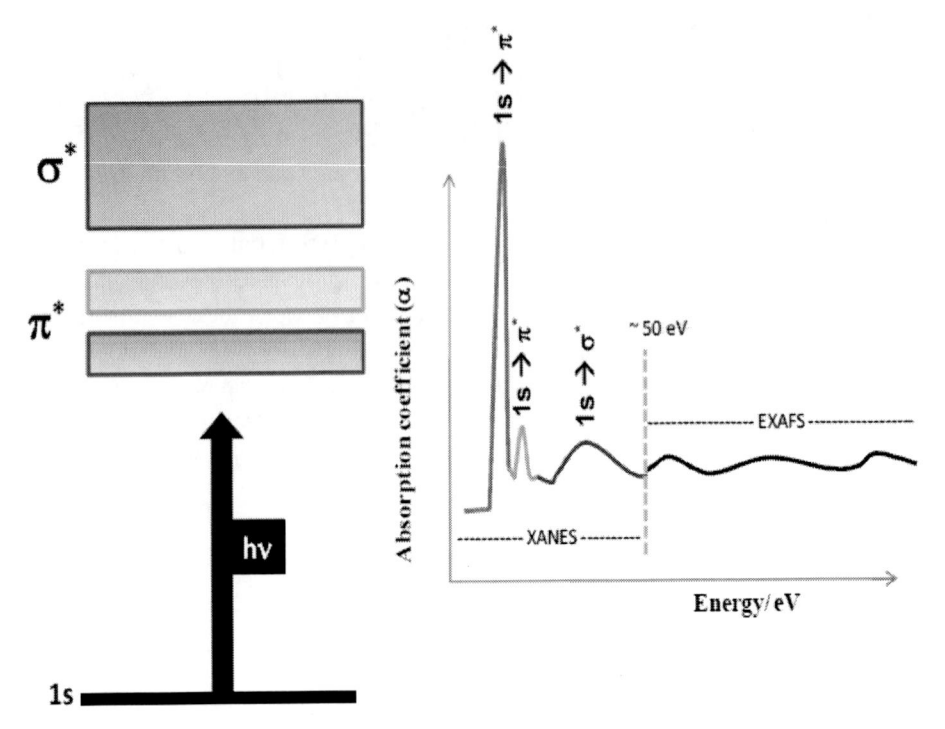

Figure 2.7. The oscillations in the absorption spectrum are owing to the different final states and contain information related to the structure around the atom that absorbed the radiation.

The XANES spectrum contains information similar to the EXAFS spectrum, but the information is more difficult to extract from the math standpoint.[87,89] This is largely due to the different possibilities of transitions that may occur in the solid in the XANES region, which in the language of scattering theory means that scattering is multiple in the XANES region.[87,89] The intensity of absorption is influenced by the number of electrons that occupy the initial state and therefore may participate in absorption, and depends on the density of unoccupied states and the transition momentum.[87,89] The unoccupied energy levels depend on the oxidation state and also the nature of the chemical bond so that this atom with its neighbors, making possible through the XANES spectra, distinguish different states of oxidation of this element. The observed modulations into the XANES spectra are also influenced by the oxidation state and nature of chemical bonds of materials under study. In the follow lines the focus will be in the analysis of XANES spectra at N K edge, because this is the most important edge for the study of PANI and its derivates.

2.2.4. Measurement of the X-Ray Absorption Spectra

For the N K XANES experiment is used a beam line having spherical grating, since this line can operate in the energy range from 250 to 1000 eV,[90] this energy range covers the K edges of carbon, nitrogen and oxygen. The absorption measurements are only possible in conditions of ultra high vacuum (the pressure inside the chamber is ca. 10^{-7} mbar. It could be imagined that the measurements could be made by the intensity ratio between the incident beam (I_o) and transmitted beam (I), similarly to the absorption experiments at visible light region. However all substances strongly absorb in the vacuum UV region, it avoids the measurement by transmittion. In order to circumvent this limitation the absorption is observed by the total electron yield detection.[91] What is measured is a signal which is directly proportional to the amount of photons absorbed. After absorption, emission of electrons (photoelectrons, electrons Auger, and secondary electrons) can occur from the sample and whose intensity is proportional to amount of photons absorbed. In order to keep the sample electrically neutral, it is measured the replacement current of the electrons in the sample (typically the current is of the order of 10^{-12} A) that is proportional to the intensity of photons absorbed.

We briefly describe this:

$$I_{(current\ of\ replacement\ of\ the\ electrons)} \propto I_{(electron\ emitted)} \propto A_{(photons\ absorbed)}$$

Another experimental concern is the number of samples that can be placed at one time in the compartment, because at each replacement of the sample it is necessary to break the vacuum and afterwards re-evacuate the chamber (typically it takes between 3 to 5 hours to reach the required pressure for samples little hygroscopic).

The arrangement used in the SGM can be placed online for solid samples at powder form up to 40 samples at once. The samples in powder form fixed on a metal rod containing carbon double-sided tape (Figure 2.8). Other grooved rods are placed on the main stem for delimit the area of sample (ca. 0.2 cm^2), and prevent mixing of one sample to another, since the measurements are made with the rod positioned vertically in the sample chamber.

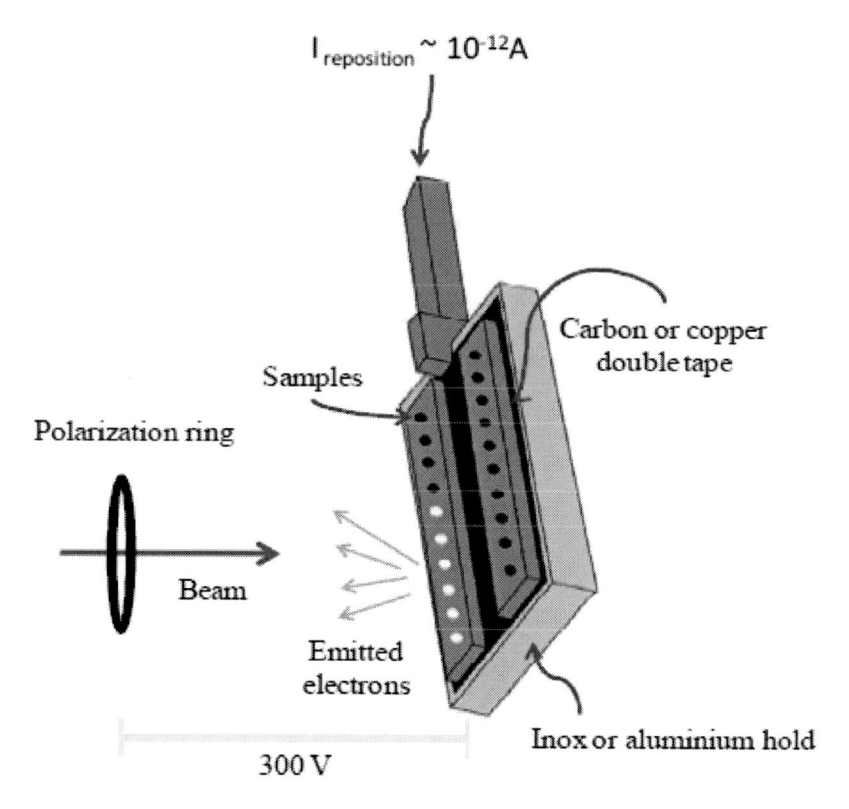

Figure 2.8. Schematic representation of the support hold used for the XANES measurements at N, C or O K edges.

2.2.5. N K XANES Spectroscopy

In the literature, mainly through the work of Henning et al.[92-94] and Hitchcock et al.[95] was possible to verify the utility of the XANES spectroscopy in the determination of the nitrogen oxidation states for different aromatic compounds, and also study effects of conjugation and the presence of quinonoid structures. More recently, this approach was employed in the study of PANI. Henning et al.[92-94] showed that the N K XANES spectrum of PANI is dominated by $1s \rightarrow 2p\pi^*$ transitions whose band position and intensities are dependent on the oxidation state of PANI. The N K XANES spectrum of PANI-EB (see Figure 2.9) is dominated by a band at 397.4 eV, that was assigned by comparison to spectrum taken for 2-hydroxy-3-methoxy-benzilaniline to the transition $1s \rightarrow 2p\pi^*$ of imine Nitrogen (=N-). It was observed that this band is hardly present in the N K XANES spectrum of PANI-LB (fully reduced form of PANI), demonstrating the consistency of the attribution. The presence of two other bands observed at 400.4 eV and 402.1 eV in N K XANES spectrum of PANI-EB, are due to effects of conjugation, and are attributed to transitions $1s \rightarrow 2p\pi^*$ of delocalized imine nitrogens and amines, respectively (see Table 2.1). Henning et al.[92-94] also obtained the N K XANES spectra of PANI synthesized electrochemically with different acids and found that the intensity of the band at 397.4 eV, which is dominant in the spectrum of PANI-EB, suffers a significant reduction in the spectrum of PANI-ES, and the intensity of this band depends on the degree of protonation of the polymer. According to these authors, a second effect of protonation is the complete N1s band suppression $2p\pi^*$ (= N-) which is replaced by a new band at 398.8 eV, attributed to the protonated imine nitrogen.

The PANI-EB, PANI-PB and PANI-ES chemically prepared [11-14] were the first samples studied by our group at the SGM beam line at LNLS/Brazil.[96-101] In addition, a very large spectral database was built in order to know the peak values for nitrogens with different types of bonds.[96-102] The spectra of PANI- EB and PANI-PB (Figure 2.9) have an intense band at 397.7 eV, which can be attributed to the imine nitrogen, similar to that proposed by Henning et al.[92-94] It is evident that this band is not present in the spectrum of PANI-ES (Figure 2.9) due to protonation of the imine nitrogen, the band, which emerges at 399.1 eV was attributed to PANI-ES nitrogen radical cation segments (see Table 2.1).

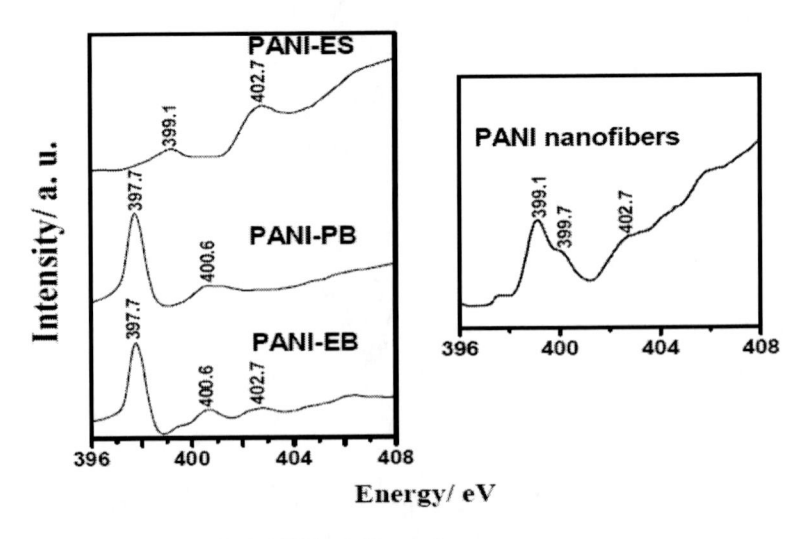

Figure 2.9. N K XANES spectra for PANI in different forms.

Table 2.1. Spectral database of 1s → π* transitions observed at N K edge for different compounds having nitrogen atoms bonded in different types of bonds and chemical environments

Samples	Energy/ eV										
	1s → π* transitions at N K edge										
	=N- Quinoid ring	=N- Phenazine or Oxazine like ring	-N=N-	=N- Pyridine like ring	$\overset{+\bullet}{-N-}$	$\overset{+}{=N-}$	=N··	$\overset{+}{=N-}$ Imidazolium ring	-NH-	-NO$_2$	$N_2H_4^{++}$
PANI-EB	397.7	--	--	--	--	--	400.6	--	402.7	--	--
PANI-PB	397.7	--	--	--	--	--	400.6	--	--	--	--
PANI-ES	--	--	--	--	399.1	--	--	--	402.7	--	--
Phenazine	--	398.2	--	--	--	--	400.6	--	--	--	--
Phenosafranine	--	398.5	--	--	--	399.4	400.7	--	~ 402	--	--
Janus Green B	--	398.3	398.7	--	--	399.6	400.4	--	~ 402	--	--
Nile Blue	--	398.3	--	--	--	399.6	--	--	~ 402	--	--
Oxazine	--	398.3	--	--	--	399.6	--	--	401.8	--	--
4-Amine azobenzene	--	--	398.7	--	--	--	--	--	402.5	--	--
4,4'-Diamine azobenzene	--	--	398.7	--	--	--	--	--	402.0	--	--
Congo red	--	--	398.7	--	--	--	--	--	--	--	--
Methyl Orange	--	--	398.7	--	--	--	--	--	401.5	--	--
Methyl red	--	--	398.8	--	--	--	--	--	--	--	--
Titan yellow	--	--	398.9	--	--	--	400.4	--	--	--	--
Sudan III	--	--	399.0	--	--	--	--	--	--	--	--
Naftol Black	--	--	398.8	--	--	--	--	--	--	--	--
Red Lake C	--	--	399.0	--	--	--	--	--	--	--	--
[N$_2$H$_4$]Cl$_2$	--	--	--	--	--	--	--	--	--	--	404.6
[N$_2$H$_4$]SO$_4$	--	--	--	--	--	--	--	--	--	--	403.5
1,10-fenantroline	--	--	--	398.8	--	--	400.9	--	--	--	--
Dinitrobenzoic acid	--	--	--	--	--	--	--	--	--	403.8	--
N,N'-difenil-1,4-phenylenediamine	--	--	--	--	--	--	--	--	402.0	--	--
Gentian violet	--	--	--	--	--	399.6	--	--	401.8	--	--
[BMIm]Cl*	--	--	--	--	--	--	--	401.9	--	--	--
[HMIm]Br**	--	--	--	--	--	--	--	401.9	--	--	--
[OMIm]Br***	--	--	--	--	--	--	--	401.9	--	--	--

* 1-butyl-3-methylimidazolium chloride.
** 1-hexyl-3-methylimidazolium bromide.
*** 1-octyl-3-methylimidazolium bromide.

After the study of PANI in different forms, our group studied nanocomposites of PANI formed in different inorganic substrates and also PANI nanofibers. The spectra of PANI nanocomposites show different profile compared to the "free" PANI, however, through the use of the spectral database previous built (see Table 2.1) it was possible to analyze the new data. Table 2.1 summarizes all spectral database, the difference between the energy values is little, and however it is possible to correlate the energy value with the type of nitrogen. It should be mentioned that the broad bands are observed at energy values larger than ca. 404 eV, and are assigned to 1s →σ* transitions, whose analysis is more complex, since these bands are much wider than the 1s → π* transitions, hindering the use of these bands in a comparative analysis. Hence, the study of the N K XANES spectra of PANI nanocomposites prove the presence of phenazinic and azo nitrogens in the PANI backbone.[96-99,102]

The study of N K XANES of PANI nanofibers (see Figure 2.9) reveal that the spectral profile is similar to the PANI-ES, however the signal of polarons are more intense and a shoulder at higher energies indicate the presence of nitrogen with delocalized bond and also protonated phenazinic rings.[103] This is a strong indication of the presence of phenazinic rings in PANI nanofibers, as suggested by our group[96-99,102] and from Stejskal's group.[65,66]

CONCLUSION

The study of the structural pattern of the polyaniline nanofibers is decisive for their applications. Nowadays, two great approaches are used to acquire the nanostructured morphology in conducting polymers derived from aniline without the use of rigid hosts: (i) polymerization of aniline in a micellar media and (ii) polymerization of aniline on the interface between two solvents. However, the morphology of PANI obtained without rigid hosts is more susceptible to the synthetic conditions (such as pH) and also post-synthesis procedures. Mainly, it is observed shifts in the vibrational frequencies of polyaniline and also variations in their intensities. The presence of Raman bands owed to phenoxazine rings is observed in PANI backbone formed in micellar media. N K XANES data also indicates the presence of phenazinic/oxazinic rings in the structure of PANI prepared by interfacial polymerization. The presence of phenoxazine units is crucial for stacking and stabilization of the nanotube wall of PANI. Probably, The π-π stacking formed by phenoxazine rings, in the PANI backbone is one of the driving forces for the formation of PANI chains with extended conformation and PANI particles with one-dimensional (needles and/or nanofibers) morphology.

REFERENCES

[1]   Shirakawa, H.; Ikeda, S. Polymer J. 1971, 2, 231.
[2]   Shirakawa, H.; Ikeda, S. J. Poylm. Sci. Chem. 1974, 12, 929.
[3]   Chiang, C. K.; Druy, M. A.; Gau, S. C.; Heeger, A. J.; Louis, E. J.; MacDiarmid, A. G.; Park, Y. W.; Shirakawa, H. J. Am. Chem. Soc. 1978, 100, 1013.
[4]   Chiang, C. K.; Fincher Jr, C. R.; Park, Y. W.; Heeger, A. J.; Shirakawa, H.; Louis, E. J.; MacDiarmid, A. G. Phys. Rev. Lett. 1977, 39, 1098.
[5]   Shirakawa, H. Angew. Chem. Int. Ed. 2001, 40, 2575.
[6]   Shirakawa, H.; Louis, E. J.; MacDiarmid, A. G.; Chiang, C. K.; Heeger, A. J. J. Chem. Soc.: Chem. Commun. 1977, 16, 578.
[7]   MacDiarmid, A. G. Angew. Chem. Int. Ed. 2001, 40, 2581.
[8]   Nigrey, P. J.; MacDiarmid, A. G.; Heeger, A. J. J. Chem. Soc.: Chem. Commun. 1979, 14, 594.
[9]   Han, C. C.; Elsenbaumer, R. L. Synth. Met. 1989, 30, 1, 123.
[10]   Heeger, A. J. Angew. Chem. Int. Ed. 2001, 40, 2591.
[11]   MacDiarmid, A. G.; Epstein, A. J. Faraday Discuss. Chem. Soc. 1989, 88, 317.
[12]   MacDiarmid, A. G.; Epstein, A. J., In Conducting polymers, emerging technologies, Technical Insights: New Jersey, 1989, page 27.

[13] MacDiarmid, A. G.; Chiang, J. C.; Richter A. F.; Sonosiri, N. L. D., Conducting Polymers, Alcácer, L. Ed., Reidel Publications: Dordrecht, 1989.

[14] MacDiarmid, A. G.; Epstein, A. J. In Frontiers of Polymers and Advanced Materials, Prasad, P. N. Ed., Plenum Press: New York, 1994, 251.

[15] Pouget, J. P.; Jozefowicz, M. E.; Epstein, A. J.; Tang, X.; MacDiarmid, A. G. Macromolecules 1991, 24, 779.

[16] Colomban, Ph.; Folch, S.; Gruger, A. Macromolecules 1999, 32, 3080.

[17] Syed, A. A.; Dineson M. K. Talanta 1991, 38, 815.

[18] Wei, Y.; Sun, Y.; Tang, X. J. Phys. Chem. 1989, 93, 4878.

[19] Sasaki, K.; Yaka, M.; Yano, J.; Kitani, A.; Kumai, A. J. Electroanal. Chem. 1986, 215, 401.

[20] Mohilner, D. M.; Argersinger, W. J.; Adams, R. N. J. Am. Chem. Soc. 1962, 84, 3618.

[21] Bacon, J.; Adams, R. N. J. Am. Chem. Soc. 1968, 90, 6596.

[22] Wawzonek, S.; McIntyre, T. W. J. Electrochem. Soc. 1967, 114, 1025.

[23] Geniés, E. M.; Boyle, A.; Lapkowski, M.; Tsintavis, C. Synth. Met. 1990, 36, 139.

[24] Do Nascimento, G. M. In Nanofibers; Kumar, A. Ed.; In-Tech; Austria, 2010, page 349.

[25] Huang, J.; Kaner, R. B. Angew. Chem. Int. Ed. 2004, 43, 5817.

[26] Huang, J.; Kaner, R. B. J. Am. Chem. Soc. 2004, 126, 851.

[27] Zhang, Z. M.; Wei, Z. X.; Wan. M. X. Macromolecules 2002, 35, 5937.

[28] Qiu, H. J.; Wan, M. X.; Matthews, B.; Dai, L. M. Macromolecules 2001, 34, 675.

[29] Wei, Z. X.; Wan, M. X. Adv. Mater. 2002, 14, 1314.

[30] Do Nascimento, G. M.; Silva, C. H. B.; Temperini, M. L. A. Macromol. Rapid Commun. 2006, 27, 255.

[31] Gao, H.; Jiang, T.; Han, B.; Wang, Y.; Du, J.; Liu, Z.; Zhang, J. Polymer 2004, 45, 3017.

[32] Rodrigues, F.; Do Nascimento, G. M.; Santos, P. S. Macromol. Rapid Commun. 2007, 28, 666.

[33] Rodrigues, F.; Do Nascimento, G. M.; Santos, P. S. J. Electron Spectrosc. Rel. Phenom. 2007, 155, 148.

[34] Davis Jr., J. H.; Gordon, C. M.; Hilgers, C.; Wasserscheld, P. In Ionic Liquids in Synthesis, Wasserscheid, P.; Welton, T., Eds.; Wiley–VCH: New York, 2002, page 7.

[35] Wasserscheid, P.; Keim, W. Angew. Chem. Int. Ed. 2000, 39, 3773.

[36] Dupont, J. J. Braz. Chem. Soc. 2004, 15, 3, 341.

[37] Earle, M. J.; Esperança, J. M. S. S.; Gilea, M. A.; Lopes, J. N. C.; Rebelo, L. P. N.; Magee, J. W.; Seddon, K. R.; Wildegren, J. A. Nature 2006, 439, 831.

[38] Nakamoto, N.; Ferraro, J. R. Introduction to Raman Spectroscopy; Academic Press, INC: London 1994.

[39] Howel, G. M. E. Molecular Spectroscopy Laboratories, Chemistry and Chemical Technology; University of Bradford: Bradford, West Yorkshire, United Kingdom, 1996.

[40] Ospitalli, F.; Sabetta, T.; Tullini, F.; Nannetti, M. C.; Di Lornardo, G., J. Raman Spectrosc. 2005, 36, 18.

[41] Smith, B. C. Fundamentals of Fourier Transform Infrared Spectroscopy, CRC Press, Inc.: USA 1996.

[42] Milosevic, M., Appl. Spectrosc. Rev. 2004, 39, 365.

[43] Batchelder, D. N. In Optical Techniques to Characterize Polymer Systems; H. Brässler, Eds.; Elsevier: Amsterdam, 1987.

[44] Batchelder, D. N.; Bloor, D. In Advances in Infrared and Raman Spectroscopy, Wiley-Heyden: London, 1984.

[45] Clark, J. H.; Dines, T. J., Angew. Chem. Int. Ed. Engl. 1986, 25, 131.

[46] McHale, J. L., Molecular Spectroscopy, Prentice-Hall: USA, 1999.

[47] Sariciftci, N. S.; Bartonek, M.; Kuzmany, H.; Neugebauer, H.; Neckel, A., Synth. Met. 1989, 29, 193.

[48] Furukawa, Y.; Ueda, F.; Hydo, Y.; Harada, I.; Nakajima, T.; Kawagoe, T., Macromolecules 1988, 21, 1297.

[49] Quillard, S.; Louarn, G.; Lefrant, S.; MacDiarmid, A. G. Phys. Rev. B 1994, 50, 12496.

[50] Berrada, K.; Quillard, S.; Louarn, G.; Lefrant, S. Synth. Met. 1995, 69, 201.

[51] Louarn, G.; Lapkowski, M.; Quillard, S.; Pron, A.; Buisson, J. P.; Lefrant, S. J. Phys. Chem. 1996, 100, 6998.

[52] Huang, W. S.; MacDiarmid, A. G. Polymer 1993, 34, 1833.

[53] McCall, R. P.; Ginder, J. M.; Leng, J. M.; Ye, H. J.; Manohar, S. K.; Masters, J. G.; Asturias, G. E.; MacDiarmid, A. G.; Epstein, A. J. Phys. Rev. B 1990, 41, 5202.

[54] Pereira da Silva, J. E.; Córdoba de Torresi, S. I.; De Faria, D. L. A.; Temperini, M. L. A. Synth. Met. 1999, 101, 834.

[55] Pereira da Silva, J. E.; De Faria, D. L. A.; Córdoba de Torresi, S. I.; Temperini, M. L. A. Macromolecules 2000, 33, 3077.

[56] Cochet, M.; Louarn, G.; Quillard, S.; Boyer, M. I.; Buisson, J. P.; Lefrant, S., J. Raman Spectrosc. 2000, 31, 1029.

[57] Cochet, M.; Louarn, G.; Quillard, S.; Buisson, J. P.; Lefrant, S., J. Raman Spectrosc. 2000, 31, 1041.

[58] Quillard, S.; Berrada, K.; Louarn, G.; Lefrant, S.; Lapkowski, M.; Pron, A. New J. Chem. 1995, 19, 365.

[59] Engert, C.; Umapathy, S.; Kiefer, W.; Hamaguchi, H. Chem. Phys. Lett. 1994, 218, 87.

[60] Niaura, G.; Mazeikiene, R.; Malinauskas, A. Synth. Met. 2004, 145, 105.

[61] Do Nascimento, G. M.; Temperini, M. L. A. J. Raman Spectrosc. 2008, 39, 772.

[62] Do Nascimento, G. M.; Pereira da Silva, J. E.; Córdoba de Torresi, S. I.; Temperini, M. L. A. Macromolecules 2002, 35, 121.

[63] Huang, J.; Wan, M. X. J. Polym. Science part A: Polym. Chem. 1999, 37, 1277.

[64] Wang, X.; Liu, J.; Huang, X.; Men, L.; Guo, M.; Sun, D. Polym. Bull. 2008, 60, 1.

[65] Trchova, M.; Syedenkova, I.; Konyushenko, E. N.; Stejskal, J.; Holler, P.; Ciric-Marjanovic, G. J. Phys. Chem. B 2006, 110, 9461.

[66] Stejskal, J.; Sapurina, I.; Trchova, M.; Konyushenko, E. M.; Holler, P. Polymer 2006, 47, 8253.

[67] Do Nascimento G. M.; Silva, C. H. B.; Izumi, C. M. S.; Temperini, M. L. A. Spectrochim. Acta Part A 2008, 71, 869.

[68] Do Nascimento, G. M.; Silva, C. H. B.; Temperini, M. L. A. Polym. Degrad. Stab. 2008, 93, 291.

[69] Wei, D.; Kvarnstrom, C.; Lindfors, T.; Ivaska, A. Electrochem. Commun. 2006, 8, 1563.

[70] Do Nascimento, G. M.; Kobata, P. Y. G.; Temperini, M. L. A. J. Phys. Chem. B 2008, 112, 11551.

[71] Jana, T.; Roy, S.; Nandi, A. K. Synth. Met. 2003, 132, 257.

[72] Lunzy, W.; Banka, E. Macromolecules 2000, 33, 425.

[73] Margaritondo, G. Introduction to Synchrotron Radiation; Oxford University Press: New York, 1988; Chapter 1 and 3.

[74] Bilderback, D. H.; Elleaume, P.; Weckert, E. J. Phys. B: At., Mol. Opt. Phys. 2005, 38, S773.

[75] Heald, S. M. In Chemical Analysis; Koningsberger, D. C.; Prins, R., Eds.; John Wiley and Sons: USA, 1988; 92, Chapter 4, page 119.

[76] More information at the official page of LNLS. http://www.lnls.br.

[77] Atkins, P. W. Physical Chemistry; Oxford University Press: Oxford-London, 1994; page 545.

[78] Margaritondo, G. Elements of Synchrotron Light for Biology, Chemistry and Medical Research; Oxford University Press: New York, 2002; Chapter 3.

[79] Margaritondo, G. Introduction to Synchrotron Radiation; Oxford University Press: New York, 1988; Chapter 2.

[80] Durham, P. J. In Chemical Analysis; Koningsberger, D. C.; Prins, R., Eds.; John Wiley and Sons: USA, 1988; 92, Chapter 2, page 53.

[81] Thompson, A. C.; Vaughan, D. X-ray Data Booklet Compiled and Edited by Lawrence Berkeley National Laboratory; Lawrence Berkeley National Laboratory University of California Berkeley: California 2001.

[82] Bianconi, A. Appl. Surf. Sci. 1980, 6, 392.

[83] Lee, P. A.; Citrin, P. H.; Eisenberger, P.; Kincaid, B. M. Rev. Mod. Phys. 1981, 53, 769.

[84] Stern, E. A. In Chemical Analysis; Koningsberger, D. C.; Prins, R., Eds.; John Wiley and Sons: USA, 1988; 92, Chapter 1 and 3.

[85] Abruña, H. D. In Modern Aspects of Electrochemistry; Bockris, J. O'M.; White, R. E.; Conway, B. E., Eds.; Plenum Press: New York, 1989, 20, Chapter 4, page 265.

[86] Parsons, J. G.; Aldrich, M. V.; Gardea-Torresdey, J. L. Appl. Spect. Rev. 2002, 37, 187.

[87] Bianconi, A. In Chemical Analysis; Koningsberger, D. C.; Prins, R., Eds.; John Wiley and Sons: USA, 1988; 92, Chapter 11, page 573.

[88] Manne, R.; Åberg, T. Chem. Phys. Lett. 1970, 7, 282.

[89] Manne, R.; Åberg, T. In Benchmark Papers in Physical Chemistry and Chemical Physics/2:X-ray Photoelectron Spectroscopy; Carlson, T. A., Ed.; Dowden, Hutchinson and Ross, Inc.: Stroudsburg-Pennsylvania 1978; Chapter 3, page 124.

[90] The beam line has a resolution of 0.1 eV. The N K XANES were calibrated considering the value 405.5 eV, according to the reference: Vinogradov, A. S.; Akimov, V. N. Opt. Spectrosc. 1998, 85, 53.

[91] Heald, S. M. In Chemical Analysis; Koningsberger, D. C.; Prins, R., Eds.; John Wiley and Sons: USA, 1988; 92, Chapter 3, page 87.

[92] Hennig, C.; Hallmeier, K. H.; Bach, A.; Bender, S.; Franke, R.; Hormes, J.; Szargan, R. Spectrochim. Acta, Part A 1996, 52, 1079.

[93] Pavlychev, A. A.; Hallmeier, K. H.; Hennig, C.; Hennig, L.; Szargan, R. Chem. Phys. 1995, 201, 547.

[94] Hennig, C.; Hallmeier, K. H.; Szargan, R. Synth. Met. 1998, 92, 161.

[95] Francis, J. T.; Hitchcock, A. P. J. Phys. Chem. 1992, 96, 6598.

[96] Do Nascimento, G. M.; Izumi, C. M. S.; Constantino, V. R. L.; Temperini, M. L. A. Activity report of Brazilian Synchrotron Light Laboratory 2002-2003, 141.

[97] Do Nascimento, G. M.; Constantino, V. R. L.; Temperini, M. L. A. J. Phys. Chem. B 2004, 108, 5564.

[98] Do Nascimento, G. M.; Constantino, V. R. L.; Landers, R.; Temperini, M. L. A. Macromolecules 2004, 25, 9373.

[99] Do Nascimento, G. M.; Temperini, M. L. A. Eur. Polym. J. 2008, 44, 3501.

[100] Sestrem, R. H.; Ferreira, D. C.; Landers, R.; Temperini, M. L. A.; Do Nascimento, G. M. Eur. Polym. J. 2010, 46, 484.

[101] Do Nascimento, G. M.; De Oliveira, R. C.; Pradie, N. A.; Gessolo Lins, P. R.; Worfel, P. R.; Martinez, G. R.; Di Mascio, P.; Dresselhaus, M. S.; Corio, P. J. Photochem. Photobio. A: Chemistry 2010, 211, 99.

[102] Do Nascimento, G. M.; Temperini, M. L. A. Quim. Nova 2006, 29, 823.

[103] Do Nascimento, G. M.; Temperini, M. L. A. submitted to publication elsewhere, 2011.

In: Polymer Synthesis
Editor: E. Kowsari

ISBN 978-1-61324-672-6
© 2012 Nova Science Publishers, Inc.

Chapter 7

SYNTHESIS OF POLYTHIOPHENE NANOPARTICLES AND MICROSPHERES BY TRANSITION METAL MEDIATED OXIDATIVE HETEROGENEOUS POLYMERIZATION

Zi Wang

National Key Laboratory of Nano/Micro Fabrication Technology,
Key Laboratory for Thin Film Microfabrication of the Ministry of Education,
Institute of Micro/Nano Science and Technology, Shanghai Jiao Tong University,
800 Dongchuan Road, Shanghai, China

1. INTRODUCTION

Although electronically conducting polymers have emerged as one of the most highlighted research fields in macromolecular science and engineering, the first preparation of conducting polymer was actually published in the 19th century. [1] In this pioneer work 'aniline black' (now we know it is polyaniline) was synthesized by the oxidation of aniline, but its electronic properties were not recognized. Natta's group first synthesized polyacetylene in 1958, and they found that its conductivity fell in the range of a semiconductor (10^{-11} to 10^{-3} S/cm). [2] Then an inorganic polymer polysulfur nitride (SN)x was produced with a high conductivity of the order of 10^3 S/cm. [3] It was not until Shirakawa, MacDiarmid and Heeger revealed metallic conductivity of p-doped crystalline polyacetylene films that the development of various organic conducting polymers became a booming research area. [4, 5] The three scientists were awarded the Noble Prize in Chemistry in 2000 for their great contribution in this field. [6-8] The conductivity of most conjugated polymers results from the presence of alternating single and double bonds along the macromolecular chain, which can delocalize the π-bonded electrons along the molecular backbone. [9] Except for polyacetylene, [10] current intensively studied conducting polymers include polyaniline, [11] polypyrrole, [12] poly(phenylenevinylene), [13] polythiophene, [14] polyfluorene, [15] polycarbazole, [16] polyphenylene, [17] poly(aryleneethynylene), [18] and their derivatives.

Polythiophene (PT) is one of the most important intrinsic conducting polymers. The unsubstituted PT is a rigid conjugated π system dominated by α-α' linkages of thiophene rings, which has excellent charge transport properties and environmental stabilities. [19] It has shown great potential in the field of organic electronics [20] especially in the production of photovoltaic modules. [13] Many conducting polymer films including PT have been prepared for the applications of photovoltaic devices, [21] chemical sensors [22] and field effect transistors. [23] Those films were usually casted from corresponding polymer solution. However, unsubstituted PT is essentially infusible and insoluble thus limiting its practical applications to a large extent.

The state-of-the-art solution for this insolubility problem is to incorporate kinds of substituents onto the PT backbone, such as alkyl, [24] alkoxyl, [25] perfluoroalkyl, [26] amine, [27] carboxyl, [28] and zwitterionic groups, [29] making PT soluble in organic solvents, water or supercritical fluids. Although this method was initially pursued to solve the processing problem, it has been widely recognized as an approach for tuning the electronic and optical properties of conducting polymers. The recent advances in molecular design for substituted PT and its copolymers aiming for various application especially photovoltaic device productions have been reported in several excellent reviews. [30-32] Furthermore, the incorporation of ionized groups on the backbone can make PT a self-doped conducting polymer, and a monograph on this topic is available. [9] Despite all these merits, the current synthetic procedures for these PT derivatives are usually complex and expensive, which generally need stringent conditions and several individual steps. Therefore they are not easily accessible for industrial scaling up. Also some toxic solvents are usually involved which are not environmentally favorable. In terms of organic photovoltaic applications, the material stability under continuous illumination has to be taken into account. [33] The soluble side chains may be passive in terms of light harvesting and charge transport, [34] and they may also make the materials soft and allow for the intrusion of moisture or other small molecules, [35, 36]. Recently, Bjerring et al prepared unsubstituted PT films by solution processing thermocleaving method. [37]

Another choice to stress the processing problem is PT dispersible micro- or nano-structures, including nanoparticles, microspheres, nanowires, and so on. Not only can these micro- or nano-structures retain the intrinsic properties of PT, but also they can introduce other bonus advantages, e.g. huge specific surface area especially favored by gas sensor application. The template synthesis has long been the primary choice for achieving this objective. [38] It is well known for its accurate morphology control, but removing the template after the synthesis is necessary. On the other hand, heterogeneous polymerization methods, including emulsion polymerization, miniemulsion polymerization, microemulsion polymerization, dispersion polymerization etc., have served decades for the preparation of conventional polymer nanoparticles and microspheres. These methods are robust, feasible, and suitable for both one-pot and continuous reaction process of industrial scale production. Considering the extensive application of conducting polymers in the future, the scaling up capability of synthetic methods seems necessary. Dispersible polyaniline and polypyrrole nanoparticles and nanofibers have been synthesized by oxidative emulsion polymerization with the doping of organic acid. [39-42] In comparison, the reports on how to prepare PT micro- or nanostructures by heterogeneous polymerization are rare.

In most cases, the oxidative polymerization of thiophene takes place in organic solvents such as chloroform and acetonitrile, [43] with the monomer soluble in continuous phase and the polymer precipitating out during the polymerization. If there is no effective stabilizing mechanism for the dispersed phase, this scenario is similar to precipitation polymerization. Trans et al pointed out that with this method one can only obtain coagulated structure with irregular morphology. [44] Instead of direct polymerization from thiophene monomer, they initiated the polymerization from thiophene oligomer and 1-D PT nanowires were formed. Nonetheless, how to achieve well-dispersed PT micro- or nanoparticles in one step by templateless oxidative polymerization starting from inexpensive thiophene monomer, which seems more favored by mass production, still remains a challenge.

The first part of this chapter will focus on the preparation of PT nanoparticles by Cu (II) catalyzed oxidative emulsion polymerization in aqueous medium, where the effect of different metal salt oxidant on the rate of reaction, the particle morphology and the properties of PT samples will be discussed. The second part will describe how to overcome the present difficulty in the preparation of well defined PT microspheres in organic solvents by oxidative dispersion polymerization. The diameter control by surfactant concentration and solvent power tuning will also be discussed.

2. POLYTHIOPHENE NANOPARTICLES BY OXIDATIVE EMULSION POLYMERIZATION

Water is an environmental friendly medium suitable for heterogeneous polymerization. However, there are few reports on the oxidative polymerization of thiophene in water. This is probably due to the very low yield caused by extremely low solubility of monomer and polymer in water. [45] Lee et al pioneered in the exploration of oxidative emulsion polymerization of thiophene in water. [46,47] In their study, hydroxyl peroxide was employed as a "co-oxidant" to regenerate the Fe^{3+} from Fe^{2+} formed during the polymerization. The results showed that high yield could be achieved by using only a catalytic amount of $FeCl_3$. The synthesized PT nanoparticles could be easily redispersed in various organic solvents, but the conductivity data were not reported. Obviously, further study on this system is necessary. The influence of the oxidants on the polymerization process and the properties of PT also need investigation, since the literatures using oxidants other than iron (III) salts are rare. [48]

In this part, we will report the experimental results of a recent study on the Cu (II) salts catalyzed oxidative polymerization of thiophene in our group [49]. The effects of different Cu (II) salts on the polymerization rate as well as the properties and morphology of PT will be discussed. The results showed that the rate of polymerization strongly depends on the types of Cu (II) salts despite their very low concentration. Also these Cu (II) oxidants could lead to PT materials with quite different morphologies and properties such as surface resistivity and thermal stability.

2.1. Synthesis of Polythiophene Nanoparticles

The oxidative emulsion polymerization of thiophene was performed in a 250mL flask equipped with mechanical stirrer at 50°C. 0.1g sodium dodecyl sulfonate (SDS) and 2g thiophene was added into 60mL deionized water, and the mixture was stirred at 300rpm for

20 min. Then 10g 30% H_2O_2 solution was introduced, and 2×10^{-4} mol Cu (II) salts as oxidants dissolved in 5mL water was charged in one portion. The reaction was allowed to proceed for 7 hours.

For yield calculation, a syringe was used to extract 5mL reacting mixture at given time intervals. The extracted mixture was diluted with 5mL deionized water, and the polymer was precipitated by adding sodium chloride followed by centrifugation. The precipitates were washed with deionized water for several times in order to remove the oxidants and surfactants, and then dried under vacuum at 50°C for 48 hours.

2.2. Effect of Copper Salts on Polymerization Process

The same amount copper salt (2×10^{-4} mol) was used in all the four polymerization runs in order to investigate the effect of different anions. The rates of polymerization for all the 4 systems were relatively high in the beginning, and they dropped gradually as the reactions proceeded. Among all the 4 oxidants, $Cu(NO_3)_2$ and $CuSO_4$ demonstrated the highest catalytic capability. The yields were above 70% at 45 min and exceeded 90% after 7 hours. Although the rate of the polymerization mediated by $CuCl_2$ was slightly lower than the above two in the beginning of reaction, the yield was pushed up to 86% after 7 hours. Whereas the rate of polymerization in $CuBr_2$ catalyzed system was significantly low, whose yield after 7 hours was 39%. Since the concentrations of Cu^{2+} were the same, the deviation in the rates of polymerization should be attributed to the different oxidant anions involved although their concentrations were low. Owning to the presence of H_2O_2 Cu^{2+} ion could be regenerated from Cu^+, so only a small amount of Cu (II) salts were used. Hence the concentrations of oxidant anions in the polymerization were also quite low. It can be inferred from above that the rate of polymerization depends strongly on the types of oxidant anions despite their low concentrations. The reaction time-yield curves can be found in ref [49].

For further investigation on the effect of oxidant anions, an extra polymerization run was performed. In addition to 2×10^{-4} mol $CuBr_2$ as oxidant, 4×10^{-4} mol HNO_3 was added to simulate the case of using $Cu(NO_3)_2$ as oxidant. However, the increase in the rate of polymerization was not observed. This indicated that the introduction of additional NO_3^- in the presence of Br^- could not lead to the same reaction rate as that in the polymerization with $Cu(NO_3)_2$ as oxidant.

In the initial state of the polymerization, some of the thiophene monomer forms micelles stabilized by SDS, and the rest (usually majority of them) forms monomer drops in water. The copper salts which initiate the polymerization are soluble in water. This scenario is similar to that of free radical emulsion polymerization, [50] but the kinetics of reaction is different from Smith-Ewart equations. Because the surface area of micelles is much larger than that of monomer drops, the reaction taking place in those drops can be omitted. [50] The procedure of polymerization is generally known as follows: [20] (1) the metal cations in aqueous phase diffuse into the micelles; (2) the oxidation of thiophene monomer by the metal cations followed by the formation of cationic radicals; (3) The coupling between the thiophene cationic radicals to form the dimers; (4) The dimers were further oxidized to form cationic radicals and continue the coupling reaction to gradually form high polymer chains. The oxidant anions probably act in two ways: they may interfere with the diffusion of metal cations from aqueous phase to micelles by electrostatic attraction; or they may interact with monomeric and oligomeric cationic radicals thus influence their propagation reaction. Although few papers directly investigated the effect of the interaction between oxidant anions and metal cations on the polymerization process, the possible influence of the interaction between surfactant ions and metal cations was realized. [46] The anionic surfactant could draw the metal cations into the micelles by electrostatic attraction and promote the initiation reaction therein. But the cationic surfactant may act in the opposite way and decrease the rate of polymerization. Since the same electrostatic attraction exists between the oxidant anions and metal cations in aqueous phase and micelles, it seems reasonable to take the possibility of this electrostatic attraction affecting on the polymerization into account. Furthermore, there has been experimental data showing that the chlorine anion in the oxidant ($FeCl_3$) acts as a dopant in the final PT materials synthesized. [48] Therefore it seems reasonable to infer that, in this research the oxidant anions such as NO_3^- might also interact with propagating cationic radicals thus interfere with the polymerization procedure. Although the complicated influence of the oxidant anions on the polymerization of thiophene needs further investigation, it is quite clear that even a low concentration of oxidant anions is enough to interfere with the polymerization process significantly.

2.3. Morphology of Polythiophene

The morphology of PT samples was observed under a Zeiss Ultra 55 field emission scanning electron microscope, with an accelerating voltage of 5kV and InLens observation mode. The samples were diluted with dionized water and coated on a silicon wafer for SEM characterization, and their morphology was shown in Figure 1. It can be seen that well dispersed PT nanoparticles with diameter ranging between 60 to 150 nm were synthesized using $CuCl_2$ as oxidant (Figure 1a). By contrast, the product with irregular morphology was obtained with $CuBr_2$ as oxidant (Figure 1b). Using $CuSO_4$ and $Cu(NO_3)_2$ could also lead to spherical PT nanoparticles with similar diameters as those prepared under $CuCl_2$. But the PT particles with $CuSO_4$ (Figure 1c) agglomerated quite some with ill-defined edges. Those by $Cu(NO_3)_2$ (Figure 1d) seem to be dispersed better. The spherical morphology of these nanoparticles indicates the formation of stable colloid phase during the polymerization proceeded. These results demonstrate that, except for the rate of polymerization, different Cu (II) salts as oxidants will also influence the morphology of PT particles.

In recent years, it has been revealed that solvent-dispersible nanostructures of PT and its derivatives can be synthesized by oxidative polymerization with the addition of surfactants. The morphology of polythiophene varied much with different combination of solvents and surfactants. Nonetheless, in most cases these obtained polythiophene particles showed agglomerated morphology more or less, thus necessitating the improvement in the stability of polymer dispersed phase during polymerization. We believe that if small molecular surfactants are used in the heterogeneous polymerization of thiophene, water as the reaction medium may be a good choice to form the stable colloid system during the polymerization. In aqueous medium, the surfactant molecules will form spherical micelles spontaneously owning to their amphiphilic nature, with the hydrophilic parts extending and the hydrophobic segments aggregating inside. These micelles can accommodate the monomers and become the main loci for polymerization.

(a)

(b)

(c)

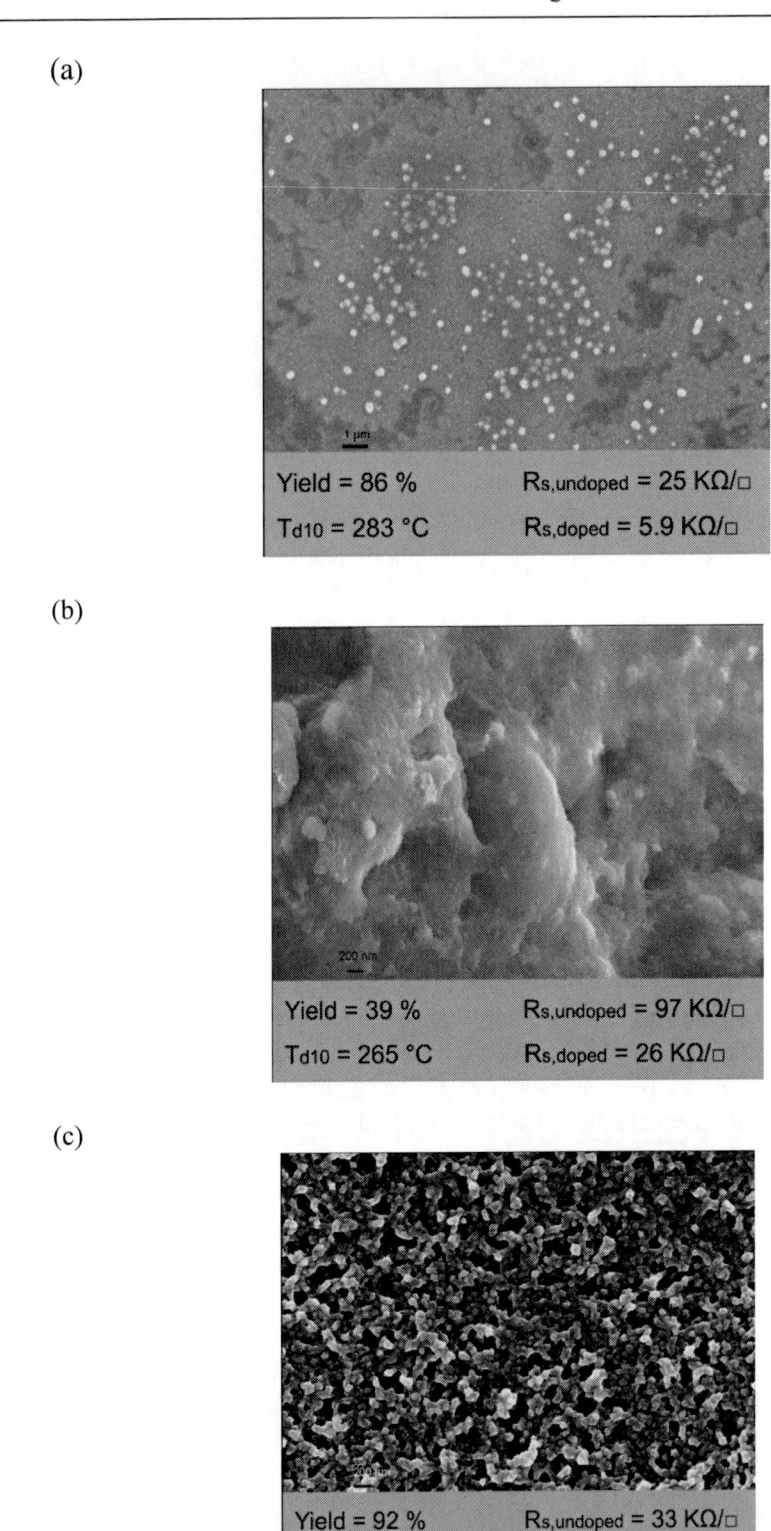

Figure 1. (Continued).

(d)

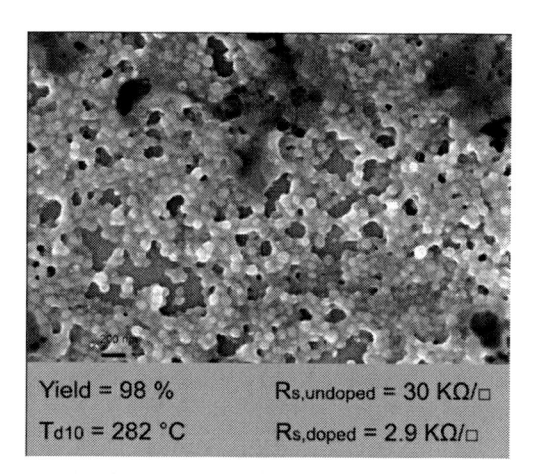

Yield = 98 % $R_{s,undoped}$ = 30 KΩ/□

T_{d10} = 282 °C $R_{s,doped}$ = 2.9 KΩ/□

Figure 1. SEM micrographs of PT samples synthesized in water with different Cu (II) oxidants: (a) CuCl$_2$; (b) CuBr$_2$; (c) CuSO$_4$; (d) Cu(NO$_3$)$_2$. $R_{s,\,undoped}$ means surface resistivity of undoped polythiophene samples; $R_{s,\,doped}$ means surface resistivity of iodine doped polythiophene samples; T_{d10} means temperature as 10% weight loss by TGA.

As the polymerization proceeds the hydrophobic segments of surfactants can absorb onto the polymer chains to form stable colloid particles. On the other hand, it is difficult for these small molecular surfactants to aggregate into micelles in organic solvents since the hydrophobic segment of surfactant is soluble in continuous phase. This might be the reason why it was quite difficult to prepare well-dispersed spherical PT particles without coagulation in organic solvents when small molecular surfactants were employed. In order to achieve stable dispersed phase in organic solvents, macromolecular surfactants may be required. This will be demonstrated later in this chapter.

After being dried, those PT nanoparticles can be redispersed in some common organic solvents. In this study, 30 mg dried PT samples were added into a vial and then 10mL of given organic solvent was introduced. Among all the 9 solvents tested, PT nanoparticles showed the highest dispersibility in DMSO and DMF without any precipitation after 1 month. They could also be easily dispersed in THF, acetone, methanol and ethanol, although some precipitation was observed after 2 weeks. By comparison, the PT particles could not be totally dissolved in dioxane under the test conditions. They could not be dispersed in ether or chloroform since no color change was observed in liquid phase of the mixture. The photos of these PT dispersions can be found in ref [49].

Many practical applications demand the conducting polymer being used as films. One of the most important properties that characterize the conducting film materials is the surface resistivity. The surface resistivity of PT samples coated on slides was measured by a RTS-8 four-probe resistivity instrument (Four Probe Tech.) at room temperature. According to the data listed in Figure 1, the surface resistivity of undoped polythiophene is quite high. Among all the 4 samples the one prepared with CuCl$_2$ shows the lowest value of 25 KΩ/□, and the two with CuSO$_4$ and Cu(NO$_3$)$_2$ gave slightly higher values of 33 KΩ/□ and 30 KΩ/□ respectively. The surface resistivity for the sample synthesized with CuBr is high as 97KΩ/□. After being doped with iodine vapor, all the surface resistivity for PT samples except that prepared with CuBr was lowered by about one magnitude. The data for the two samples with CuSO$_4$ and Cu(NO$_3$)$_2$ become slightly lower than that with CuCl$_2$. Although a significant drop of surface resistivity was observed on doping for the sample with CuBr, it is still on the order of $10^4 \Omega/\square$. Such a high surface resistivity for this sample may be related with its low yield and irregular morphology.

According to the UV-vis absorption spetra for DMSO dispersion of PT nanoparticles, all the PTh samples showed absorption around 400 nm due to the π-π^* transition for large π-conjugation structure, compared with the absorption peak around 320 nm attributed to the π-π^* transition for thiophene monomer. These results are similar to those reported by Lee et al in FeCl$_3$ catalyzed polymerization. [46] However, the UV absorption at 400 nm indicated somewhat limited effective conjugated length of the synthesized polythiophene which needs to be improved in future study. It is also interesting to note that, the wavelength of absorption peak for polythiophene prepared with CuBr (406 nm) was slightly higher than the other three (398 nm), despite its low yield and conductivity. The UV-vis absorption spectra can be found in Ref [49].

The PT prepared with CuSO$_4$ exhibited the best thermal stability with the temperature for 10% weight loss (T_{d10}) of 292°C and the high residual weight of 25%. The thermal stability for the two samples synthesized with CuCl$_2$ and Cu(NO$_3$)$_2$ is almost the same, whose T_{d10} are 283°C and 282°C respectively. The T_{d10} for the sample prepared with CuBr are the lowest, being 265°C, and the residual weight is only 10%. For the oxidative polymerization of thiophene, the molecular weight rises with the monomer conversion. The molecular weight of the polythiophene synthesized with CuBr is probably low due to the low yield, thus decreasing its thermal stability. In addition, the TGA test for the polythiophene synthesized without surfactant (with Cu(NO$_3$)$_2$ as oxidant) was performed in order to check if the surfactant will affect the thermal stability. The result showed that T_{d10} for this sample is 271°C, lower than that using SDS as surfactant. In the case of aqueous polymerization of thiophene, it seems adding anionic surfactant SDS can improve the thermal stability of the PT synthesized. The TGA curves for all the samples can be found in Ref [49].

3. POLYTHIOPHENE MICROSPHERES BY OXIDATIVE DISPERSION POLYMERIZATION

As previously described, most of the oxidative polymerizations of thiophene were performed in organic solvents, and it has been difficult to obtain well dispersed PT particles with conventionally used small molecular surfactants. We postulate that the problem encountered in organic solvents can be solved by using macromolecular stabilizers instead of small molecule surfactants. [51] The experimental results showed that, by using polyvinylpyrrolidone (PVP) as a macromolecular stabilizer in the oxidative polymerization of thiophene, we obtained well-dispersed PT microspheres with tunable diameters ranging between 180 nm and 1.1 μm.

3.1. Two Different Scenarios in Aqueous Medium and Organic Solvent

We first made a comparison between two oxidative polymerizations of thiophene with SDS in water and ethanol respectively, in order to describe how our idea is generated. In aqueous medium, the initial state of reaction is similar to that of emulsion polymerization as previously described. The micelles are the main loci for polymerization owning to their high monomer concentration compared to the monomer in solution. Therefore micelle nucleation dominates, and the formation of stable colloid particles is straightforward. In this way well-dispersed PT particles can be obtained in aqueous medium with SDS as surfactant (see Part 2). But when water is replaced by organic solvents (e.g. ethanol used here), the reaction begins as a homogeneous system because both thiophene and SDS are soluble. When the molecular weight of the propagating oligomers exceeds the critical chain length for precipitation, there is not enough driving force for the hydrophobic segments of surfactant to absorb onto the oligomer chains (namely "over soluble"). Then these oligomers coagulate and precipitate instead of continuing the propagation reaction in the colloid particles stabilized by surfactants. In this case similar to conventional precipitation polymerization, it is difficult to obtain discrete PT particles (Figure 2a). But if appropriate macromolecular stabilizers with large excluded volume and good compatibility with oligomer chains are used, the coalescence of unstable particles will stop when there are enough stabilizer chains absorbed on the particle surface to provide steric stabilization. [52, 53] This procedure is similar to conventional dispersion polymerization.

3.2. Effect of Macromolecular Surfactant in Organic Solvents

Then we performed the oxidative polymerization of thiophene in ethanol in the presence of macromolecular surfactant. The experiments were similar to those of emulsion polymerization described above, except that PVP ($M_v \sim 40,000$) dissolved in the mixture of 50g ethanol and 10g 30% H_2O_2 solution was employed as surfactant. 1.5g $Cu(NO_3)_2 \cdot 3H_2O$ was used as oxidant. The reaction was allowed to proceed for 24 hours. For yield calculation, the reaction mixtures were added with 200 mL deionized water, and purified by five centrifuging-washing cycles. Finally all the samples were dried under vacuum at 50°C for 48 hours.

In the presence of 5 wt% PVP, well-dispersed PT particles with quite regular morphology were obtained (Figure 2b). The numerical average particle diameter (D_n) is 181 nm with a quite narrow distribution ($D_w/D_n = 1.04$, D_w represents the weight average particle diameter). This result shows that PVP is an effective stabilizer for dispersion polymerization of thiophene in ethanol/H_2O_2 solution mixture. In order to see if the particle size could be controlled in the same way as in conventional dispersion polymerization, we next decreased the stabilizer concentration to 2.0 wt%. As expected, D_n increased to 556 nm (Figure 2c). With lower stabilizer concentration, more unstablized particles have to coalesce before sterically-stable particles can form. [52] Although the particle size distribution is somewhat broad ($D_w/D_n = 1.09$), stable dispersed phase still manage to form with decreased surfactant concentration. The dependence of particle size on monomer concentration was also checked. Figure 2d shows that when monomer mass was increased from 6g to 9g, the particle size increased as expected ($D_n = 231$ nm). Nonetheless, particle size distribution also increased considerably ($D_w/D_n = 1.19$) and some agglomeration was observed.

In order to check if it was possible to extend current methodology to other organic solvents with quite different solvent power, we performed one extra polymerization run in acetone/H_2O_2 solution mixture. Although PVP is not well soluble in acetone alone, it can be dissolved in acetone/H_2O_2 solution mixture. It should be stressed that 1.5g $FeCl_3 \cdot 6H_2O$ rather than $Cu(NO_3)_2 \cdot 3H_2O$ was used here, because the rate of Cu (II) mediated polymerization in acetone was found very slow (less than 15% yield after 24 hours). Figure 2e shows that well dispersed PT microspheres can still be obtained by the same procedure. No agglomeration is observed, but the size of PT microspheres is much larger than those prepared in ethanol/H_2O_2 solution mixture (D_n = 1125nm, D_w/D_n = 1.08). The conformation of PVP chains should be less extended than that in ethanol due to the anti-solvent effect of acetone, so the excluded volume for each PVP chain is also decreased. Therefore more unstablized particles have to coalesce before sterically-stable particles can form, and such a large increase in particle size is actually expected. Moreover, the relatively high yield (81%) should also be taken into account.

(a)

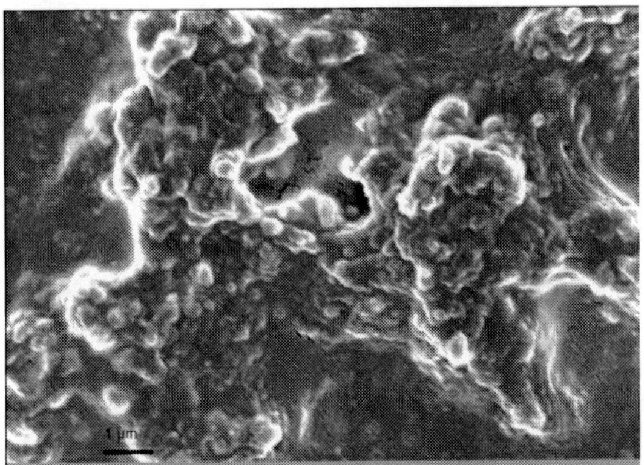

6 g monomer with 5 wt% SDS in ethanol

Yield = 46 %   Dn = -   Dw/Dn = -

(b)

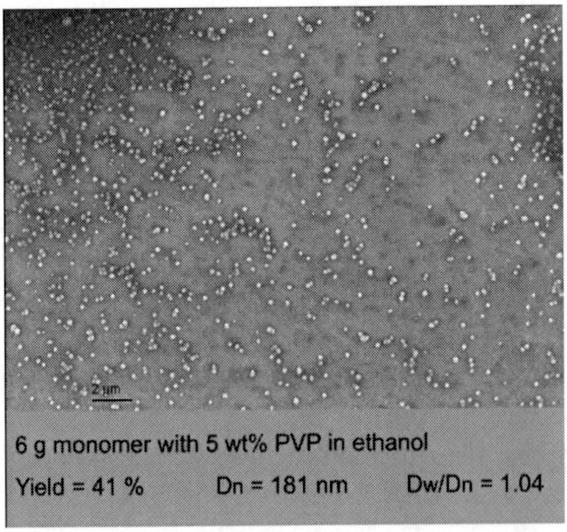

6 g monomer with 5 wt% PVP in ethanol

Yield = 41 %   Dn = 181 nm   Dw/Dn = 1.04

(c)

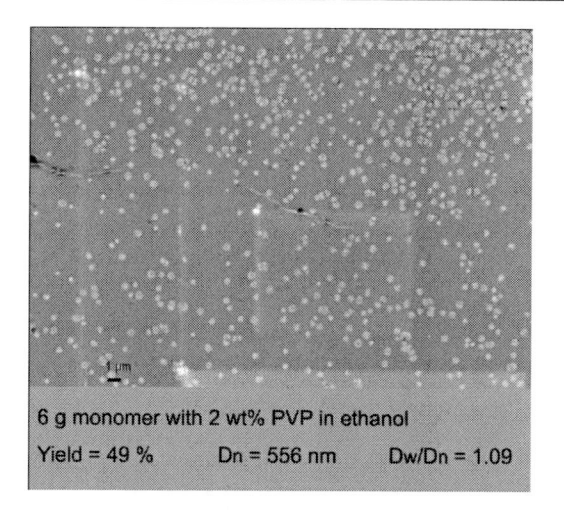

(d)

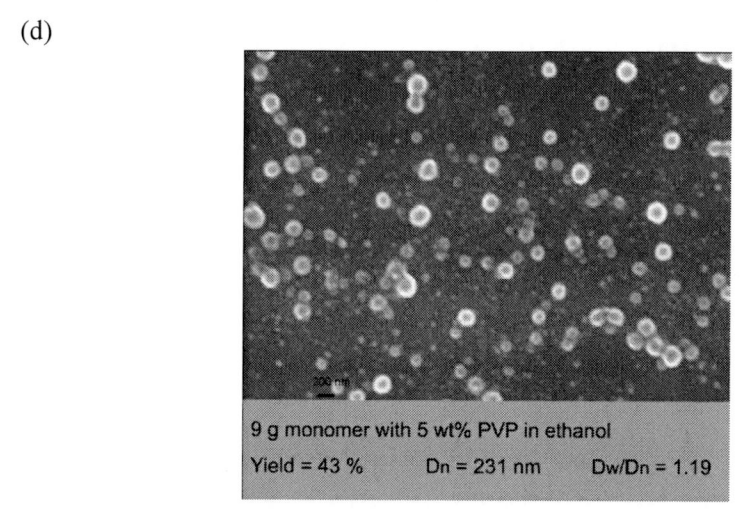

(e)

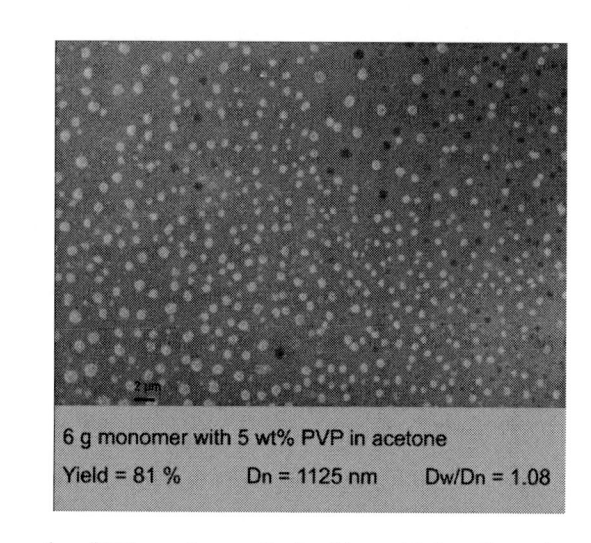

Figure 2. SEM micrographs of PT samples synthesized by oxidative dispersion polymerization.

It is well-known that one of the most important applications of dispersion polymerization is to prepare polymer microsphere whose size (0.1-15 μm) falls in-between those of suspension polymerization (50-1000 μm) and emulsion polymerization (0.06-0.7 μm). [52] Indeed, the diameter of PT particles prepared by emulsion polymerization is found smaller than 150 nm. [46, 49] In comparison, the dispersion polymerization of thiophene reported here is capable of generating PT microsphere sizing between 180 nm and 1.1 μm. Owning to their unique properties, conducting polymer microspheres are expected to show significant potential in multiple fields especially for medical and electronic applications. [12] Recently, high efficiency polymer solar cells based on unsubstituted PT has been reported, showing its capability for photovoltaic applications. [34] In the production of photovoltaic devices, stable PT dispersion is especially suitable for the fabrication of polymer inks for spin-coating and roll-to-roll process. [54] While other conducting polymer microspheres (e.g. polyaniline and polypyrrole) have been synthesized successfully by dispersion polymerization, realizing that PT microspheres can also be prepared by this facial methodology may considerably expand their capability for practical applications.

Although PT microspheres can be prepared through this way, quite high amount of metal salts is required compared with that in the oxidative emulsion polymerization. [46, 49] The rate of polymerization in organic solvents is also lower than that in water. This is probably because the metal salts may complex with solvent molecules thus reducing its oxidative capability, and similar phenomenon was observed in Fe (III) catalyzed oxidative polymerization of pyrrole in ethanol. [55] The relatively low local monomer concentration compared to emulsion polymerization in water is also one of the factors which reduce the rate of polymerization. Additionally, the relatively broad particle size distribution upon the increased monomer concentration indicates that the efficiency of stabilizer still needs to be improved. Regarding to the previous studies on conventional dispersion polymerization, [56-63] elaborately designed amphiphilic block copolymer stabilizers or reactive macromonomers may show higher stabilizing efficiency and allow better control on PT microspheres morphology. They may also be used for the preparation of conducting polymer composite microspheres and the surface functionalization of conducting polymer particles.

ACKNOWLEDGMENTS

The author appreciates the financial support by National Natural Science Foundation of China and Shanghai Science and Technology Grant.

REFERENCES

[1]   Letheby, H. J. Chem. Soc. 1862, 15, 161-163.
[2]   Natta, G.; Mazzanti, G. Atti accad. Nazl. Lincei Rend. 1958, 25, 3-12.
[3]   Walatka, V. V. J.; Labes, M. M.; Perlstein, J. H. Phys. Rev. Lett. 1973, 31, 1139-1142.
[4]   Chiang, C. K.; Fincher, C. R. J.; Jr. Park, Y. W.; Heeger, A. J.; Shirakawa, H.; Louis, E. J.; Gau, S. C.; MacDiarmid, A. G. Phys. Rev. Lett. 1977, 39, 1098-1101.
[5]   Shirakawa, H.; Louis, E. J.; MacDiarmid, A. G.; Chiang, C. K., Heeger, A. J. Chem.

Commun. 1977, 578-580.

[6]   Shirakawa, H. Angew. Chem. Int. Ed. 2001, 40, 2574-2580.

[7]   MacDiarmid, A. G. Curr. Appl. Phys. 2001, 1, 269-279.

[8]   Heeger, A. J. Angew. Chem. Int. Ed. 2001, 40, 2591-2611.

[9]   Freund, M. S.; Deore, B. Self-doped Conducting Polymers, John Wiley and Sons: Chichester, 2007.

[10] Masuda, T. J. Polym. Sci. Part A: Polym. Chem. 2007, 45, 165-180.

[11] Bhadra, S.; Khastgir, D.; Singha, N. K.; Lee. J. H. Prog. Polym. Sci. 2009, 34, 783-810.

[12] Guimard, N. K.; Gomez, N.; Schmidt, C. E. Prog. Polym. Sci. 2007, 32, 876-921.

[13] Gune, S.; Neugebauer, H.; Sariciftci, N. S. Chem. Rev. 2007, 107, 1324-1338.

[14] Osaka, I.; McCullough, R. D. Acc. Chem. Res. 2008, 41, 1202-1214.

[15] Knaapila, M.; Winokur, M. J. Adv. Polym. Sci. 2008, 212, 227-272.

[16] Heeger, A. J. Chem. Soc. Rev. 2010, 39, 2354-2371.

[17] Grimsdale, A. C.; Müllen, K. Adv. Polym. Sci. 2006, 199, 1-82.

[18] Thomas, S. W.; Joly, G. D.; Swager, T. M. Chem. Rev. 2007, 107, 1339-1386.

[19] Schenning, A. P. H. J.; Meijer, E. W. Chem. Commun. 2005, 3245-3258.

[20] Roncali, J. Chem. Rev. 1992, 92, 711-738.

[21] Mikroyannidis, J. A.; Stylianakis, M. M.; Dong, Q.; Zhou, Y.; Tian, W. Synth. Met. 2009, 159, 1471-1477.

[22] Liu, W.; Pink, M.; Lee, D. J. Am. Chem. Soc. 2009, 131, 8703-8707.

[23] Lim, B.; Baeg, K.; Jeong, H.; Jo, J.; Kim, H.; Park, J.; Noh, Y.; Vak, D.; Park, J.-H.; Park, J.-W.; Kim, D.-Y. Adv. Mater. 2009, 21, 2808-2814.

[24] McCullough, R. D.; Lowe, R. D.; Jayaraman, M.; Anderson, D. L. J. Org. Chem. 1993, 58, 904-912.

[25] Sheina, E. E.; Khersonsky, S. M.; Jones, E. G.; McCullough, R. D. Chem. Mater. 2005, 17, 3317-3319.

[26] Li, L.; Counts, K. E.; Kurosawa, S.; Teja, A.S. Adv. Mater. 2004, 16, 180-183.

[27] Ogawa, K.; Stafford, J. A.; Rothstein, S. D.; Tallman, D. E.; Rasmussen, S. C. Synth. Met. 2005, 152, 137-140.

[28] Bao, Z.; Lovinger, A. J. Chem. Mater. 1999, 11, 2607-2612.

[29] Nilsson, K. P. R.; Inganas, O. Nat. Mater. 2003, 2, 419-424.

[30] Bundgaard, E.; Krebs, F. C. Sol. Energy Mater. Sol. Cells 2007, 91, 954-985.

[31] Kroon, R.; Lenes, M.; Hummelen, J. C.; Blom, P. W. M.; De Bore, B. Polym. Rev. 2008, 48, 531-582.

[32] Thompson, B. C.; Fréchet, J. M. J. Angew. Chem. Int. Ed. 2008, 47, 58-77.

[33] Jørgensen, M.; Norrman, K.; Krebs, F. C. Sol. Energy Mater. Sol. Cells 2008, 92, 686-714.

[34] Gevorgyan, S. A.; Krebs, F. C. Chem. Mater. 2008, 20, 4386-4390.

[35] Krebs, F. C.; Norrman, K. Prog. Photovolt: Res. Appl. 2007, 15, 697-712.

[36] Norrman, K.; Krebs, F. C. Sol. Energy Mater. Sol. Cells 2006, 90, 213-227.

[37] Bjerring, M.; Nielsen, J. S.; Nielsen, N. C.; Krebs, F. C. Macromolecules 2007, 40, 6012-6013.

[38] Fu, M. X.; Zhu, Y. F.; Tan, R. Q.; Shi, G. Q. Adv. Mater. 2001, 23, 1874-1877.

[39] Han, M. G.; Cho, S. K.; Oh, S. G.; Im, S. S. Synth. Met. 2002, 126, 53-60.

[40] Zhang, X.; Goux, W. J.; Manohar, S. K. J. Am. Chem. Soc. 2004, 126, 4502-4503.

[41] Zhang, X.; Manohar, S. K. J. Am. Chem. Soc. 2004, 126, 12714-12715.

[42] Zhong, W.; Liu, S.; Chen, X.; Wang, Y.; Yang, W. Macromolecules 2006, 39, 3224-3230.

[43] Gok, A.; Omastova, M.; Yavuz, A. G. Synth. Met. 2007, 157, 23-29.

[44] Tran, H. D.; Wang, Y.; D'Arcy, J. M.; Kaner, R. B. ACS Nano 2008, 2, 1841-1848.

[45] Li, X.-G.; Li, J.; Huang, M.-R. Chem. Eur. J. 2009, 15, 6446-6455.

[46] Lee, S. J.; Lee, J. M.; Cheong, I. W.; Lee, H.; Kim, J. H. J. Polym. Sci. Part A. Polym. Chem. 2008, 46, 2097-2107.

[47] Lee, J. M.; Lee, S. J.; Jung, Y. J.; Kim, J. H. Curr. Appl. Phys. 2008, 8, 659-663.

[48] Masuda, H.; Asano, D. K.; Kaeriyama, K. Synth. Met. 2001, 119, 167-168.

[49] Wang, Z.; Wang, Y.; Kong, E. S. W.; Dong, X.; Zhang, Y. Synth. Met. 2010, 160, 921-926.

[50] Odian, G. in Principles of Polymerization. 4th Ed. John Wiley and Sons: New York 2004, p351

[51] Wang, Z.; Wang, Y.; Hu, N.; Wei, L.; Chen, S.; Zhang, Y. J. Polym. Sci. Part A. Polym. Chem. 2010, 48, 5265-5269.

[52] Kawaguchi, S.; Ito, K. Adv. Polym. Sci. 2005, 175, 299-328.

[53] Paine, A. J. Macromolecules 1990, 23, 3109-3117

[54] F. C. Krebs, S. A. Gevorgyan, J. Alstrup, J. Mater. Chem. 2009, 19, 5442.

[55] Pich, A.; Lu, Y.; Adler, H-J. P.; Schmidt, T.; Arndt, K-F. Polymer 2002, 43, 5723-2729.

[56] Riess, G.; Labbe, C.; Macromol. Rapid Commun. 2004, 25, 401-435.

[57] Oh, J. K.; J. Polym. Sci. Part A: Polym. Chem. 2008, 46, 6983-7001.

[58] Tomita, K.; Ono, T. J. Polym. Sci. Part A: Polym. Chem. 2009, 47, 762-770.

[59] Tomita, K.; Ono, T. J. Polym. Sci. Part A: Polym. Chem. 2009, 47, 2281-2288.

[60] Wang, Z.; Yang, Y. J.; Liu, T.; Dong, Q. Z.; Hu, C. P. Polymer, 2006, 47, 7670-7679.

[61] Mumtaz, M.; Labrugère, C.; Cloutet, E.; Cramail, H.; J. Polym. Sci. Part A: Polym. Chem. 2010, 48, 3841-3855.

[62] Bousquet, A.; Ibarboure, E.; HÉRoguez, V.; Papon, E.; Labrugere, C.; Rodríguez-Hernández, J. J. Polym. Sci. Part A: Polym. Chem. 2010, 48, 3523-3533.

[63] Muranaka, M.; Ono, T. J. Polym. Sci. Part A: Polym. Chem. 2009, 47, 5230-5240.

In: Polymer Synthesis
Editor: E. Kowsari

ISBN 978-1-61324-672-6
© 2012 Nova Science Publishers, Inc.

Chapter 8

Facile Synthesis of Conductive Polymers and Their Nanocomposites Using Ultrasound Radiation

E. Kowsari[*]

Department of Chemistry, Amirkabir University of Technology,
Tehran, Iran

Abstract

In materials science, sonochemistry is mostly used for the fabrication of nanomaterials, but it has also been used for the polymerization of monomers. Recently, great efforts are being made to improve chemical and electronic properties of conductive polymers, especially via the organization of polymeric conductive chains as fibers, hollow tubes, and spheres. Conductives such as polyaniline with organized morphology can be used for advanced applications in electronic devices and molecular sensors as a result of its high surface area and low cost. In order to achieve these objectives, a few authors have been using ionic liquids (ILs), organic salts that are liquid at low temperature (typically lower than 100 °C), as synthetic media to prepare nanostructured conductive polymer, especially IL derived from the imidazolium cation. In this chapter, an IL-assisted sonochemical method is reported for the synthesis of conductive polymer and nanocomposites showing controlled conductivity. This method avoids the use of conventional oxidants and metal complexes. The effect of the ultrasonic irritation time and frequency on the morphology, conductivity, and yield are discussed.

1. Introduction

Sonochemistry is the research area in which molecules undergo chemical reaction due to the application of powerful ultrasound radiation (20 KHz–10 MHz) [1]. The physical

[*] Department of Chemistry, Amirkabir University of Technology, No. 424, Hafez Avenue, 1591634311, Tehran, Iran, E-mail address: kowsarie@aut.ac.ir. Corresponding author: Fax: +98 (21)64542762, Tel.: +98 (21) 64542769.

phenomenon responsible for the sonochemical process is acoustic cavitation. Ultrasonics induces chemical changes due to the phenomena of acoustic cavitation, which involves the formation, growth, and instantaneously implosive collapse of bubbles in liquid, which can generate local hot spots having a temperature of roughly 5000 °C, pressure of about 500 atm, and a lifetime of a few microseconds [2]. With growing concerns over the environmental impact and health hazards of traditional volatile organic solvents, researchers continue to search for greener alternatives. Room-temperature ionic liquids (RTILs), as a new solvent system, is increasingly arousing worldwide interest due to their high mobility, low melting points, negligible vapor pressure, thermal stability, low toxicity, large electrochemical window, non-flammability, and ability to dissolve a variety of chemicals [3–6]. Ionic liquids have favorable intrinsic properties that make them of interest as solvents and catalyst for various chemical reactions. The same properties that make the liquids effective solvents or catalysts also make them interesting liquids for studies involving sonochemistry, acoustic cavitation, and sonoluminescence. Recently, a few papers are reported to use ultrasound as a tool to improve conducting polymers synthesis process [7–13]. As a consequence, polyaniline nanoparticles have been prepared with more uniform size distribution. Also, polyaniline colloids preparation has been improved, where morphologies and conductivity have been modified and improved, respectively.

In this chapter, A sonochemical-assisted method and IL-assisted sonochemical were reviewed for the synthesis of conductive polymer and nanocomposites showing controlled conductivity. The IL-assisted sonochemical avoids using conventional oxidants and metal complexes. The effect of the ultrasonic irritation time and frequency on the morphology, conductivity, and yield are discussed.

2. DISSCUSSION

2. 1. Ultrasound and Ionic-Liquid-Assisted Synthesis of Polyaniline-Y_2O_3 Nanocomposite

A sonochemical method has been employed to prepare polyaniline-Y_2O_3 nanocomposite with controlled conductivity with the assistance of an ionic liquid (IL) by Kowsari and Faraghi [14] Ultrasound energy and the IL replace conventional oxidants and metal complexes in promoting the polymerization of aniline monomer for the first time. Structural characterization has revealed that the resulting nanocomposite consists of microspheres of average diameter 3–5 μm. The products were found to consist of regular solid microspheres covered with some 40 nm nanoparticles. Under certain polymerization conditions, polyaniline nanofibers and nanosheet were obtaine. Figure 1 shows the morphology of product.

The method may open a new pathway for the preparation of nanoscale conducting polymer nanocomposites with the aid of ILs. The conductivity of the product varies with the mass ratio of aniline monomer to Y_2O_3 and IL. Figure 2 shows the the conductivities of PANI/Y_2O_3 composites at different concentrations of ILs.

The reaction conditions have been optimized by varying parameters such as the aniline/Y_2O_3 ratio and the type and amount of IL used. The effect of the ultrasonic irritation time and frequency on the morphology, conductivity and yield were discussed.

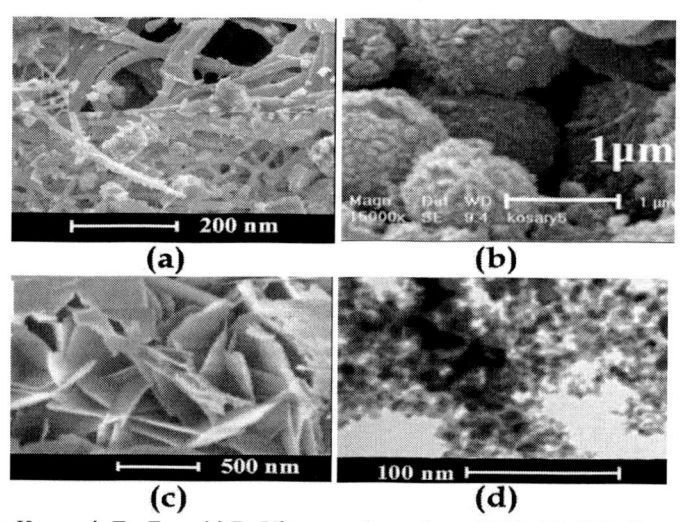

(a)   (b)
(c)   (d)

(Reproduced from Kowsari, E.; Faraghi F. *Ultrason. Sonochem.* 2010, *17*, 718, Copyright (2010), with permeation from Elsevier).

Figure 1. SEM images of PANI/ Y_2O_3 composite at different type of IL: (a) PANI/ Y_2O_3 composite (Y_2O_3 = 30%, aniline = 0.2 M, and IL = [hepmim]•H_2PO_4 = 0.6 M), (b) PANI/ Y_2O_3 composite (Y_2O_3 = 30%, aniline = 0.2 M, and IL = [hepmim]•HSO4 = 0.6 M), (c) PANI/ Y_2O_3 composite (Y_2O_3 = 30%, aniline = 0.2 M, and IL = [hepmim]•NO_3 = 0.6 M). (d) TEM image of Y_2O_3.

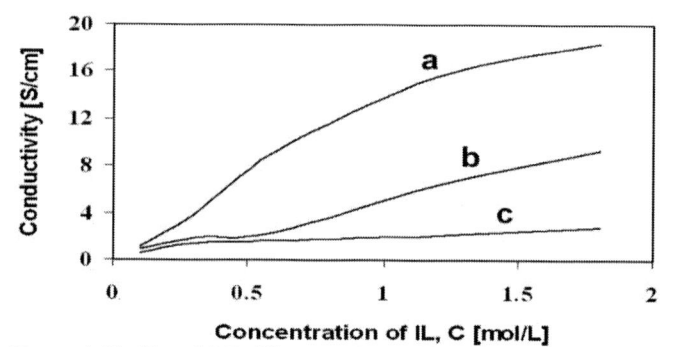

(Reproduced from Kowsari, E.; Faraghi F. *Ultrason. Sonochem.* 2010, *17*, 718, Copyright (2010), with permeation from Elsevier).

Figure 2. The conductivities of PANI/Y_2O_3 composites at different concentrations of ILs: (a) PANI/Y_2O_3 composite (Y_2O_3 = 30%, aniline = 0.2 M, and IL = [hepmim]•H_2PO_4 = 0.6 M), (b) PANI/Y_2O_3 composite (Y_2O_3 = 30%, aniline = 0.2 M, and IL = [hepmim]•HSO_4 = 0.6 M), (c) PANI/ Y_2O_3 composite (Y_2O_3 = 30%, aniline = 0.2 M, and IL = [hepmim]• NO_3 = 0.6 M.

2.2. Synthesis of the PANI Nanofibers by the Ultrasonic Irradiation

Conventionally, micro-sized irregular polyaniline (PANI) particles were synthesized by dropwise addition of the ammonium persulfate (APS) solution into the aniline (ANI) solution with mechanical stirring. By replacing the mechanical stirring with an ultrasonic irradiation, PANI nanofibers in diameters of ~50 nm and lengths of 200 nm to several micrometers were prepared by Jing and coworkers [15].

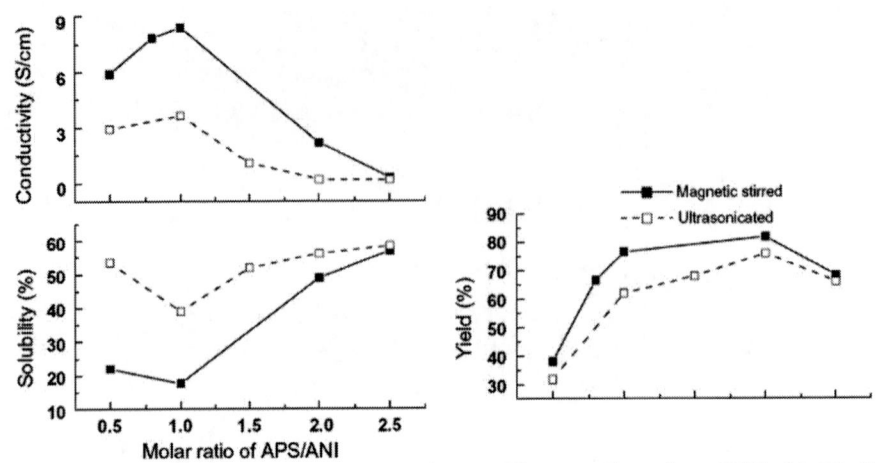

(Reproduced from Jing , X.; Wang,Y.; Wu,D.; Qian, J. *Ultrason. Sonochem* **2007**, *14*, 75, Copyright
(2007), with permeation from Elsevier).

Figure 3. Effect of APS/ANI molar ratios on the yield, conductivity and solubility of the PANI.

Transmission electron microscopy (TEM) and scanning electron microscopy (SEM)
showed that at the early stage of polymerization, the polymers formed in both the mechanical
stirred and ultrasonicated systems are in the form of nanofiber. However, with continuing of
the reaction, these primary nanofibers grow and agglomerate into irregular shaped PANI
particles in the mechanical stirred system, while in the case of the ultrasonic irradiation, the
growth and agglomeration are effectively prevented, preserving thus the PANI nanofibers in
the final product. By increasing the APS/ANI molar ratio from 0.5 to 2.5, the aspect ratios of
the PANI nanofibers decreased. The PANI nanofibers exhibit higher solubility than the
irregular shaped PANI particles. Although the yield, as well as the conductivity of the
ultrasonic synthesized PANI nanofibers, was slightly lower than the irregular shaped PANI
particles, the ultrasonic synthesis approach is one of the facile and scalable approaches in
synthesizing PANI nanofibers in comparison with other ones without use of templates (e.g.,
the interfacial polymerization and rapid mixing polymerization). Effect of APS/ANI molar
ratios on the yield, conductivity and solubility of the PANI is shown in Figure 3.

2.3. Synthesis of Polyaniline–Silver (Polyaniline/Ag) Nanocomposite by an Ultrasound Assisted in Situ Mini Emulsion Polymerization

Polyaniline–silver (polyaniline/Ag) nanocomposite was prepared by an ultrasound
assisted in situ miniemulsion polymerization of aniline along with different loading of silver
nanoparticles by Barkade and coworkers [16]. Colloidal silver nanoparticles were synthesized
by reduction of silver nitrate ($AgNO_3$) with sodium borohydride ($NaBH_4$) using sodium
dodecyl sulfate(SDS). Films of Polyaniline/Ag were casted using1-methyl- 2-pyrrolidone
(NMP) by spin coating, which were further tested for ethanol vapor sensing.

2.4. Synthesis of Polythiophene Nanofibers Using Ionic Liquids and Ultrasound Radiation for High Performance Redox Supercapacitors

Supercapacitor is a kind of new storage energy devices that can provide higher specific power and longer cycle life than batteries, and higher specific energy than conventional dielectric capacitors because of the high specific capacitance of its electrode materials [17−19]. Recent interests in supercapacitors have been stimulated by their potential applications such as power storage devices operating in parallel with batteries in hybrid electric and electric vehicles [20−22].

The electrode materials used for electrochemical capacitor applications include carbon and carbon nanotubes [23], metal oxides [24], and conducting polymers [25]. But the redox supercapacitors are constructed by metal oxides and conducting polymers are the promising electrode materials. Among these two electrode materials, conducting polymers are widely studied for redox supercapacitors. They are generally offer three important properties: high specific capacitance, due to the involvement of whole polymer mass in the charging process; high conductivity in the doped state and fast charge/discharge electron-transfer kinetics.

Hence, in the present investigation, various morphologies of polythiophene (PT) have been designed and successfully prepared by IL, sonochemical assisted polymerization in aqueous medium and used for the first time to use as electrode material for symmetric type redox supercapacitors. The morphologies of PT could be controlled in ribbons, fibers and sheet by changing the concentrations of FIL, as an oxidant and templeting agent. The structure, thermal stability and the conductivity have been characterized, and a mechanism for the transformation of the morphology of polythiophene has been proposed. The PT produced has good stability and is free from contaminants. It have been demonstrated that the interaction of polythiophenes with the FIL results in a drastic increase of the electrical conductivity and a strong suppression of inter band optical absorption of these polymers. The effect is tentatively attributed to the protonic doping of the polymer chains by sulphate groups of the hydrolyzed hydrogensulphate available within FIL. These effects point to an attractive new route of doping of conjugated polymers that might be interesting for applications in molecular sensors, transparent conductors, and organic electronics. The overall capacitor performances were found to be good, so that the ionic liquid assisted polymerization method is the easiest and more convenient method for the synthesis of polythiophene nanofiberss for high performance supercapacitor applications.

2.4.1. Morphology Investigation

Fanctinal ionic liquids (FIL)s, with different counter ions (Figure 4), were used for shaping the PT morphology tend to agglomerate into block.

Figure 5a–d exhibits the morphology of PT-FIL1-1, PT–FIL1-1.5, PT– FIL1-1.75, and PT– FIL1-2 samples. To compare the effects of FIL1 on the basis of the properties of the resulting polythiophene, the monomer concentration was kept constant. PT–FIL1-1.0 (Figure 5a) has a sheet structure. The differences in the structure of PT samples prepared in the presence of different concentrations of FIL are clearly visible. At the same magnification, PT–FIL1.50 reveals an interesting nanosheet structure (Figure 5b). The nanosheet structure of PT– FIL-1.0 is shown in Figure 5c. In Figure 5b, the smooth surface of PT–FIL1-1.75. with small deformed nanosheet is visible. Smooth surfaces of PT– FIL1-1, PT– FIL1-1.5, and PT– FIL1-1.75 can be a reason for the better conductivities of these samples compared with PT–

FIL-2.0. The SEM study shows that the presence of FILs in polymerization strongly affects the morphology of PTs. FILs plays a key role in tailoring the resultant conducting PT structures. Globular PT nanostructures are obtained in the presence of FIL with amount of 2 mmol. As it is known, the morphology and electrical properties are the most important relations for conductive polymers. FIL additives change the morphology of PTs and this is reflected in the conductivity changes.

FIL is an effective template guiding the PT chain conformation by forming the aggregates or micelles during the polymerization. The small aggregates of thiophene or oligo thiophene act as nucleation centers for PT growth. It is believed that the structure of the associated nucleation centers guided the growth and determined the final morphology of the PT. Also, IL plays both the roles of the acid, instead of HCl, and oxidant, intead of APS, in polymerization thiophen. The former compounds were used to be utilized in tradditional methods.

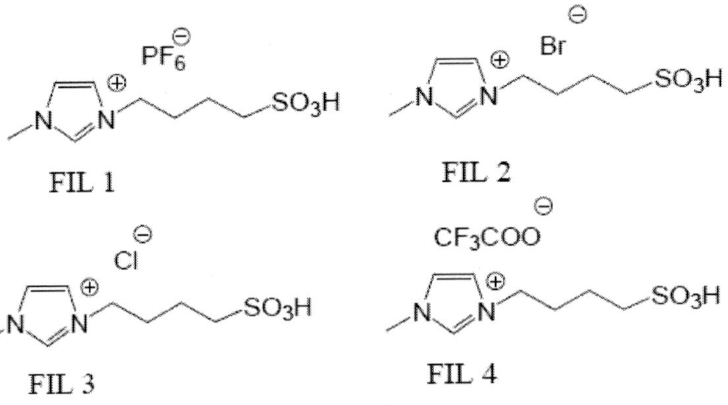

Figure 4. Structures of the FILs used in this study.

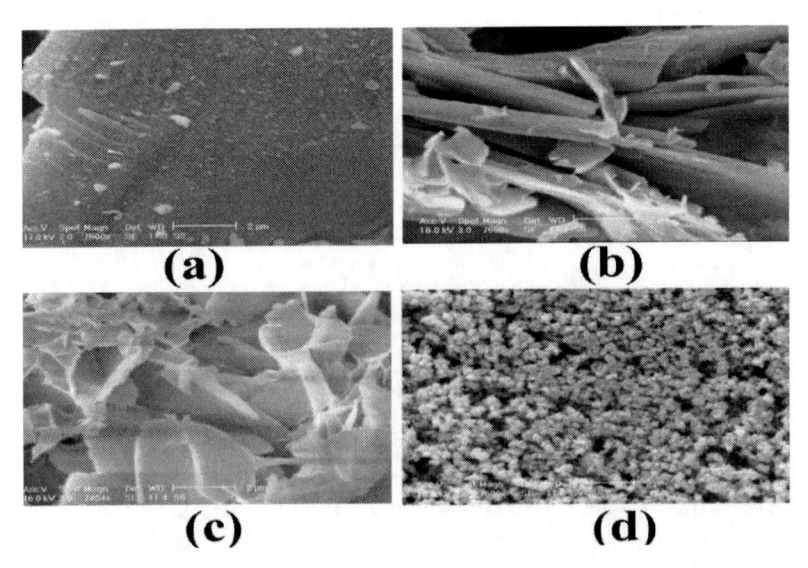

Figure 5. SEM images showing the morphological evolution of PT samples with different concentrations of FIL1: (a) 1.0 mmol, (b) 1.50 mmol, (c) 1.75 mmol, and (d) 2 mmol. (thiophene = 0.2 M, ultrasonic frequencie = 20 kHz, ultrasonic irradiation time = 60 min, The ultrasonic power was kept at 100 W.

2.4.2. Effect of the FILs Concentrations on PT Morphology

The effect of the concentrations of FILs on PT (thiophene = 0.2 M) morphologies is illustrated in Figure 6.

(a)   **(b)**

(c)   **(d)**

Figure 6. SEM images of PT at different type of FILs: (a) FIL1, (b) FIL2, (c) FIL3,, and (d) FIL4 (FIL = 3 mmol, thiophene = 0.2 M, ultrasonic frequencie = 20 kHz, ultrasonic irradiation time = 60 min, The ultrasonic power was kept at 100 W).

In order to compare the effects of FIL additives on the properties of the resulting PT, the monomer concentration ratio was kept constant. The differences in the structure of PT prepared in the presence of the different FILs are clearly visible. At certain constant amount of thiophene (with FIL1) the mixture of nanotube and nanoparticle is revealed (Figure 6a). The nanofiber structures of PT in the presence of FIL2 is shown in Figure 5b. It was found that in the presence of FIL3, the products were regular nanoporous (Fig 6c). As Fig 6d depicts, in the presences of FIL4 the diameter of nanofibers decrease and nanoparticles can be observed atop them.

2.4.3. Initial Concentration of Thiophene

PTs were prepared with different concentrations of thiophene and fixed amounts of FIL1 (2mmol mmol). The conductivities of the PTs were found to increase as concentration of the used thiophene increased.

2.4.4. Effect of the Ultrasonic Irradiation Time on the Morphology of PT

The effect of the ultrasonic irradiation time on PT morphology (thiophene = 0.2 M) is shown in Figure 7 .

As is shown in Figure 7, for FIL4, as the ultrasonic irradiation time increases, the diameters of the fibers decrease and are attached to each other. (Figure 7c). As a result, an increase in ultrasonic irradiation time (FIL4), for instance at 20, 40, and 60 min, increases the yield of the polymer increase to 45 wt %, 78 wt %, 92 wt %, respectively. The polymer conductivity increased with ultrasonic irradiation time and reached a constant value with ca. 60 min of ultrasonic irradiation.

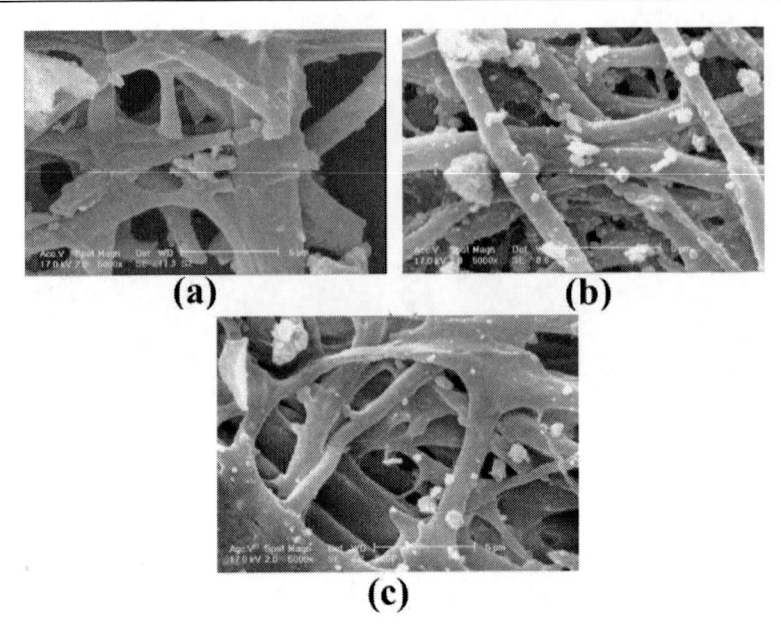

Figure 7. SEM images of PT at different ultrasonic irradiation times (a) 20 min, (b) 40 min and (c) 60 min. (FIL4 = 3 mmol, thiophene = 0.2 M, ultrasonic frequencie = 20 kHz, ultrasonic irradiation time = 60 min, The ultrasonic power was kept at 100 W).

2.4.5. Effect of Ultrasonic Frequency on PT Morphology and Yield

The effect of the ultrasonic frequency on PT morphology is investigated by Li and coworkers [25]. In the present study, as is shown in Fig 8, when there is an increase in a frequency from 20 to 40 kHz, the morphology of PT transforms from attached nanoparticles into nanoporous. In additions, the yield would be 65 wt %, 77 wt %, and 90 wt % at 20, 30, and 40 kHz for frequency.

Figure 8. SEM images of PT at different ultrasonic frequencies (a) 20 kHz, (b) 30 kHz and (c) 40 kHz. (thiophene = 0.2 M, ultrasonic irradiation time = 60 min, FIL3 = 3 mmol, the ultrasonic power was kept at 100 W).

Ultrasonic irradiation at 40 kHz yields constantly higher degradation efficiencies compared with that at 20 kHz for all FILs. Since, in the present study, the FIL replaces conventional oxidant and metal complexes for polymerization, the increase of FILs degradation leads to the increase of the concentration of alkyl radicals and the increase of thiophene polymerization and, therefore, the increase of yield of product.

2.4.6. Effect of the Ultrasonic Irradiation Time on the Conductivity of PT

The value of conductivity increases with reaction time and attains a maximum. Reasonably, high conductivity (3.9 Scm^{-1}) was obtained after 60 min; at longer times, the conductivity decreases. As ultrasonic irradiation time increases to 60, conductivity increases and remains constant afterwards. This process is true for all three types of FILs.

2.4.7. The Mechanism for the Formation of PT

The effects of acoustic cavitation on a number of room-temperature FILs, including imidazolium cation, have previously been examined [27]. It was observed that sonication led to varying degrees of decomposition of the ILs studied. Sonicated imidazolium FIL produces various alkyl halides, imidazole, and radicals products. Although the mechanistic details of the reaction are not clearly known, it is reasonable to assume that during sonication, radical cations are produced from the decomposition of the FILs. Monomers are oxidized by them to the thiophene radical cations, which can form the dimers through a head-to-tail coupling. Then, the dimers are sequentially oxidized to quinoid units which could be deprotonated to afford nitrenium ions, and react with a thiophene monomer to produce the trimers. This process is repeated, leading to form oligomers, and eventually to the formation of PT.

2.4.8. Thermal Stability

The thermal stability of the PT samples prepared by the FIL polymerization pathway was studied by TGA (Figure 9).

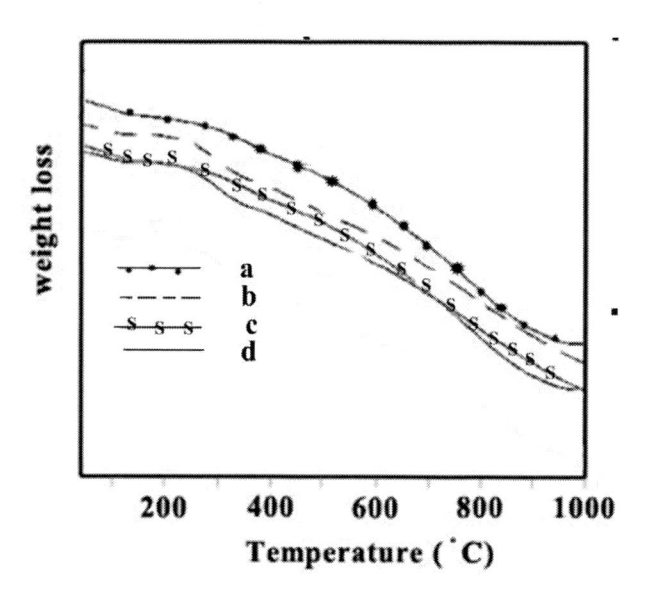

Figure 9. TG curves (inset) of PT samples with different concentrations of FIL: (a) 1.5 mmol, (b) 1.75 mmol, (c) 1.0 mmol, and (d) 2 mmol.

PT exhibits a three-stage decomposition pattern. In the first stage, the weight loss observed up to 110 °C (4.2-5 wt. %) is due to the loss of water molecules present in the polymer matrix. The second-stage weight loss observed from 230 to 370 °C (16-18 wt. %) can be attributed to the loss of the dopants group from the polymer chain. The third-stage weight loss from 370 °C onwards is due to the degradation of polymer backbone.

The polymer nanofibers (FIL2, = 3 mmol, thiophene = 0.2 M, ultrasonic frequencies = 20 kHz, ultrasonic power = 100 W) were used as electrode materials for symmetric type high performance redox supercapacitor studies with PVdF-co-HFP based microporous polymer electrolyte containing 1M LiPF$_6$ in 1:1 EC/PC electrolyte. Its specific capacitance was found to be 138 Fg^{-1}. This capacitance was decreased slowly on continuous cycling due to both polymer degradation and mechanical stress on the PT. The overall capacitor performances were found to be satisfactory, so the IL-ultrsonic assisted polymerization method is the easiest and more convenient method for the synthesis of PT nanofibers for high performance supercapacitor applications.

2.5. Synthesis of Polypyrrole (PPy) and Gold Nanoparticles (Au-NPs) or Platinum Nanoparticles (Pt-NPs) by a Sonochemical Method

Park and coworkers [28] prepared colloidal dispersions of hybrid nanocomposite composed of polypyrrole (PPy) and gold nanoparticles (Au-NPs) or platinum nanoparticles (Pt-NPs) by a sonochemical method, in which Au ion and pyrrole monomer in an aqueous solution were reduced and oxidized, respectively, by ultrasonic irradiation. Figure 10 shows the TEM images of Au-NPs/PPy nanocomposite suspensions corresponding to 4 h ultrasonic irradiation.

The reaction step for the formation of Me-NPs (Au, Pt)/PPy nanocomposite under ultrasonic irradiation can be expected as shown in Figure 11, which is a sequence of reactions for the formation of polypyrrole on the Au-NPs or Pt-NPs surface (core–shell structure)

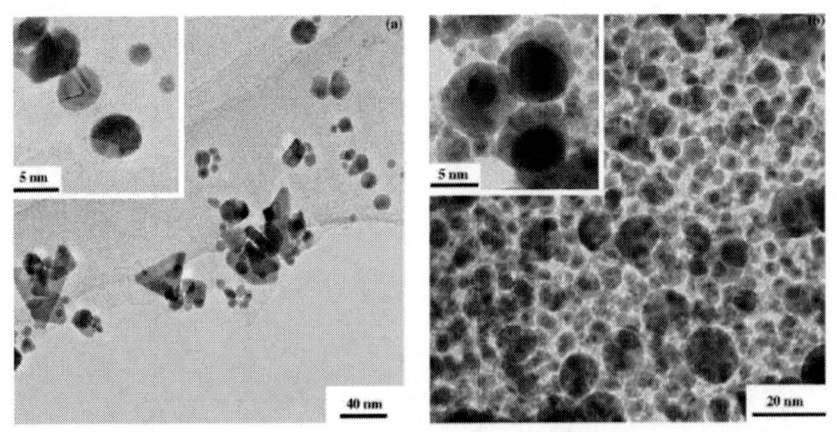

(Reproduced from Park, J. E.; Atobe, M.; Fuchigami,T. *Electrochimica Acta* 2005, *51*,849, Copyright (2005), with permeation from Elsevier).

Figure 10. TEM image obtained for (a) Au-NPs and (b) Au-NPs/PPy nanocomposite prepared by ultrasonic irradiation for 4 h.

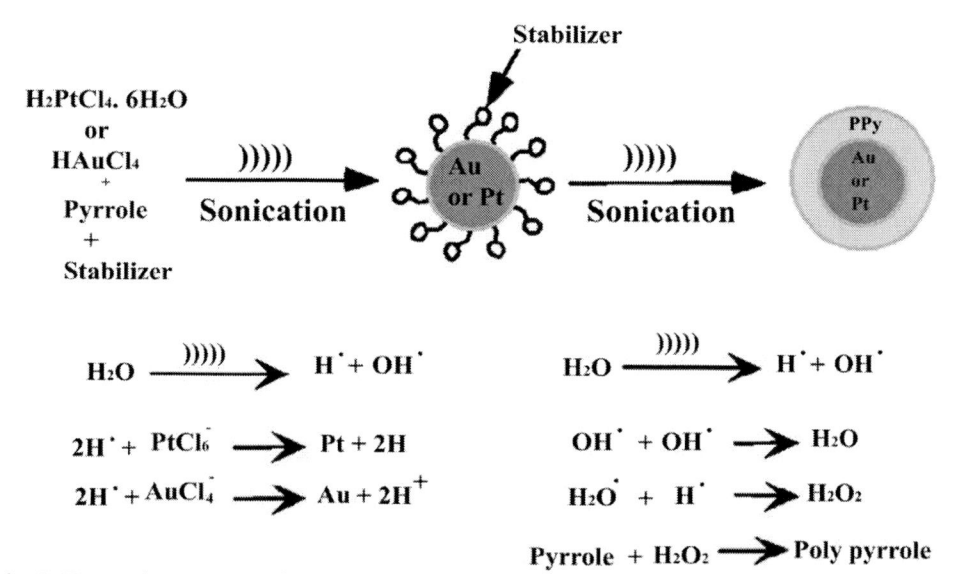

(Park, J. E.; Atobe, M.; Fuchigami,T. *Electrochimica Acta* 2005, *51*,849, Copyright (2005), with permeation from Elsevier).

Figure 11. Plausible reaction pathway for the formation of Me-NPs (Au, Pt)/PPy nanocomposite.

2.6. The Polypyrrole Flower-Like Nanostracture Synthesized in the Biphasic IL/Water System

2.6.1. Morphology

The PPy flower-like nanostracture was synthesized in the biphasic IL/water system and ultrasound radiation Morphology studies of the PPy flower-like were performed by applying SEM (Figure 12).

IL seems to act as a template of liquid phase. IL is an isotropic organic liquid composed entirely of ions, but it is also anisotropic conductors due to its self-organized structures. The macroscopic orientation of self-organize dmonodomains in IL may play a key role in the enhancement of properties because the boundary in randomly oriented domains highly disturbs 1anisotropic transportation of charges and ions. The IL may result in the formation of uniform flower-like PP nano structure.

Figure 12. SEM images showing the morphological evolution of the PPy flower-like nanostracture synthesized in the biphasic IL/water system and ultrasound radiation.

2.6.2. Micelle Formation

Figure 13 shows formation of IL micelle by IL and the orientation of PPy in formation of nano flower. First, hydrophobic IL forms micelle. Since IL CTC is lower than traditional surfactant, lower concentration of IL is required for the formation of micelle. Then, pyrrole is added to the media and polypyrrole nanoparticles are formed in the micelle where some radicals, produced from decomposition of IL by sonic radiation, are present and act as oxidants. Then, some of these nanoparticles aggregate and produce flower-like polypyrrole.

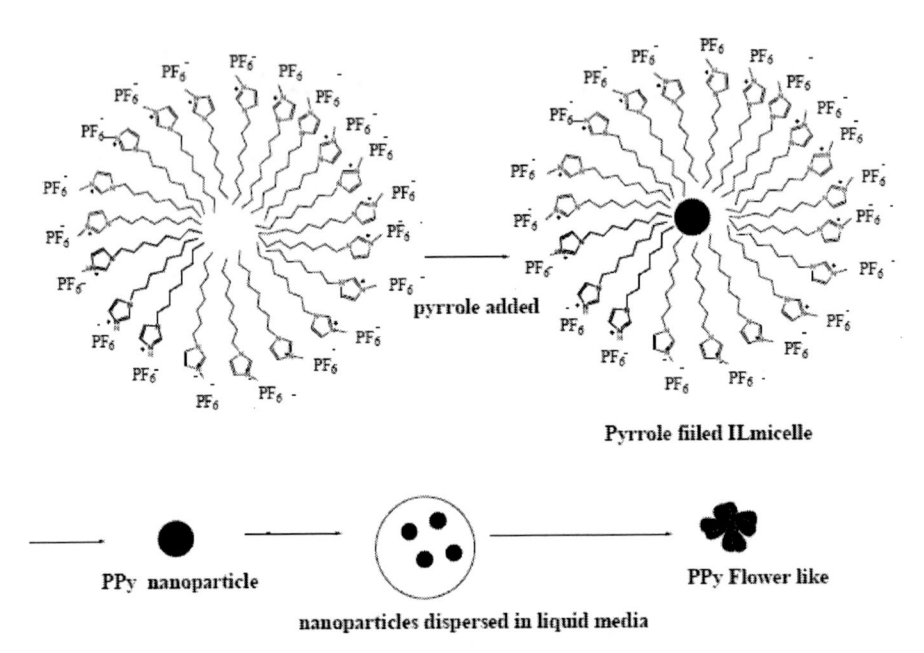

Figure 13. Schematic representation of IL micelle and the orientation of of PPy.

CONCLUSION

Both size and composition control are important to prepare conducting polymer nanocomposites with desirable properties. Some interesting results, using ultrasound, is in the preparation of organic/inorganic composite and the enhancement of intercalation of organic molecule in inorganic lattices. Functional conducting polymer nanocomposites synthesized by this method exhibit many intriguing properties, which have been extensively explored in the applications of electronic nanodevices, chemical and biological sensors, catalysis and electrocatalysis, energy devices, microwave absorption and biomedicine.

ACKNOWLEDGMENT

The author wishes to express her gratitude to National Elite Foundation for the financial support.

REFERENCES

[1]   Ramesh, S.; Koltypin, Y.; Prozorov, R.; Gedanken, A. Chem. Mater. 1997, 9, 546.

[2]   Suslick,K. S.; Science 1990, 247,1439.

[3]   Welton, T. Chem. Rev. 1999, 99, 2071.

[4]   Wasserscheid, P.; Keim,W. Angew. Chem., Int. Ed. Engl. 2000, 39, 3772.

[5]   Dupont, J.; de Souza,R. F.; P.A.Z. Suarez, P.A. Chem. Rev. 2002,102, 3667.

[6]   Xu, W.; Cooper, E. I.; Angell, C. A. J. Phys. Chem. B. 2003, 107, 6170.

[7]   Kwon, J. Y.; Kim, E. Y.; Kim, H. D. Macromol. Res. 2004, 12, 303.

[8]   Wang, J. Z.; Hu, Y.; Song, L. Solid State Ionics 2004, 167, 425.

[9]   Wang, J. Z.; Hu, Y.; Tang,Y. Mater. Res. Bull. 2003, 38, 1301.

[10]   Atobe, M.; Chowdhury, A. N.; Fuchigami, T.; Nonaka, T. Ultrason. Sonochem. 2003, 10, 77.

[11]   Lim,S. T.; H.J. Choi,H. J.; M.S. Jhon, M. S. J. Ind. Eng. Chem. 9 (1) (2003) 51.

[12]   Xia, H. S.; Wang, Q. Chem. Mater. 2002, 14, 2158.

[13]   Xia, H. S.; Wang, Q. J. Nanopart. Res. 2001, 3, 401

[14]   Kowsari, E ; Faraghi,G.Ultrason. Sonochem. 2010, , 4, 718.

[15]   Jing , X.; Wang,Y.; Wu,D.; Qian, J. Ultrason. Sonochem. 2007, 14, 75.

[16]   Barkade, S. S.; Naik, J. B.; Sonawane, S. H. Colloids Surfaces A 2011, doi.org/10.1016/j.colsurfa.2011.02.002

[17]   Kökötz, R., Carlen M. Electrochimica Acta, 2000, 45, 2483.

[18]   Lewandowski, A, Gallnski, M. J. Power Sources, 2007, 173, 822.

[19]   Vix-Guterl, C.; Saadallah, S.; Jurewicz, K.; Frackowiak, E.; Reda, M.; Paementler, J.; Patarin, J., Beguin, F. Mat. Sc.i Eng. B. 2004, B108, 148.

[20]   Thomas, E. R.; Demisa, H. J.; Zhu, Z. H., Lu, G. Q. Electrochem. Comm., 2008, 10, 1594.

[21]   Payman, A.; Pierfederici, S.; Meibody-tabar, F.; Energ. Convers. Manage. 2008, 49, 1637.

[22]   Tlan, Y. M.; Song Y., Tang, Z.; Guo, Q.; Liu, L. J. Power Sources. 2008, 184, 675.

[23]   Morimoto, T.; Hiratsuka, K.; Sanada, Y.; Kurihara, K. J. Power Sources. 1996, 60, 239.

[24]   Rinzler, A. G.; Liu, J.; Dai, H.; Nickolaev, P.; Huffman, C. B.; P.J. Boul, P. J.; Lu, A. H.; Heymann, D.; Rao, A. M. Appl. Phys. A 1998, 67, 29.

[25]   (25). Baughman, R. H.; Cui, C.; Zakhidov, A. A.; Iqbal, Z.; Barisci, J. N.; Spinks, G. M.; Wallace, G. G.; Mazzaldi, A.; Roth, S.; Kertesz, M. Science 1999, 284, 1340.

[26]   Li, Y.; Wang, Y.; Wu, D.; Jing, X. J. Appl. Polym. Sci.2009, 113, 868.

[27]   Flannigan,J. D.; Hopkins, D. S. D.; Suslick, K. S. J. Organomet Chem. 2005, 690, 3513.

[28]   Park, J. E.; Atobe, M.; Fuchigami,T. Electrochimica Acta 2005, 51,849.

In: Polymer Synthesis
Editor: E. Kowsari

ISBN 978-1-61324-672-6
© 2012 Nova Science Publishers, Inc.

Chapter 9

MECHANOCHEMICAL POLYMERISATION: A FACILE GREENER SYNTHETIC ROUTE TO PREPARE NANOSTRUCTURED CONDUCTING POLY (2-AMINO DIPHENYLAMINE)

S.P. Palaniappan and P. Manisankar[*]

Department of Industrial Chemistry, School of Chemistry
Alagappa University, Karaikudi, Tamil Nadu, India

ABSTRACT

Room-temperature mechanochemical route to produce different materials is widely regarded as a simple and efficient synthetic method. The deployment of this synthetic strategy to produce conducting polymers was first realized only in 1960. The extension of this method to prepare some polyaniline type conducting polymers began in late 1990. However, utilization of this method for preparing nanostructured conducting polymers was discovered very recently. In this chapter, we intend to discuss some nuances behind the mechanochemical synthesis of poly(2-amino diphenylamine). The influence of oxidant namely ammonium persulphate and ferric chloride, inorganic doping acids on the properties of poly(2-amino diphenylamine) will be briefly explained on a comparative basis. A comprehensive account on the physicochemical properties of as prepared polymers and newer research findings from our research results would also be vividly presented. The promising areas of application of the mechanochemically prepared polymers are to be suggested.

Keyword: Mechanochemical synthesis, conducting polymer, polyaniline, poly(2-amino diphenylamine), nanostructure

---

[*] Corresponding author Email: pms11@rediffmail.com (or) pmsankarsiva@yahoo.com.

1.1. INTRODUCTION

Polymeric materials are widely employed in various fields. Ionic conducting polymers are used in battery application such as high capacity lithium batteries. Emulsion polymerization was used for production of binders for paints and adhesives and for coatings in the paper industry. Blends of different polymers generally form two-phase systems and the properties of the blends depend on the mutual compatibility between the polymers and the interactions across the phase boundaries. For polymers used in medical applications the surface properties are of high importance. The free volume of a polymer, which corresponds to the unoccupied regions accessible to segmental motions, plays an important role in understanding many characteristic properties such as time and frequency dependent electrical and mechanical properties and transport properties of low molecular species.

Conducting polymers have potential applications at all levels of microelectronics. Conjugated polymers in the undoped and doped conducting state have an array of utilities in microelectronics industry. Conducting polymers finds application in electron beam lithography, in metallization (electrolytic and electroless) of plated through-holes for printed circuit board technology, in excellent electrostatic discharge protection for packages and housings of electronic equipment, in excellent corrosion protection for metals and in electromagnetic interference shielding.

Conducting polymers such as polyacetylene, polypyrrole, etc. have been the subjects of numerous investigations in the past two decades. These materials have properties that make them suitable for several applications in sensor, biosensor, rechargeable batteries, molecular electronic devices, electrochromic display devices, corrosion inhibitors, electromagnetic shielding materials, etc. Among electrically conducting polymers, polyaniline and its derivatives have been the focus of attention due to their environmental stability, ease of synthesis, exciting electrochemical, optical and electrical properties. Polyaniline can be doped to highly conducting state by protonic acids or by electrochemical oxidation. They show moderate conductivity upon doping with protonic acids and have excellent stability under ambient conditions.

Mechanochemistry of inorganic materials is a well-developed part of the science and is in the existence for quite a long period of time. Organic mechanochemistry concerns with conversion of mechanical energy into the driving force for molecular or structural phase transitions and in some cases results in newer molecules. There are a many instances where a mechanical forces introduced in the system gets transformed into a chemical driving force. The reactivity of the mechanically activated solids varies depending on the nature of the reactants employed for the mechanochemical reaction.

A wide variety of methodologies for the synthesis of conducting polymers particularly polyaniline have been reported. These studies are basically aimed at achieving superior properties by fine tuning the particle size and shape of the targeted polymeric material. Even though there are few reports [1-5] on the synthesis of conducting polymers by this novel route, they are mostly confined to polyaniline and its o-methyl [6,7] and o-methoxy [6,8] derivatives. O.Yu.Posudievsky et.al. have reported the preparation of polypyrrole and oligomers of thiophene and paraphenylene employing this method [9] We have recently reported the preparation of polypyrrole nanospheres through this greener route [10]. In this perspective, we intended to apply this novel method for the preparation of poly(2-amino diphenyl amine) (P2ADPA). Meanwhile, doping process results in dramatic changes in the electronic, electrical, magnetic, optical and structural properties of the polymer [11]. Oxidants are also found to play an important role in the determining some key properties of conducting polymers. Hence, the effect of inorganic dopants and oxidants on the properties of mechanic-hemically prepared P2ADPA is also presented in this chapter.

PANI is nowadays one of the most promising conducting polymers from a technologic perspective. Both its electrical conductivity and its environmental stability are quite good but, unfortunately, the electrical conductivity is strongly pH dependent and, in addition, PANI is almost insoluble in common organic solvents. In particular, the lack of solubility limits the processability of this material and, consequently, the practical applications are quite restricted as well. This has encouraged the researchers in the field to explore the possibility of obtaining new conducting polymers with chemical structures similar to or derived from that of polyaniline. It is thought that a polymeric material showing a slightly modified PANI structure could show better solubility in solvents keeping, at the same time, most of the chemical stability and electrical conductivity of the original polymer. In fact, during the last few years many attempts towards the synthesis of this kind of conducting polymers have been accomplished. In general, the insertion of some ionizable groups in the polymer backbone such as the carboxylic and sulfonic among others resulted in modified polyaniline structures with better processability and pH dependence and worse electrical properties [12-14]. It has also been proposed the use of some aniline dimers such as the ortho substituted one, 2-aminodiphenylamine, or its *para* isomer 4-aminodiphenylamine as monomer species for either the electrochemical homopolymerization [15-17] or their co-polymerization with aniline [17-21]. In this way, we could obtain polymeric materials having structures similar to PANI but with some specific properties enhanced. To our knowledge, despite the extensive study on the polymer obtained during the co-polymerization of 2-amino diphenylamine (2ADPA) and aniline, the homopolymer synthesized from the electrochemical oxidation of the ortho dimer of aniline has not been well characterized [17]. The subsequent sections in this chapter will shed more light on the physicochemical properties of mechanochemical homo polymerization of 2ADPA in the presence of different doping acids using two different oxidants namely APS and FC.

1.2. MECHANOCHEMICAL SYNTHESIS AND CHARACTERISATION OF POLY (2-AMINO DIPHENYLAMINE)

2ADPA was subjected to mechanochemical treatment along with two different oxidants namely ammonium peroxydisulfate (APS) and ferric chloride (FC) separately in the presence of different doping acids viz., HCl, H_2SO_4 and H_3PO_4. The resultant powder material was found to be polymerised product of 2ADPA. The effect of afore-mentioned oxidants and doping agents on the physicochemical properties of as prepared P2ADPAs is briefly discussed in this chapter.

1.2.1. Synthesis of Doped/Undoped P2ADPA

1.64 g of solid 2-ADPA was taken in a glass mortar and was hand-ground for 5 minutes using a pestle. To this finely grounded 2-ADPA, 2.2 g of solid APS or 2.7 g of solid FC was added. After the addition of the oxidant, the solid phase mixture was further hand-ground immediately for 20 minutes until the colour of the product turned dark reddish brown (See scheme 1.1). The formed polymeric product was washed thoroughly with water, methanol and diethyl ether. After repetitive washings, the polymer was dried in vacuum oven at 40°C for 12 h. The purified dry P2ADPA powders prepared in the presence of APS and FC individually was used for further characterization.

The preparation of doped P2ADPA involves the addition of 0.5 ml of doping agent (37 wt.% HCl / 96 wt.% H_2SO_4 / 87 wt.% H_3PO_4) to 1.64 g of solid 2ADPA in a glass mortar. This monomer-doping agent reactant mixture was thoroughly hand-ground for 30 minutes in order to achieve homogeneity. The rest of the polymerization procedure was followed as given above. The purified P2ADPAs obtained employing APS or FC as oxidants were used for performing further studies. Herein, undoped P2ADPA is termed as P2ADPA, P2ADPA doped with HCl as P2ADPA-HCl, P2ADPA doped with H_2SO_4 as P2ADPA-H_2SO_4 and P2ADPA doped with H_3PO_4 as P2ADPA-H_3PO_4.

1.2.2. Elemental Analysis, Yield and Processability

The percentage composition of C, H and N is consistent with the repeat unit composition (open chain phenazine like structures) in all of the prepared P2ADPA samples. The presence of Cl, S and P in P2ADPA-HCl, P2ADPA-H_2SO_4 and P2ADPA-H_3PO_4 respectively (table 1.1) suggests the doping of respective anions has taken place in the polymeric backbone due to the usage of different acids during their preparative process. Interestingly, it could be noticed that undoped P2ADPA prepared using APS has 0.40% sulphur which might be due to the self-doping during the course of mechanochemical polymerisation with APS. However, S to N ratio of 0.03 indicates negligible doping of P2ADPA. In a similar fashion, undoped P2ADPA prepared using FC has 0.45% chlorine due to the self-doping of chorine present in FC. From the highest S to N ratio observed for P2ADPA-H_2SO_4, it could be reiterated that doping level is pronounced in P2ADPA-H_2SO_4 rather than the other salts in both cases of APS and FC.

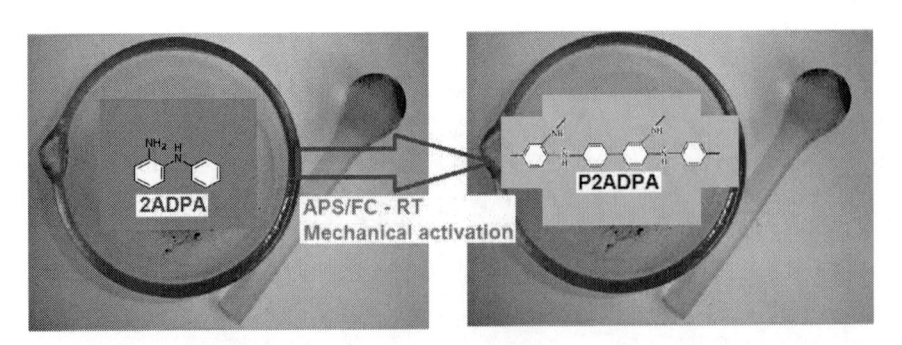

Scheme 1.1. Scehmatic representation of the mechanochemical polymerisation of P2ADPA.

Table 1.1. Elemental composition of P2ADPA and its salts

P2ADPAs prepared using APS							
Polymer	%C	%H	% N	% Cl	% S	% P	Cl/N or S/N or P/N ratio
P2ADPA	67.87	4.68	13.17	--	0.40	--	0.03
P2ADPA-HCl	66.63	4.52	13.02	1.59	--	--	0.12
P2ADPA-H_2SO_4	66.50	4.32	12.89	--	2.02	--	0.15
P2ADPA-H_3PO_4	66.58	4.36	13.08	--	--	1.40	0.10
P2ADPAs prepared using FC							
P2ADPA	67.01	4.70	13.14	0.45	--	--	0.03
P2ADPA-HCl	66.82	4.54	13.12	1.62	--	--	0.12
P2ADPA-H_2SO_4	66.42	4.26	12.85	--	2.51	--	0.19
P2ADPA-H_3PO_4	66.48	4.30	12.96	--	--	1.53	0.11

This may be attributed to the fact that polyprotic nature and strong oxidisability of H_2SO_4 favours the doping and polymerization process more than the other protic acids employed for solid-state polymerization. Moreover 2-ADPA is found to form salt more readily with 96 wt.% H_2SO_4 rather than 37 wt.% HCl and 87 wt.% H_3PO_4 due to the strong protonating capability of H_2SO_4. Unlike aniline, 2ADPA does not form salt with HCl easily and is also found to form least doped H_3PO_4 salt. This argument goes in agreement with the elemental composition analysis data. The order of doping level as in both APS and FC system is P2ADPA-H_2SO_4 > P2ADPA-HCl > P2ADPA-H_3PO_4. It can be seen from table 1.2 that the yield obtained for all P2ADPA salts are comparable to each other. As expected, P2ADPA prepared in the absence of protonating agent in both APS and FC has resulted in a comparatively lesser amount of polymer than the other polymers. This may be due to the creation of lesser number of radical centers for the polymerization reaction to proceed. All of the as prepared P2ADPA salts are found to be highly dispersible in common solvents like ethanol, acetone and double distilled water. These observations indicate the processable nature of all the doped P2ADPA powders. From the above experimental observations, it could be asserted that the facile solid-state method does assists in the preparation of pure and processable P2ADPA salts in appreciable quantity in both of the tested oxidants.

Table 1.2. Yield of P2ADPA and its salts

P2ADPAs prepared using APS	
Polymer	Yield in %
P2ADPA	62
P2ADPA-HCl	66
P2ADPA-H_2SO_4	70
P2ADPA-H_3PO_4	65
P2ADPAs prepared using FC	
P2ADPA	63
P2ADPA-HCl	69
P2ADPA-H_2SO_4	72
P2ADPA-H_3PO_4	67

1.2.3. FTIR spectra

Figure 1.1A (a-d) shows the FTIR spectra of P2ADPA salts using APS as oxidant. The characteristic peak at ~3213-3391 cm^{-1} is caused by the N-H stretching mode of the secondary amine. The peak at ~1521-1523 cm^{-1} is characteristic of the C-C multiple bond stretching mode of benzene ring. The C-N stretching mode of the aromatic secondary amine causes the peak at ~1301-1304 cm^{-1}. The peak at ~1240-1242 cm^{-1} can be assigned for the C-H bending vibration of diphenoquinone.

The sharp peaks at ~746-750 and ~690-695 cm^{-1} is assigned for the terminal phenyl groups and C-H out-of-plane phase bending vibration of mono-substituted benzene rings at the ends of the polymer chains respectively. The vibrational bands located in the range of 612, 568 and 492 cm^{-1} is associated with $(PO_4)^{3-}$ and $(SO_4)^{2-}$ groups. The results indicate that the backbone structures of P2ADPA-HCl, P2ADPA-H$_2$SO$_4$, and P2ADPA-H$_3$PO$_4$ obtained in this solid state synthesis method are identical to each other.

Figure 1.1B (a-d) shows the FTIR spectra of P2ADPA salts synthesized by solid-state polymerization method using FC as oxidant. All the bands observed in Figure 6.1A are found to appear in these spectra suggesting the polymer backbone is very similar to that of P2ADPA prepared using APS oxidant (table 1.3). From the characteristic peaks noticed in the FTIR spectrum, it can be concluded that the mechanochemical route has successfully yielded pristine P2ADPA and its doped forms.

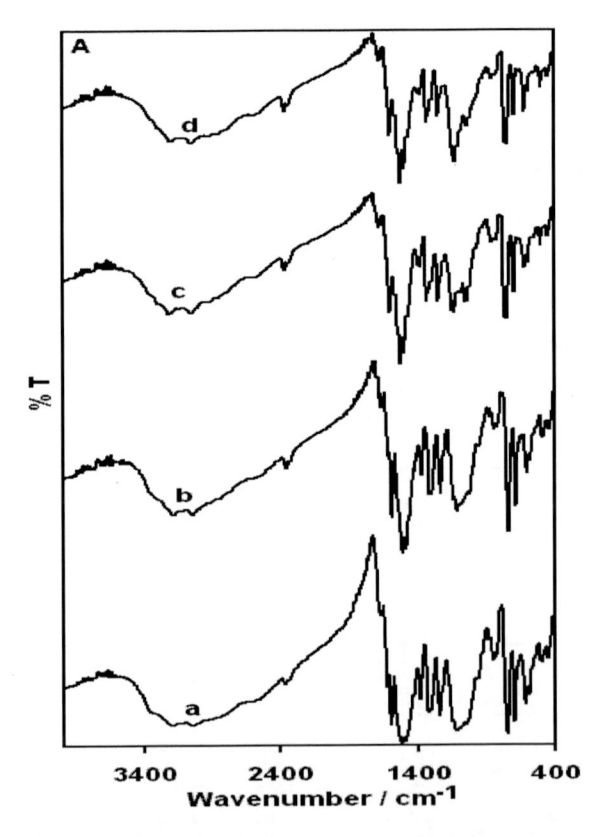

Figure 1.1A. FTIR spectra of P2ADPA and its salts prepared using APS: (a) P2ADPA (b) P2ADPA-HCl (c) P2ADPA-H$_2$SO$_4$ and (d) P2ADPA H$_3$PO$_4$.

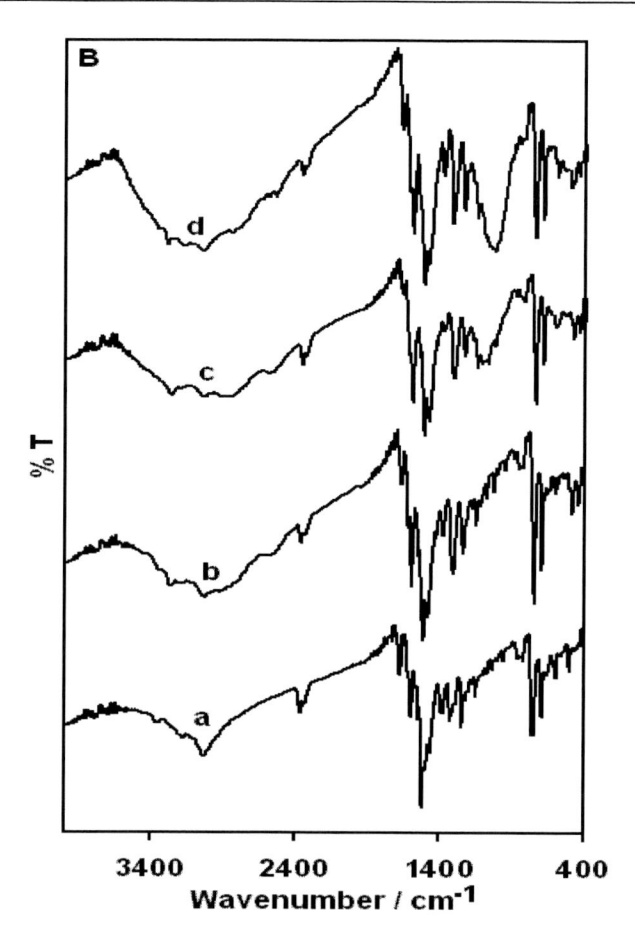

Figure 1.1B. FTIR spectra of P2ADPA and its salts prepared using FC: (a) P2ADPA. (b) P2ADPA-HCl (c) P2ADPA-H_2SO_4 and (d) P2ADPA H_3PO_4.

Table 1.3. Assignment of bands found in FTIR spectra of P2ADPA and its salts

P2ADPAs prepared using APS				
Polymer	$(N-H)_s$	$(C-C)_s$	$(C-N)_s$	$(C-H)_b$
P2ADPA	3213	1523	1301	1242
P2ADPA-HCl	3215	1520	1302	1241
P2ADPA-H_2SO_4	3216	1522	1302	1240
P2ADPA-H_3PO_4	3213	1521	1304	1242
P2ADPAs prepared using FC				
P2ADPA	3223	1521	1309	1248
P2ADPA-HCl	3220	1520	1306	1246
P2ADPA-H_2SO_4	3218	1524	1305	1245
P2ADPA-H_3PO_4	3216	1522	1306	1247

1.2.4. UV–Vis Absorption Spectra

Figure 1.2A (a-d) represents the UV–Vis absorption spectra of P2ADPA and its salts prepared using APS dispersed in NMP solution. The spectral profile show two characteristic absorption peaks at ~274-279 nm, ~446-452 nm (see table. 1.4). The well-defined peak at ~274-279 nm can be ascribed to $\pi–\pi^*$ electronic transition of the benzenoid rings in the polymer backbone, while the broad peak at ~446-452 nm can be attributed to polaron–π^* transition or bipolaronic state of P2ADPA.

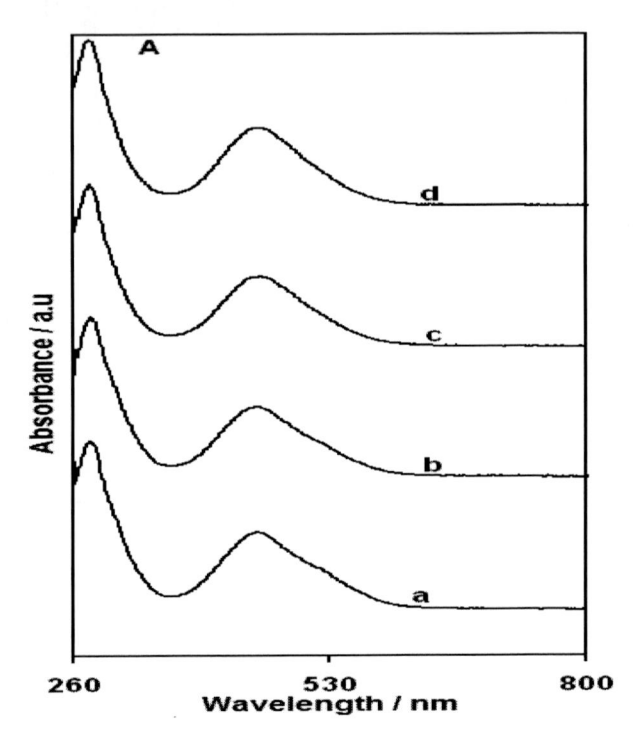

Figure 1.2A. UV-Vis spectra of P2ADPA and its salts prepared using APS: (a) P2ADPA (b) P2ADPA-HCl (c) P2ADPA-H_2SO_4 and (d) P2ADPA H_3PO_4.

Table 1.4. Assignment of UV-Vis absorption peaks of P2ADPA and its salts

P2ADPAs prepared using APS		
Polymer	$\pi–\pi^*$ (nm)	polaron–π^* (nm)
P2ADPA	274	452
P2ADPA-HCl	276	446
P2ADPA-H_2SO_4	279	450
P2ADPA-H_3PO_4	276	450
P2ADPAs prepared using FC		
P2ADPA	279	442, 512
P2ADPA-HCl	279	450, 507
P2ADPA-H_2SO_4	284	504, 545
P2ADPA-H_3PO_4	282	444, 504

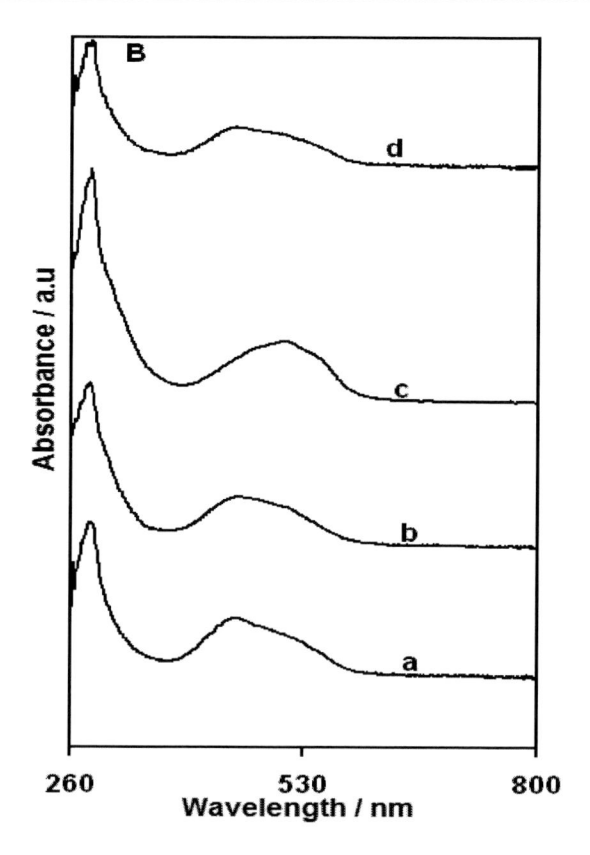

Figure 1.2B. UV-Vis spectra of P2ADPA and its salts prepared using FC: (a) P2ADPA. (b) P2ADPA-HCl (c) P2ADPA-H_2SO_4 and (d) P2ADPA H_3PO_4

Figure 1.2B (a-d) shows the UV–Vis absorption spectra of P2ADPA and its salts prepared using FC dispersed in NMP solution. Here too, one sharp peak at ~279-282 nm could be noticed for all the polymers. In addition to this peak there is a polaronic band at 442-504 nm along with a shoulder at around 504-512 nm, characteristic feature of phenazine type structure present in the polymer backbone. The intensity of polaron–π^* peak roughly indicates the amount of polaronic chain present in the polymer backbone. It could be observed that irrespective of oxidant used for polymerization, P2ADPA prepared using H_2SO_4 as doping agent shows more intense polaron–π^* band. This may be due to the more oxidative ability and polyprotic nature of 96 wt.% H_2SO_4 in comparison with the other doping acids. From these results, it can be said with conformity that H_2SO_4 serves as better doping agent among the tested protonating acids for preparing P2ADPA with better electronic properties.

1.2.5. XRD Pattern

The powder diffractograms of the different samples of P2ADPA prepared by APS and FC are shown in figs.1.3A (a-d) and 1.3B (a-d). The two major diffraction peaks in the range of 2θ = 19.58°-20.93° and 22.43°-23.68° found in the X-ray diffraction patterns of doped P2ADPA prepared using APS [Figure1.3A (a-d)] could be attributed to the periodicity parallel to the polymer chain and periodicity perpendicular to polymer chain respectively. The latter peak is stronger in the case of P2ADPA-H_2SO_4 [Figure 1.3A (c)] indicative of the formation of P2ADPA in a highly doped conducting state compared to the other polymers. There were no or very less intense peaks in the diffraction pattern of P2ADPA and P2ADPA-H_3PO_4 at around $2\theta = 30.63°$. This shows the formation of polymeric backbone with no or very less doping in these cases.

The XRD profile of the mechanochemically synthesized P2ADPA and its salts using FC possess ordered systems. The relatively sharp and stronger diffraction peaks at $2\theta = \sim11-13°$, 18°, 23.6° and 30° can be found in all the XRD spectrum of P2ADPA and its salts. It can be followed from Figure 1.3B (b-d) that there is a medium intense reflex around 11-13° in all of the doped P2ADPA. These reflexes could be due to the presence of closely packed benzene rings. The reflexes at around 23.6° is rendered by the periodicity parallel to the polymer chain and that around 30° is caused by the periodicity perpendicular to the polymer chain. On close examination of the peak at around 23° and 30°, it could be noticed that the same peaks for P2ADPA-H_2SO_4 is stronger than that of the remaining polymers indicating that it is in a highly doped crystalline state.

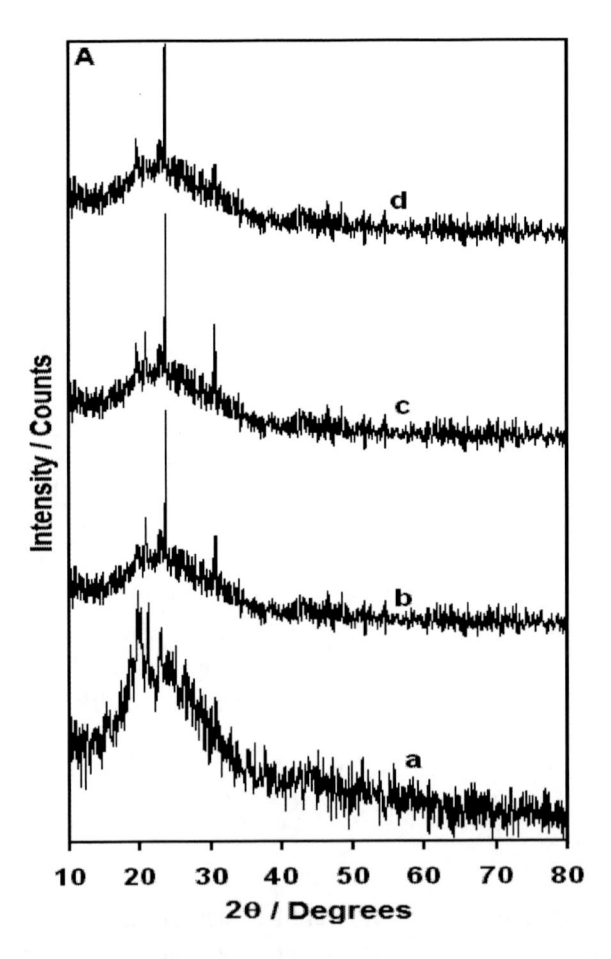

Figure 1.3A. XRD pattern of P2ADPA and its salts prepared using APS: (a) P2ADPA. (b) P2ADPA-HCl (c) P2ADPA-H_2SO_4 and (d) P2ADPA H_3PO_4

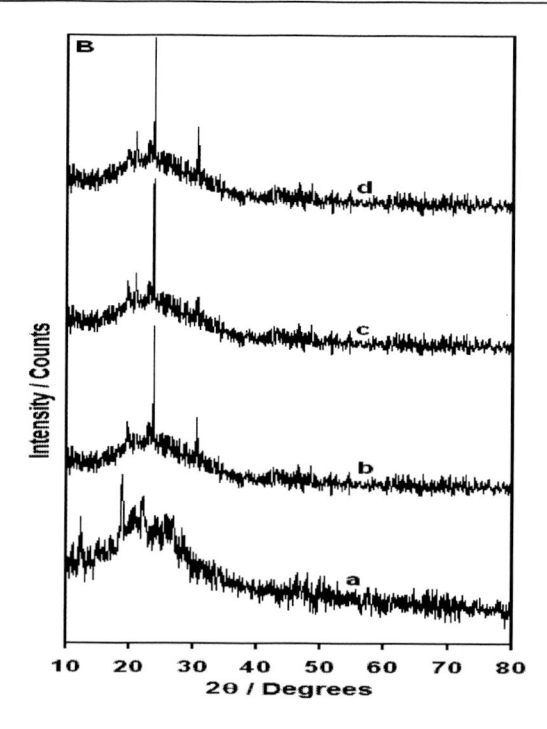

Figure 1.3B. XRD pattern of P2ADPA and its salts prepared using FC: (a) P2ADPA. (b) P2ADPA-HCl (c) P2ADPA-H$_2$SO$_4$ and (d) P2ADPA H$_3$PO$_4$.

1.2.6. Morphology

The greener synthetic route adopted to synthesize P2ADPA yielded polymeric products with distinctly different morphologies. Figure 1.4A (a-d) displays the FESEM image of P2ADPA and its salts prepared using APS. Since the mechanochemical polymerization is a surface confined process, the nature of reacting species plays a vital role in influencing the morphology of the formed polymeric particles. As seen in the FESEM images of P2ADPA and its salts, formation of aggregates of microparticles could be noticed. Figure 1.4A (a) shows aggregated P2ADPA-HCl particles with an average size of 250 nm. Figure 1.4A (b) displays P2ADPA-H$_2$SO$_4$ nanoparticles with an average size of 100 nm. The morphology of P2ADPA-H$_3$PO$_4$ (Figure1.4A (c)) shows bigger non-uniform micro-structured polymeric particles. Similarly the image of undoped P2ADPA [Figure1.4A (d)] also highlights the presence of particulates with average size more than 300 nm. The morphological difference is due to the difference in interaction between the reactants employed in the polymerization with APS. The influence of mechanical forces on the parent reagents and reaction products greatly influence the morphological feature of P2ADPA and its salts.

FESEM image of P2ADPA-HCl [Figure 1.4B (a)] is found to possess interconnected polymeric fibers along with few spherical shaped polymeric particles. The majority of P2ADPA-HCl particles were found as fused nanorods with sizes ranging from 35 to 50 nm. Figure 1.4B (a) indicates the presence of nanorods with random dimensions. Figure 1.4B (b) presents the FESEM image of P2ADPA-H$_2$SO$_4$ that has belt-like structures with sizes ranging from 65-80 nm. This particular morphology is quite distinct from that of P2ADPA-HCl. Even though the methodology adopted to produce the polymers is same, depending on the nature of the reacting species and the surface that is on offer for the mechanochemical reaction to precede the morphology differs. The surface morphological feature of P2ADPA-H$_3$PO$_4$ [Figure 1.4B (c)] consists of bulkier P2ADPA-H$_3$PO$_4$ particulates. Meanwhile, Figure 1.4B (d) showing the P2ADPA particles witnesses the formation of irregular aggregates along with the presence of small sized particles with indefinite shape. From these FESEM pictures, it is very easy to comprehend the fact that the role of doping agent and oxidizer plays a crucial role in the formation of different nanostructures while adopting this greener route for preparing P2ADPAs.

Figure 1.4A. FESEM image of P2ADPA and its salts prepared using APS: (a) P2ADPA-HCl (b) P2ADPA-H$_2$SO$_4$ (c) P2ADPA H$_3$PO$_4$ and (d) P2ADPA.

Figure 1.4B. FESEM image of P2ADPA and its salts prepared using FC: (a) P2ADPA-HCl (b) P2ADPA-H$_2$SO$_4$ (c) P2ADPA H$_3$PO$_4$ and (d) P2ADPA.

1.2.7. Electrochemical Activity

The electrochemical activity of the P2ADPA and its salts was investigated with cyclic voltammetry. As expected, CV with no characteristic peak is obtained for plain GCE in 0.5 mol/L H_2SO_4 [Figure 1.5A (a)]. Figure 1.5A (b-e) shows the cyclic voltammograms (CVs) of the polymer films (fabricated from P2ADPA prepared using APS) on GCE in 0.5 mol/L H_2SO_4. The anodic and cathodic potentials are listed in table 1.5. A redox couple is noticed in the CVs of P2ADPA and its salts. This redox peak in the positive sweep at 0.08-0.23 V and at -0.14 to -0.18 V is well-known as the formation of N,N'-diphenyl benzidine type radical cation (polaronic form of P2ADPA) (see scheme 1.2). The redox peaks are relatively sharper and the peak separation potential is lesser in CVs of P2ADPA-HCl and P2ADPA-H_2SO_4 than P2ADPA-H_3PO_4. However, little broader peak and a positive potential shift in the redox potential are observed in the CV of P2ADPA may be due to the absence of dopants in the polymer backbone.

The redox behaviour of P2ADPA and its doped forms prepared by using FC is displayed in Figure 1.5B (b-e). The CV shows an oxidation peak at around 0.03-0.06 V and reduction peak at around -0.09 - -0.12 V characterizing the formation of N,N'-diphenyl benzidine type radical cation (polaronic form of P2ADPA). The CVs recorded for P2ADPA and its salts prepared through the mechanochemical route suggest the formation of redox-active P2ADPA salts.

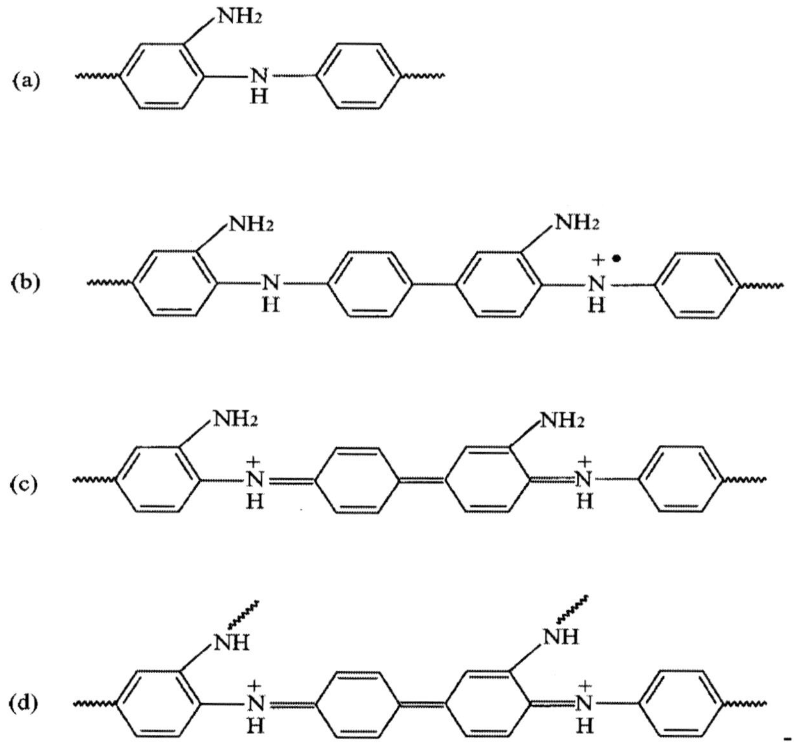

Scheme 1.2. Schematic structures of P2ADPA: (a) Structure of poly-2-aminodiphenylamine; (b) *N,N*-diphenyl benzidine- type radical cations (c) *N,N*-diphenylbenzidine-type dications.and (d) Branching of 2ADPA units.

Table 1.5. Redox potentials of P2ADPA and its salts

P2ADPAs prepared using APS		
Polymer	E_{pa}/V E_{pc}/V	
P2ADPA	0.23	-0.14
P2ADPA-HCl	0.05	-0.16
P2ADPA-H_2SO_4	0.19	-0.18
P2ADPA-H_3PO_4	0.08	-0.18
P2ADPAs prepared using FC		
P2ADPA	0.06	-0.12
P2ADPA-HCl	0.03	-0.19
P2ADPA-H_2SO_4	0.06	-0.16
P2ADPA-H_3PO_4	0.06	-0.09

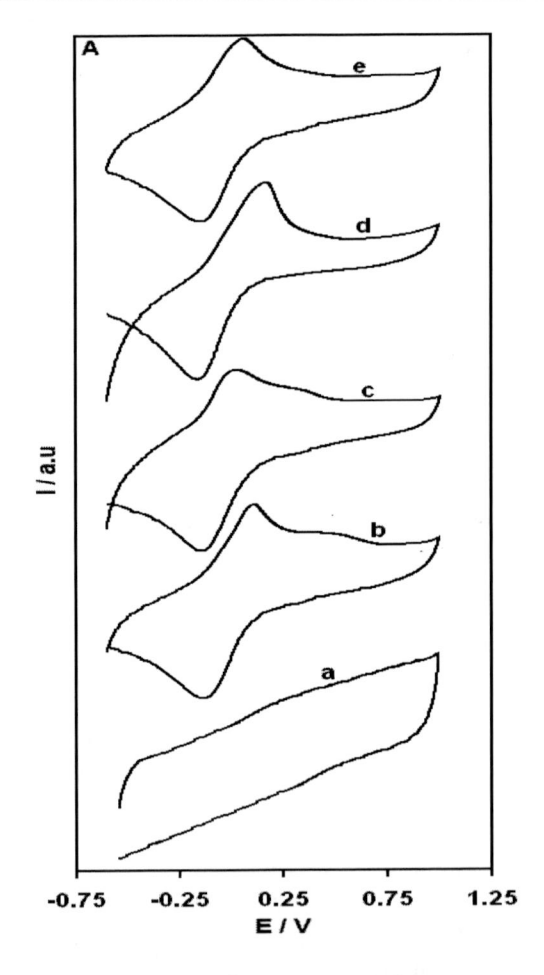

Figure 1.5A. Cyclic voltammograms of P2ADPA and its salts prepared using APS: (a) Plain GCE (b) P2ADPA/GCE (c) P2ADPA-HCl/GCE (d) P2ADPA-H_2SO_4/GCE and (d) P2ADPA -H_3PO_4/GCE.

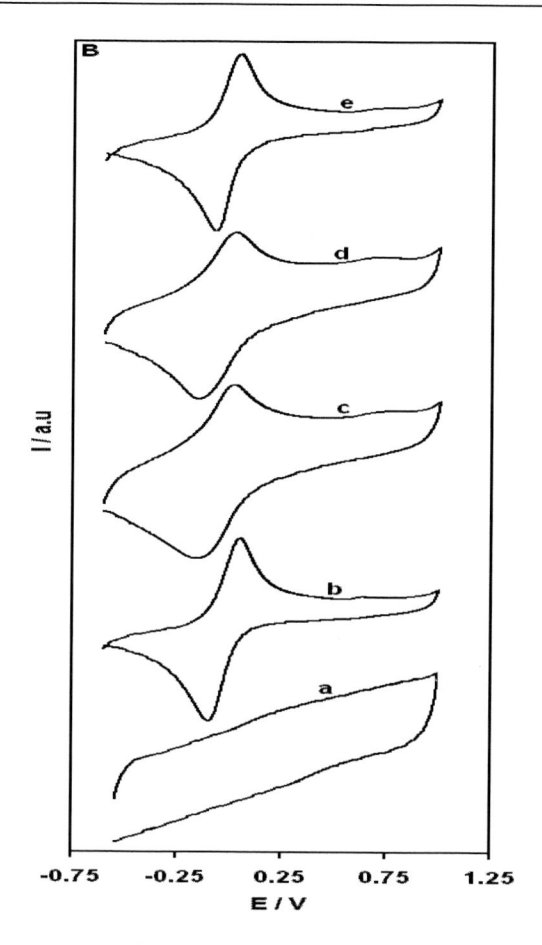

Figure 1.5B. Cyclic voltammograms of P2ADPA and its salts prepared using FC: (a) Plain GCE (b) P2ADPA/GCE (c) P2ADPA-HCl/GCE (d) P2ADPA-H_2SO_4/GCE and (d) P2ADPA -H_3PO_4/GCE.

1.2.8. Conductivity

The conductivity of P2ADPA and its salts prepared by APS and FC is listed in table 1.6.

Table 1.6. Conductivity values of P2ADPA and its salts

P2ADPAs prepared using APS	
Polymer	Conductivity (S/cm)
P2ADPA	0.68×10^{-3}
P2ADPA-HCl	1.56×10^{-2}
P2ADPA-H_2SO_4	1.2×10^{-1}
P2ADPA-H_3PO_4	0.23×10^{-2}
P2ADPAs prepared using FC	
P2ADPA	0.86×10^{-3}
P2ADPA-HCl	1.92×10^{-2}
P2ADPA-H_2SO_4	6.5×10^{-1}
P2ADPA-H_3PO_4	0.38×10^{-2}

As expected P2ADPA prepared by employing APS or FC is found to be less conductive rather than its salts. P2ADPA-H_2SO_4 is found to possess higher conductivity than other salts irrespective of the oxidant employed for solid-state polymerization. From the results of FTIR spectra, UV–Vis spectra and cyclic voltammogram studies, the conductivity differences of P2ADPA salts is found to rely on the characteristics of doping acids used viz., HCl, H_2SO_4 and H_3PO_4. Since 96 wt.% H_2SO_4 is more oxidative, polyprotic and strong acid and since it forms salt more readily with DPA compared to other acids used, P2ADPA-H_2SO_4 is found to show better conductivity.

1.2.9. Thermal Stability

The thermal curves obtained from thermogravimetric analysis (TGA) and differential thermal analysis (DTA) of P2ADPA, P2ADPA-HCl, P2ADPA-H_2SO_4 and P2ADPA-H_3PO_4 are presented in Figure 1.6A (a-d) respectively.

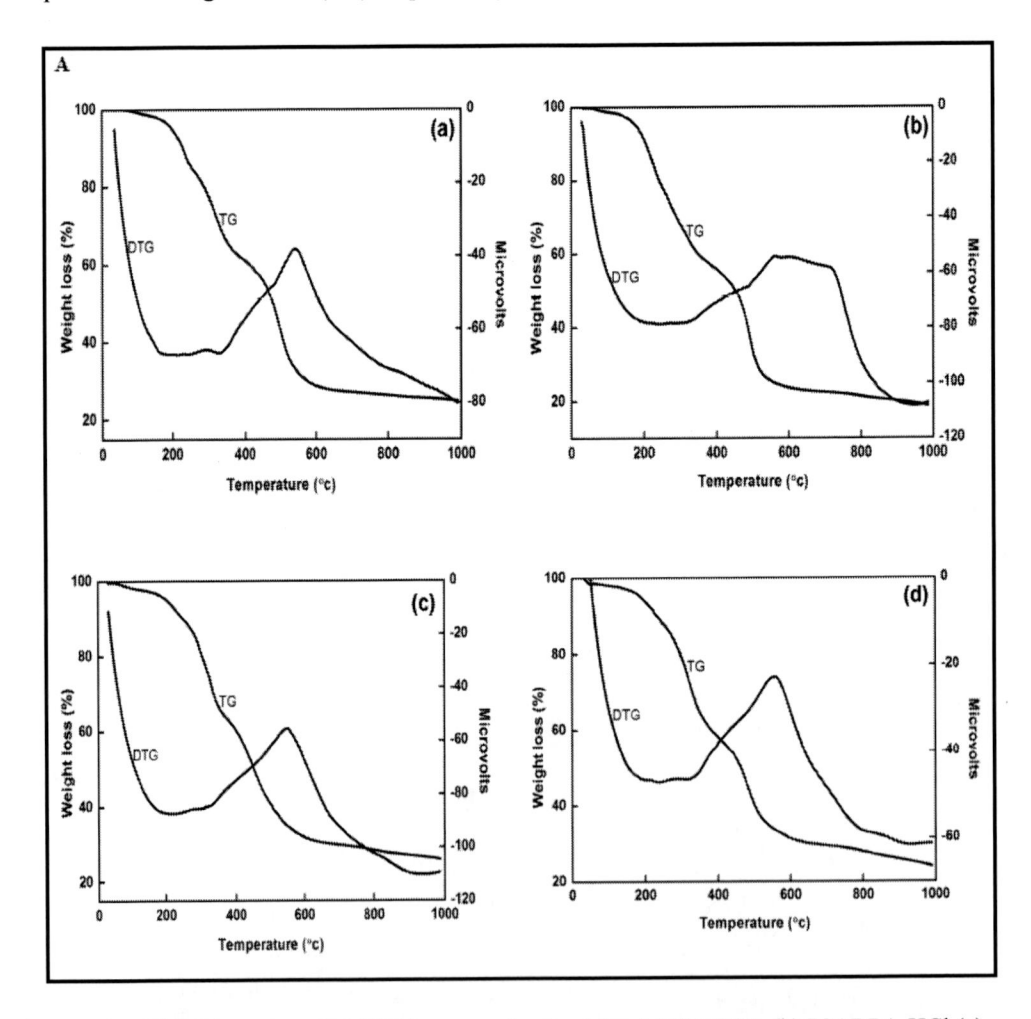

Figure 1.6A. TG/DTA curves of P2ADPA prepared using APS: (a) P2ADPA (b) P2ADPA-HCl (c) P2ADPA-H_2SO_4 and (d) P2ADPA H_3PO_4.

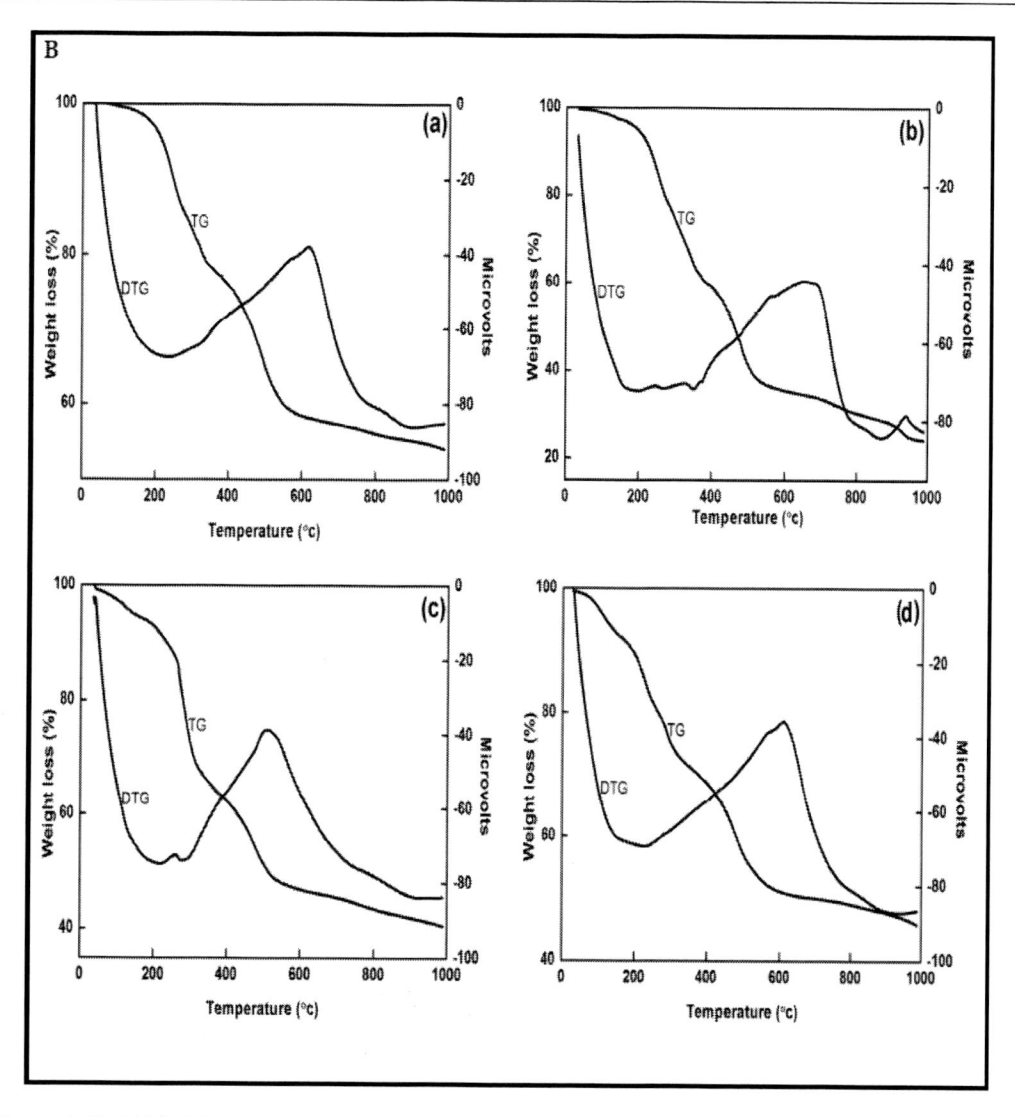

Figure 1.6B. TG/DTA curves of P2ADPA prepared using FC: (a) P2ADPA (b) P2ADPA-HCl (c) P2ADPA-H_2SO_4 and (d) P2ADPA H_3PO_4.

The first step loss (6-9%) in all polymers occurred at 60°C due to the loss of trapped water molecules present in the polymer matrix. The second step loss (14-18%) observed at 220°C-300°C is due to the loss of dopants and low molecular weight fragments of the polymer. The absence of such thermal event for P2ADPA (Figure 1.6A (a)) confirms the conjecture that HCl/H_2SO_4/H_3PO_4 ion, which are present as dopant ions, are removed in this temperature range. Degradation of P2ADPA main chain starts from 300°C as observed for both doped and undoped samples. Doped P2ADPA shows similar weight losses at temperature beyond 400°C. DTA curves of doped polymer shows two exothermic peaks at 485-500°C and 600-700°C. The former exotherm is due to the thermal expulsion of dopants present in the polymer chain and latter is due to the thermal degradation of polymer.

CONCLUSION

P2ADPA doped with inorganic acids (e.g., HCl, H_2SO_4 and H_3PO_4) were synthesized by solid-state polymerization by using APS or FC. Elemental analysis highlighted the presence of corresponding dopant anions (Cl, S, P) in the polymeric backbone of P2ADPA prepared using APS or FC. The yield obtained from this reaction route is found to be convincing. All of the prepared P2ADPA salts were processable using common solvents like water and acetone. Spectroscopic studies showed the presence of quinoid and benzenoid units along with phenazine like structure in all doped forms of P2ADPA. These results were further supported by conductivity measurements and cyclic voltammetry. More crystalline P2ADPA salt was obtained by doping with H_2SO_4 in comparison with other protonating acids in both the cases of APS and FC. FESEM pictures depicted the formation of nanostructures for P2ADPA-H_2SO_4 prepared using FC. All the as-prepared P2ADPA salts are found to be thermally stable up to 300°C. Among these inorganic acids, H_2SO_4 was found to be more suitable as a protic acid media in order to prepare P2ADPA salt with high conductivity through this mechanochemical route irrespective of the oxidants (APS or FC) employed for the polymerization process. These differences mainly depended on the chemical character of inorganic acid employed for mechanochemical insitu doping.

REFERENCES

[1] Huang, J.; Moore, J.A.; Henry Acquaye, J.; Kaner, R.B. *Macromolecules* 2005, *38,* 317-321.

[2] Abdiryim, T.; Xiao-Gang Z.; Jamal R. *Mater. Chem. Phys.* 2005, *90,* 367-372.

[3] Abdiryim, T.; Jamal, R.; Nurulla I. *J. Appl. Polym. Sci.* 2008, *107,* 3864-3870.

[4] Du, X-S.; Zhou, C-F.; Wang, G-T.; Mai, Y-W. *Chem. Mater.* 2008, *20,* 3806-3808.

[5] Zhou, C-F.; Du, X-S.; Liu, Z.; Ringer, S.P.; Mai Y-W. *Synth. Met.* 2009, *159,* 1302-1307.

[6] Jamal, R.; Abdiryim, T.; Nurulla, I. *Polym. Advan. Technol.* 2008, *19,* 1461-1466.

[7] Abdiryim, T.; Xiao-Gang, Z.; Jamal, R. *J. Appl. Polym. Sci.* 2005, *96,* 1630-1634.

[8] Jamal, R.; Abdiryim, T.; Ding, Y.; Nurulla, I. *J. Polym. Res.* 2008, *15,* 75-82.

[9] Posudievsky, O.Yu.; Goncharuk, O.A.; Pokhodenko, V.D. *Synth. Met* 2010, *160,* 47-51.

[10] Palaniappan, SP.; Manisankar, P. *Mater. Chem. Phys.* 2010, *122,* 15-17.

[11] Sinha, S.; Bhadra, S.; Khastgir, D. *J. Appl. Polym. Sci.* 2009, *112(5),* 3135-3140.

[12] Karyakin, A.K.; Strakhova, A.K.; Yatsimirski, J. *Electroanal. Chem.* 1994, *371,* 259-265.

[13] Thiemann, C.; Brett, M.A.; *Synth. Met.* 2001, *123,* 1-9.

[14] Wei, X.-L.; Wang, Y.Z.; Long, S.M.; Bobeczko, C.; Epstein, A.J. *J. Am.Chem. Soc.* 1996, *118,* 2545-2555.

[15] Kitani, K.; Yano, J.; Kunai, A.; Sasaki, K. *J. Electroanal. Chem.* 1987, *221,* 69-82.

[16] Genies, E.M.; Penneau, J.F.; Lapkowski, M.; Boyle, A. *J. Electroanal. Chem.* 1989, *269,* 63-75.

[17] Chen, W.C.; Wen, T.C.; Gopalan, A. *J. Electrochem. Soc.* 2001, *148*, E427-E434.

[18] Sasaki, K.; Kaya, M.; Yano, J.; Kitani, A.; Kunai, A. *J. Electroanal. Chem.* 1986, *215* 401-407.

[19] Wei, Y.; Sun, Y.; Jang, G.W.; Tang, X. *J. Polym. Sci. Part C Polym. Lett.* 1990, *28*, 81-87.

[20] Zimmermann, A.; Kunzelmann, U.; Dunsch, L. *Synth. Met.* 1998, *93*, 17-25.

[21] Cotarelo, M.A.; Huerta, F.; Quijada, C.; Cases, F.; Vazquez, J.L. *Synth. Met.* 2004, *148* , 81-85.

PART 3. SYNTHESIS OF BIODEGRADABLE POLYMER

In: Polymer Synthesis
Editor: E. Kowsari

ISBN 978-1-61324-672-6
© 2012 Nova Science Publishers, Inc.

Chapter 10

SYNTHESIS OF BIODEGRADABLE ELASTOMERS WITH TRIBLOCK AND MULTIBLOCK STRUCTURES: MAXIMIZATION OF ELASTICITY AND SHAPE RECOVERY

Vitali T. Lipik[1] and Marc J. M. Abadie[1,2]

[1]School of Materials Science and Engineering, MSE, Division of Materials Technology,
Nanyang Technological University, NTU, 50 Nanyang Avenue, Singapore,
[2]ICGM – AIME, UMR CNRS 5253, Institut Charles Gerhardt de Montpellier and
Laboratory of Polymer Science and Advanced Organic Materials, LEMP/MAO,
Université Montpellier 2, Montpellier, France

ABSTRACT

In this work, we discuss the relationship between the structure of biodegradable polymers and their ability to recover their shape. Polymers were synthesized using ε-caprolactone and L(DL)-lactide with different architectures. Experiments and simulations were run to discover the correlations among the crystallinity of polymers, polycaprolactone and polylactide PCL, PLA segment lengths, the molecular weight of blocks and shape recovery.

The elastic properties of biodegradable thermoplastic polymers are based on physical crosslinking, which appears to be caused by the separation of amorphous and crystalline phases. Therefore, macromolecules of biodegradable copolymers should contain both amorphous and crystalline phases. Elasticity can be achieved using block or multiblock structures of the polymer.

Initially, we present the results of cyclic behavior in shape recovery for random copolymers and copolymers with multiblock structures, which can be described by the hypothetical structures (PLA-co-PCL)-b-(PCL-co-PLA)-b-(PLA-co-PCL) and (PLA-co-PCL)-b-(PCL-co-PLA), respectively. Analyzing the constructed model, we found that shape recovery was positively correlated with PLA-co-PCL molar mass (hard block) but inversely correlated with PLA crystallinity and block length. However, the molar mass, crystallinity and block length of PLA were directly connected. Therefore, we attempted to increase PLA molar mass but decrease crystallinity. A small amount of ε-caprolactone as the comonomer was added during the PLA synthesis for this purpose. In this case, the block structure was lost, and the formation of complex multiblocks occurred. The best results were achieved for diblock copolymers with the following theoretical structure: (PLA-co-PCL)-b-(PCL-co-PLA) with 30% ε-caprolactone and 30% L-lactide in the PCL block. Each block has molar mass of 20 kDa. This multiblock polymer had 89% shape recovery.

Several attempts were made using transesterification reactions, which allow access to high levels of irregular structures in macromolecules that should help improve shape recovery. To increase the influence of transesterification reactions, a hard PLA block was synthesized, followed by PLA-co-PLA random block synthesis. A polymer with the targeted structure PCL-PLA-PCL 20-40-20 kDa, which has 35% ε-caprolactone in the middle block and 25% L-lactide in the PCL block, was revealed to have the best shape recovery at 90%. It is worth noting that this polymer retained this property at up to 300% elongation. One additional positive feature of this sample was the high speed of shape recovery estimated by the size of the cyclic test loop. Furthermore, the received recovery characteristics were similar or even surpassed the properties of industrial biodegradable elastomers.

Next, we tried to synthesize the triblock structure PLA-b-(PCL-co-PLA)-b-PLA, in which the central soft block is represented by an amorphous copolymer with a random structure and side hard blocks of macromolecules synthesized from PLA homopolymers. To accomplish this, a series of polymers were synthesized where the molar mass of the middle block was included as the variable in the design of the experiment. The molar mass of the PLA hard block and the ε-caprolactone:L-lactide molar ratio was also varied. The best shape recovery was recorded for a polymer with the structure PLA-PCL-PLA 5-100-5 kDa with 50% L-lactide in the middle block, reaching 92% after 100% sample stretching. We further discovered that the molar mass of the middle block significantly influenced shape recovery. However, the closest relationship was revealed between shape recovery and PLA segment length.

Reducing the influence of the middle block crystallinity on shape recovery by replacing L-lactide with DL-lactide in triblock copolymers was also investigated. In this case, three parameters of the model were varied: the molar mass of the middle block, molar ratio (ε-caprolactone:DL-lactide) in the middle block and molar mass of the hard block, which was synthesized from PLA homopolymers. Two resulting polymers revealed good shape recovery: polymers with structures PLA-b-(PCL-co-PDLA)-b-PLA 5-40-5 kDa (50 %LA in PCL) and 5-20-5 kDa (with 75 % LA in PCL) attained 93 and 92% shape recovery after 100% sample stretching. Moreover, the polymer with the optimized structure PLA-b-(PCL-co-PDLA)-b-PLA 10-40-10 kDa (50 %LA in PCL) reached 96% shape recovery.

1. INTRODUCTION

Synthetic thermoplastic elastomers (TPEs) were first developed in the 1970s based on styrene and butadiene. These polymers were found to be useful for many applications, from snowmobile tracks to shoe soles. The primary advantages of TPEs are their ease of processing using conventional methods (e.g., extrusion, injection molding) and the range of properties achievable through structure manipulation. In addition, biodegradable TPEs have potential use as implantable devices because of their controllable elasticity and the possibility of changing their chemistry and structure. [1] Elastomeric polymers can also be used to displace all implant applications where Nitinol is currently used. Such applications include cardiovascular stents and occluders for atrial septal defects (ASD). The advantage of using biodegradable elastomers over Nitinol is that the presence of scaffolding material in these cases is usually redundant once the device is covered with tissue, and furthermore, the presence of nickel can lead to hypersensitivity in some individuals. These elastomers also have other potential applications in the biomedical area, especially in the area of controlled drug delivery and tissue engineering. [2]

In recent years, biodegradable polymers have attracted great attention in the biomedical field because scientists are looking for more ways to exploit their capabilities for use in the human body. Theoretically, any material will degrade in the bioenvironment. However, the terms of degradation can be very different, even within a single class of materials such as polymers. For example, the estimated time of degradation for polyolefins varies from several years up to several hundred years depending on the environment and polymer structure. Other polymers such as proteins degrade very rapidly (a couple of weeks). When biodegradable polymers are discussed, it is usually in reference to poly(glycolic acid), polycaprolactone, poly(ethylene glycol) and poly(lactic acid) (scheme 1).

In addition to the above-mentioned polymers, there is a large variety of biodegradable copolymers made both by copolymerization of glycolide, caprolactone, lactide, dioxanon and other monomers and by block copolymerization. The introduction of the biodegradable portion (monomer units or polymers) occurs through polyurethanes or even polyethylene terephthalate. [3,4,5,6] Their commercial applications in the biomedical field include heart stents, surgical sutures and long-term drug delivery. [7] Biodegradable polymers can also be used as temporary scaffolds that facilitate the regeneration of tissues in the field of tissue engineering. The main advantages of using biodegradable polymers as implants are as follows: first, a second surgical procedure to remove the implants after they have served their purpose is not required; second, they are useful for short-term applications; and third, they overcome problems caused by permanently implanted devices.

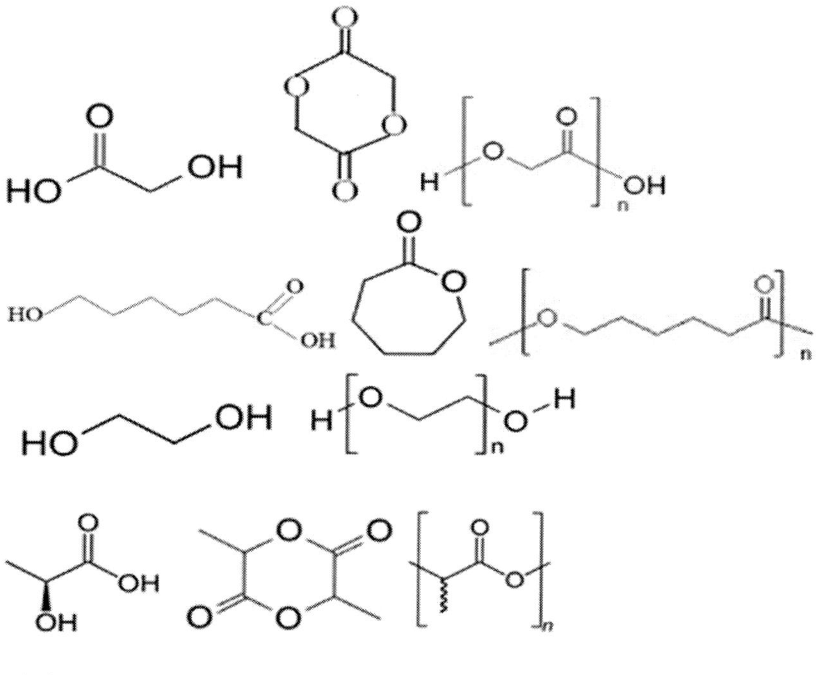

Scheme 1. Biodegradable polymers and their primary monomers.

Polymers are classified as biodegradable when they can be subjected to the destructive action of different agents. Degradation can be induced by oxygen and high temperature, by hydrolytic process or by enzymatic action. [8,9] For example, ester bonds in macromolecules of biodegradable polyesters are the first to degrade in these degradation processes.

Thermoplastic elastomers, or TPEs, are a type of polymeric materials that do rely on physical crosslinking to achieve high degrees of elastic deformation. They exhibit elastomeric (or rubbery) behavior at ambient conditions but yet are thermoplastic in nature. One of the best known and widely used TPE is a block copolymer consisting of block segments of a hard and rigid thermoplastic that alternates with block segments of a soft and flexible elastic material. Normally, the hard segments are located at the chain ends while the soft segments are located in the middle. One common example would be the styrene-butadiene-styrene (SBS) copolymer. At ambient temperatures, the soft, amorphous, central segments grant the material elastomeric behavior and the hard segments prevent sliding of the chains. Furthermore, below the melting temperature of the hard component, hard chain-end segments from numerous adjacent chains aggregate to form rigid crystalline domain regions. These domains are pseudo-crosslinked or physically crosslinked and act as anchor points to restrict soft-chain segment motions; they function in much the same way as chemical crosslinking for thermoset elastomers. When the thermoplastic elastomers are heated to elevated temperatures above the melting temperature of the hard component, however, the physical crosslinking is destroyed, and the polymer melts just like any other thermoplastic, which makes fabrication easy. This is in contrast to thermoset elastomers, which do not experience melting. Since the melting-solidification process is reversible and repeatable for thermoplastic elastomers, they can be recycled and reshaped; thus, production costs and time are significantly reduced compared to thermoset elastomers, which are generally not recyclable. [10]

2. SYNTHESIS OF POLYMERS

Several types of copolymers have been synthesized based on ε-caprolactone, L-lactide and DL-lactide. Random PCL-PLA copolymers were synthesized in one step by combining two types of monomer in a flask and maintaining certain conditions for 24 hours.

When a block structure was the target of synthesis, the first block was polymerized followed by the polymerization of the second block. All copolymers were synthesized by means of a coordinated ring opening polymerization, which is a type of "living polymerization". For this type of synthesis, after the complete exhaustion of one monomer or a mixture of monomers, a second type of monomer or mixture was added and the growth of the macromolecule resumed. This feature allowed for the successful synthesis of block structures.

Several types of polymers with different structures were synthesized. First, we synthesized diblock copolymers where one block possessed a random structure: (PCL-co-PLA)-b-PLA. The PLA hard block was designed as a homopolymer. Second, copolymers in which each block was partly randomized by the addition of a small amount of another monomer: diblock − (PCL-co-PLA)-b-(PLA-co-PCL) and triblock (PLA-co-PCL)-b-(PCL-co-PLA)-b-(PLA-co-PCL). Copolymers with inverse structures were also synthesized. For this copolymer, a PLA block was first synthesized, followed by the addition of ε-caprolactone or a mixture of two monomers (ε-caprolactone and L-lactide). This type of synthesis introduces a high level of irregular structures into the macromolecules due to the high level of transesterification. We named these types of copolymers 'block-like structures' with the understanding that transesterification may significantly destroy molecules of macroiniciator (in our case PLA) and distort the structure from the targeted block architecture. For the preparation of this type of polymer, however, we describe it as a block copolymer (e.g., PCL-PLA-PCL).

In addition to copolymers, two types of common triblock structures were obtained: PLA-b-(PCL-co-PLA)-b-PLA and PLA-b-(PCL-co-PDLA)-b-PLA. The middle, or soft, block was aimed to obtain a maximum degree of randomization.

Schematically, the coordination-insertion ring opening polymerization could be represented as follows (Scheme 2):

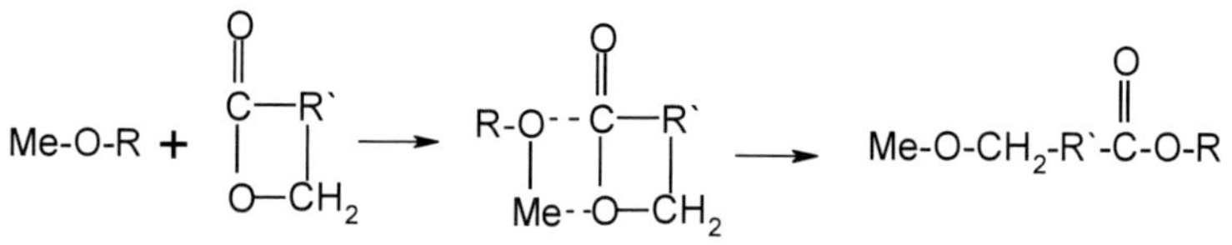

Scheme 2. Coordination-insertion ring opening polymerization.

In this reaction, there is an insertion of the opened cycle of a monomer between the metal and oxygen of the catalyst or inside of an initiator (as a rule, this insertion occurs between a carbon atom and a hydroxyl group of an alcohol). Varying the monomer:initiator molar ratio according to the following equation, a polymer with a defined molar mass can be obtained (Equation 1):

$$Mn(polymer) = \frac{[M] \times Mn(monomer)}{[I]} \times \chi \qquad (1)$$

where Mn(polymer) and Mn(monomer) are molecular masses of the polymer and monomer, respectively, in g/mol; M and I are the quantities of monomer and initiator, in moles; and χ the functionality of the alcohol used as initiator.

The syntheses were carried out in a three-neck round bottom flask (100 mL) equipped with a thermometer, a condenser and a magnetic stirrer. The flask was purged with argon, evacuated twice and then kept under an argon atmosphere. The argon was dried using a water absorption system containing silica gel. The synthesis temperature was maintained by immersing of flask into silicone oil bath. Toluene, monofunctional initiator for diblock synthesis - hydroxyl butyl vinyl ether (HBVE) or 1,4-butanediol (difunctional initiator for triblock synthesis) and Sn(Oct)$_2$ (catalyst) were added to the flask at 90°C and stirred for 30 minutes. The quantity of HBVE or 1,4-butanediol and monomers used was based on the desired degree of polymerization. The amount of catalyst (Sn(Oct)$_2$) was calculated based on the ratio of initiator to catalyst, which was constant for all syntheses at 10:1. Next, the necessary quantity of monomers (ε-caprolactone and L-lactide) was added to create the soft, middle block of the triblock copolymer. The temperature was then increased to 110°C, and the reaction was left for 24 hours to polymerize.

For the synthesis of the 2nd block, the defined quantity of L-lactide was added to the flask and the reaction was left for another 24 hours at 110°C. After a total of 48 hours, the toluene solution was poured into cold methanol. The precipitated polymer was then filtered, washed several times with methanol and dried in a vacuum oven for 48 hours at 40°C.

3. METHODS OF ANALYSIS

3.1. Nuclear Magnetic Resonance Spectrometry (NMR)

^{13}C-NMR spectra were obtained with a Bruker 400 spectrometer using $CDCl_3$ as a solvent. The samples were dissolved in deutero-chloroform in 5-mm sample tubes. The solution concentration was adjusted to 200 mg of polymer in 1 mL of solvent.

The average lengths of poly(ε-caprolactone) (L_{CL}) and poly(L-lactide) blocks (L_{LA}) were calculated from the intensities of the peaks of the carbonyl signals [11,12] using Equations 2a and 2b:

$$L_{PCL} = \frac{I_{CCC}}{I_{LCC}} + 1, \quad L_{PLA} = \frac{I_{LLL}}{I_{LLLC}} + 1 \qquad (2a, 2b)$$

where I_{CCC} and I_{LLL} are the intensities of ε-caprolactone - ε-caprolactone and L-lactide –L-lactide triads, respectively, and I_{LCC} and I_{LLLC} represent the intensities of ε-caprolactone – L-lactide and L-lactide - ε-caprolactone triads and tetrads, respectively.

3.2. Differential Scanning Calorimetry (DSC)

The glass transition temperatures and melting enthalpies of the polymer samples were measured using a TA Instruments Model Q10 DSC machine equipped with a DSC Refrigerated Cooling System to achieve low temperatures. The software used was TA Instruments Control. Polymer samples were weighed (4–6 mg) and placed into hermetic aluminum pans. DSC analysis was performed by first heating the sample to 200°C to eliminate internal stresses and then equilibrating the samples at -80°C, followed by ramping the temperature to 200°C at a rate of 10°C per minute. The crystallinity of the polymers (C) was calculated using Equation 3:

$$C = \frac{\Delta H}{\Delta H_{100}} \times 100 \qquad (3)$$

where ΔH is the experimental melting enthalpy of polymer, in J/g, and ΔH_{100} is the melting enthalpy of the polymer with 100% crystallinity (ΔH_{100}=139 J/g for polycaprolactone [13] and ΔH_{100}=93 J/g for polylactide [14]).

3.3. Shape Recovery of Elastomers. Types of Shape Recovery Measurement

3.3.1. Creep and Recovery Test

Dynamic mechanical analysis (DMA) was applied to analyze the shape recovery of polymers. Specifically, ideal elastomers are able to recover almost completely regardless of the level of stress applied during the creep test. The elastomeric properties of the triblock copolymers were therefore evaluated and quantified using creep and recovery tests. We subjected each polymer to low and high stresses (corresponding to 30% and 80% of their tensile yield stress, respectively) for creep and recovery measurements. In this test, we then estimated recovery using Equation 4:

$$\mathrm{Re\,cov}\,ery = \frac{L_M - L_R}{L_M - L_0} \times 100\ \%, \tag{4}$$

where L_M is the maximum sample length at elongation, in mm; L_R is the sample length after recovery, in mm; and L_0 is the initial length of the sample, also in mm.

3.3.2. Cyclic Test

The shape recovery of polymers was measured by the tensile cyclic test using standard tensile testers. Dumbbell-shaped samples were pulled in a vertical direction at a rate of 10 mm/min at room temperature or at a desired temperature if the machine is equipped with a temperature-controlled chamber. The degree of elongation is chosen based on the polymer's elastic properties. Data for shape recovery calculation can be taken from the machine software; or shape recovery can be calculated using the initial and final (after recovery) distance between the marks which are put manually on the part of sample subjected to the elongation before test. The sample with marks can be pulled up to 100%, for example. After the desired elongations were achieved, the microtester device was stopped, and the sample was removed. The sample then regained its original size to different degrees. The final size of the sample or the distance between marks was measured immediately after the recovering time. The degree of shape recovery could then be calculated using Equation (4).

The cyclic test can also be used to reveal the shape recovery behavior of the sample. Firstly, samples are pulled several times at the same magnitude of elongation. This test is performed to demonstrate the resistance of a polymer to stress and its ability to keep initial shape under dynamic conditions. Secondly, the sample can also be pulled several times at different levels of elongation. This is useful to show the limit of deformation when the sample starts to lose its elasticity. For comparison, the elastic behavior of polyurethane (left) and a PLA-PCL copolymer (right) during the cyclic test is shown in Figure 1.

It is evident that polyurethane possesses much better elastic properties than the PCL-PLA copolymer, nearly recovering its initial shape even after elongation of 300%.

The shape of the loop of the cyclic test can be used to estimate the rate of recovery. A sharp shape of the loop, as in the case of polyurethane, indicates a high rate of recovery. The oval-like shape of the cyclic loop of the PCL-PLA copolymer, however, corresponds to a very low rate of recovery. Furthermore, when the portion of the curve corresponding to recovery is vertical, there is no recovery.

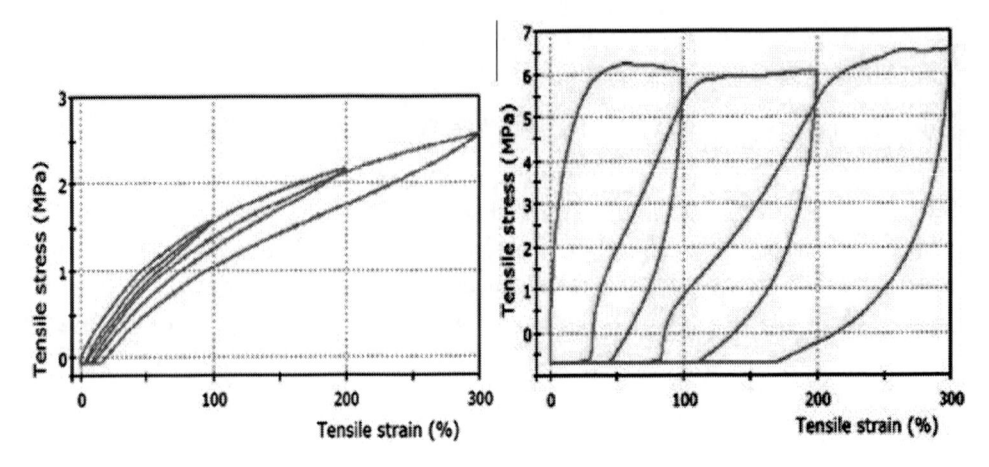

Figure 1. Cyclic test of polyurethane (left) and a PCL-PLA copolymer (right).

3.3.3. Cyclic Shape Recovery Behavior

Shape recovery behavior of polymers can be better visualized using multicyclic and multitensile tests. In these analyses, the sample is stretched several times to the same strain and released after each elongation. A machine is used to fix the modification in the sample shape during elongation and after removing the stress (Figure 2).

A good elastomer will have nearly identical paths in these tests and the ratio stres:strain will be very similar for every run. For example, 9 curves in Figure 2 (right) have very similar overlapping routes that are difficult to distinguish. When testing a copolymer with poor shape recovery, the curve of each new cycle moves to the right compared with the previous one (Figure 2, left).

3.3.4. Shape Memory Test

The shape memory test measures the ability of a material to remember its initial form and return to this initial form after deformation. In the case of polymer shape memory, the glass transition temperature is the basic point for measuring this property. The basic premise for measuring shape memory is presented in Figure 3.

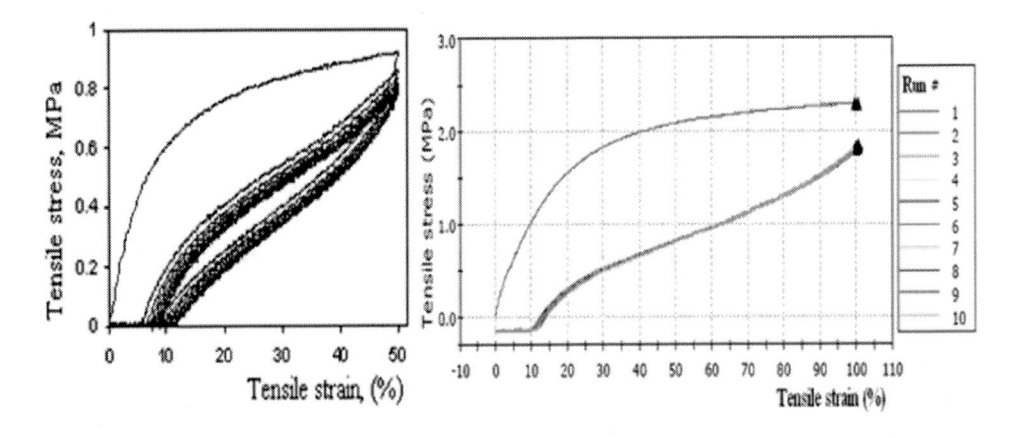

Figure 2. Shape memory behavior in cyclic (left) and tensile (right) tests.

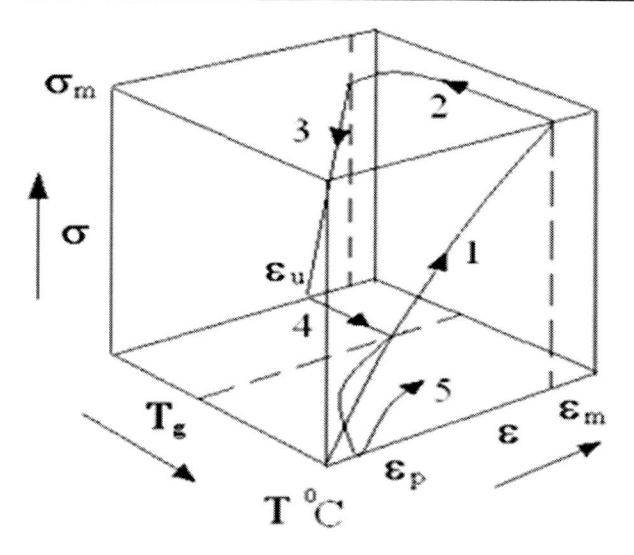

Figure 3. Path of the shape memory test of polymeric materials.

The first step is to subject the sample to the desired maximum deformation ε_m (path 1 in Figure 3). Next, the temperature is reduced below the glass transition temperature of the polymer (path 2 in Figure 3). Then, the load of the sample is removed, and the sample moves to point ε_u. The final step is the increase of the temperature back to the initial at a certain rate. The sample strain achieves the point ε_p due to the heating. The difference between the initial shape and the strain ε_p is then the unrecoverable length. [15]

Shape recovery is calculated by two equations considering two parameters: R_f and R_r, which are the ability to fix mechanical deformation ε_m resulting in a temporary shape ε_u and the ability to restore the mechanical deformation of the permanent shape ε_p after application of a certain deformation ε_m, respectively (Equations 5 and 6) [16]:

$$R_f(N) = \frac{\varepsilon_u(N)}{\varepsilon_m} \tag{5}$$

$$R_r(N) = \frac{\varepsilon_m - \varepsilon_p(N)}{\varepsilon_m - \varepsilon_p(N-1)} \ \text{or} \ R_r = \frac{\varepsilon_m - \varepsilon_p}{\varepsilon_m} \tag{6}$$

An ideal shape recovery value is equal or close to 100%. Some biodegradable copolymers have R_f and R_r equal to 99.0 and even 99.6% (e.g., the multiblock copolymer based on macroblocks of PLA and blocks of random copolymers of poly[glycolide-co-ε-caprolactone]). [17]

4. SHAPE RECOVERY OF RANDOM AND MULTIBLOCK COPOLYMERS

We will now consider shape recovery of copolymers with different synthesized structures based on two types of monomers: ε-caprolactone and lactide (D- and L-lactide). Based on our previous experience, we found that the molar ratio caprolactone:lactide in the range from

70:30 to 50:50 provides a random copolymer structure without significant crystallinity of PCL or PLA. This can clearly be seen in the comparison of the DSC data of PCL-PLA copolymers with a targeted molar mass of 100 kDa and different quantities of L-lactide (Figure 4).

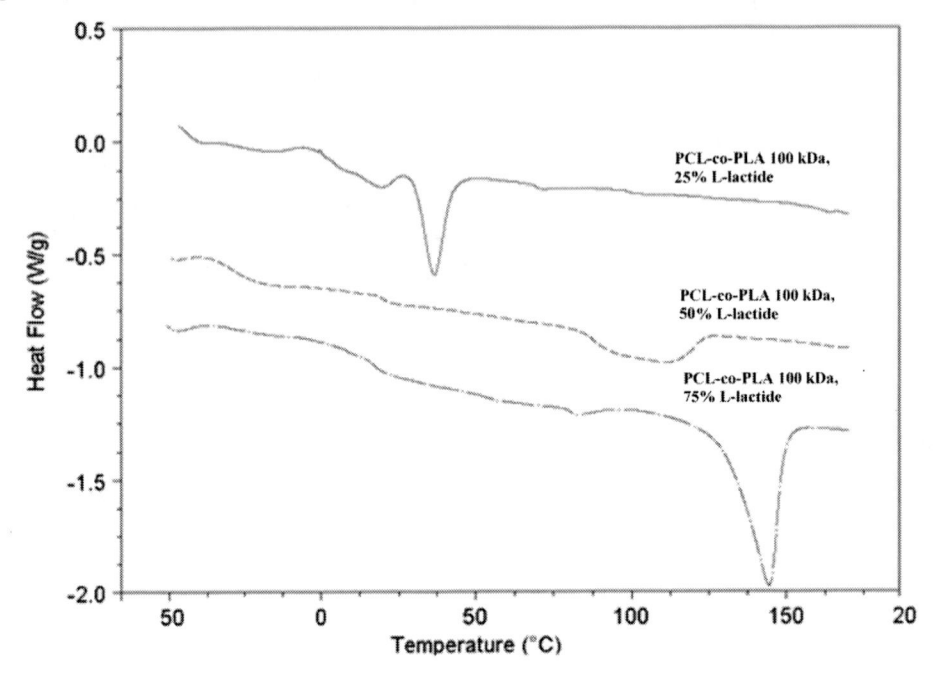

Figure 4. DSC data of PCL-PLA copolymers with different amounts of L-lactide (from the top – 25, 50 and 75%).

As can be seen, the molar mass of the copolymer influences mechanical properties, particularly shape recovery. It is interesting that the PCL-PLA copolymer with 25% L-lactide with a molar mass up to 50 kDa behaved like a viscous liquid. When the molar mass of the copolymer was then increased to 70 kDa, it became solid but still quite soft. Above 70 kDa, the copolymer acquires a rubbery-like behavior. For example, the PCL-PLA copolymer with 25% L-lactide and a molar mass 100 kDa had elongation at break 572% at 25°C (modulus = 1.26 MPa). The shape recovery of this copolymer is presented in Table 1.

Table 1. Shape recovery of a PCL-PLA copolymer with a CL:LA molar ratio of 75:25

	Recovery of polymer after elongation		
Elongation, %	100	200	300
Shape recovery, %	91	89	78

It can be seen that a polymer with low crystallinity has a better ability to recover at low deformation, but this amorphous polymer loses this ability at high levels of deformation because the polymer starts to flow.

Therefore, it is evident that some level of crystallinity in a polymer is necessary for good elasticity and shape recovery. This degree of crystallinity can be obtained by the synthesis of block copolymers or polymers with a multiblock structure. If a PCL-PLA copolymer does not have a block structure, then the best recovery rate and recovery strain are achieved with PCL content in the range of 30-40%. [18]

4.1. PCL-PLA Diblock with a Multiblock Structure

The multiblock structure of a polymer is even more important than block architecture when considering shape recovery because of the better control of segment length and the subsequent control of crystallinity allowed by this structure. Therefore, a series of PCL-PLA copolymers with diblock-like structures were synthesized to estimate the influence of PCL and PLA segment length on elasticity. The basic idea was to synthesize diblock PCL-b-PLA with the modification of each block by the addition of a co-monomer. Specifically, this meant that some quantity of L-lactide was added at the first step of the synthesis of PCL and some quantity of ε-caprolactone was added during the synthesis of the second block (PLA). The elastomeric nature of the resulting copolymers was then quantified by the elongation at break. Elongations of all of the synthesized diblocks with multiblock structures are summarized in Table 2 (molecular weight of diblock PCL-b-PLA was 20-20 kDa for all copolymers).

The best results were achieved for polymers with small amounts of L-lactide added to the PCL (5-10%). If we increased the quantity of lactide added, increasing the random polymer structure and reducing the length of multiblocks, their mechanical properties deteriorated. PCL and PLA segment lengths with the highest elongations at break were calculated by means of ^{13}C NMR spectra and were equal to 4.7 and 8.1 monomer units, respectively. From Table 2, it can be seen that the crystallinity, polymer ratio and segment length played important roles in the mechanical properties of a copolymer.

Table 2. Elongation at break of PCL-PLA copolymers with multiblock structures, %

% LA in PCL block	% CL in PLA block			
	5	10	20	30
5	353	368	248	141
10	400	1038	955	937
20	83	101	155	130
30	41	49	88	122

Remembering this influence of molar mass on mechanical properties and leaning on our previous result [19], a new series of diblock polymers, such as PCL-PLA copolymers with multiblock structures, were synthesized. The molar mass of each block in the PCL-b-PLA structure was kept constant at 20-20 kDa. The synthesis for these multiblock structures was made similarly as for the copolymers presented in Table 2. But to disrupt PLA crystallinity and keep untouched the first PCL-co-PLA block decreasing the influence of transesterification, an amount of D,L-lactide was added during the synthesis of the second PLA block. Targeted copolymer composition, structure characteristics and shape recovery after 100% elongation of the obtained polymers are presented in Table 3.

**Table 3. Structural characteristics and shape recovery
of synthesized PCL-PLA copolymers**

% LA in PCL (mol)	% D,L-lactide in PLA (mol)	PCL/PLA crystallinity, %	PCL/PLA block length, monomer units	Recovery after the elongation of sample to 100%
10	10	5.2/25.3	4.7/27.0	44
10	20	4.3/20.6	4.5/25.1	56
10	30	4.4/11.8	3.9/21.6	65
20	10	2.3/22.8	3.5/25.2	61
20	20	2.0/20.9	3.1/25.0	76
20	30	2.1/9.8	3.4/20.8	83
30	10	0/23.2	3.2/26.3	71
30	20	0/18.9	3.0/23.6	79
30	30	0/7.6	3.3/19.8	92
40	10	0/18.0	3.2/24.7	80
40	20	0/14.8	3.1/23.3	90
40	30	0/7.3	3.5/17.2	86

The shape recovery of the synthesized copolymers was also tested at 200 and 300% elongation. Experiments showed that the ability to recover initial shape decreased with increased deformation. Additionally, the shape recovery of thermoplastic biodegradable elastomers was better expressed for small deformations and small strains. Therefore, these materials could not resist high deformations, as can polyurethanes or other crosslinked polymers. At high elongation, the displacement of the amorphous and crystalline phases occurs, leading to yielding of the polymer and its inability to return to its initial state. [15]

The shape recovery of synthesized copolymers differs considerably depending on the obtained crystallinity and on the shape of the crystals, and PCL and PLA segment lengths. We could therefore conclude that the shape recovery of polymers clearly depends on the polymer composition and its structure. On one hand, the polymer should not be completely amorphous and flexible because such consistence cannot be restored after deformation. On the other hand, the polymer should not be excessively crystalline; otherwise, it will not be elastic.

Semicrystalline polymers have high elongation at break but do not return to their initial state or shape after stress removal. For example, in a previous work, [18] the shape recovery of a copolymer became worse with increasing PLA crystallinity. Conversely, the strain fixity rate depends on PLA segment content, as it increases with the augmentation of PLA quantity in copolymer. [20]

It is worth noting that some characteristics of a polymer structure correlate with shape recovery. Calculated coefficients of correlation revealed the closest correlation between PCL and PLA block length and shape recovery (-0.81 and -0.65, respectively). Logically, strong correlation between shape recovery and PCL/PLA crystallinity takes place (R = -0.84 and -0.81 respectively). Because the coefficients of correlation were negative, shape recovery should improve when the block lengths of PCL and PLA and PLA crystallinity decrease.

To more clearly visualize the dependence of shape recovery on polymer structure, we used the program TableCurve 3D v4 for simulations of this relationship. The shape recovery of the obtained copolymers with constant targeted molar masses of 20-20 kDa could be described by Equation 7 with the coefficient of correlation R = 0.986.

$$R=15.83849+3.13777\times\%LA-0.92594\%\times\%DL-9.83333\times10^{-2}\times\%LA^2+7.50516\times10^{-2}\times(\%DL)^2+$$
$$0.117\times\%LA\times\%DL+1.38888\times\%LA^3-1.2508\times10^{-3}\times\%DL^3-10^{-3}\times\%LA\times\%DL-2\times10^{-3}\times\%LA^2\times\%DL \qquad (7)$$

where R is the degree of recovery, in %; %LA is the quantity of L-lactide added to the PCL in the first step of synthesis, in % mol; and %DL is the quantity of D,L-lactide added to the PLA in the second step of synthesis, in % mol. A simulated graphical function of the relationship between shape recovery and the structure of the synthesized copolymer is presented in Figure 5.

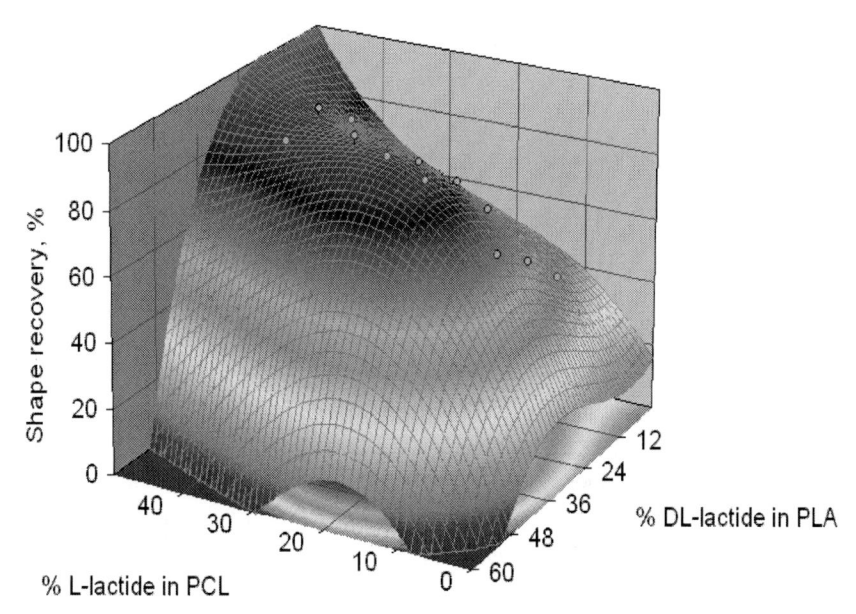

Figure 5. Simulated shape recovery values of multiblock copolymers (degree of recovery after 100% elongation).

Considering the graph of Figure 5, it can be seen that the best shape for recovery of the PCL-PLA copolymer was obtained with 10-20% D,L-lactide added to the PLA and 40-50% L-lactide added to the PCL. According to the simulation, maximum shape recovery property will be observed at minimum crystallinity in PCL block which should be almost amorphous; but some PLA crystallinity is necessary to keep good mechanical properties. Large quantity of amorphous D,L-lactide added to PLA block leads to the decrease of points of physical crosslinking and gives soft or even flowing material.

For comparison, the worst shape recovery among all of the synthesized copolymers was found for the polymer with 10% L-lactide in PCL and 10% D,L-lactide in PLA (recovery equal to 44%). The best reached shape recovery (92%) was found at polymer with 30% L-lactide in PCL and 30% D,L-lactide in PLA The curves for cyclic tests of both copolymers are presented in Figure 6.

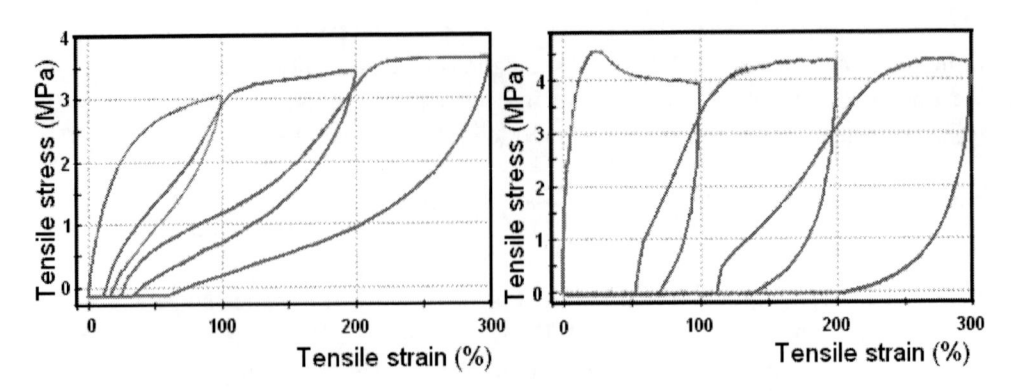

Figure 6. The best (left) and worst (right) results for shape recovery in PCL-PLA copolymers.

It can be clearly seen that one material returned to its initial state (left graph) and the other yielded under the applied stress with almost no recovery.

4.2. Shape Recovery of Copolymers with a Triblock-Like Structure

Using a bifunctional initiator (1,4-butanediol) and carrying out the synthesis over two steps, a triblock structure of polymer could be obtained. A series of elastomeric ABA triblock copolymers were thus synthesized using stannous octoate and 1,4-butanediol as the catalyst and initiator of polymerization, respectively, with sequential addition of the comonomers in toluene. The aim of this experiment was to reveal the influence of polymer structure on mechanical properties and to achieve triblock copolymers with good elastic properties.

Because a triblock structure gives more space to vary the properties of a polymer, we attempted to create triblock-like PLA-PCL-PLA copolymers where each block had a multiblock structure. Larger segments in the soft block (B) were made mainly of PCL, and larger segments in the hard blocks (A) were made mainly of PLA. To disrupt the crystallinity of the PCL and PLA, we added L-lactide during ε-caprolactone polymerization and ε-caprolactone during L-lactide polymerization.

Syntheses were planned according to the design of Galois field. The targeted molar mass of the soft block (B) was 40,000 Da for all syntheses. In addition, the quantity of L-lactide added to the soft block, the quantity of ε-caprolactone added to the hard block and the molar weight of the hard block (PLA) were varied in the experiment. Furthermore, the quantity of L-lactide added to the first step of polymerization was 10, 25 or 50 mol %, and the quantity of ε-caprolactone added to the second step of synthesis was 0, 10 or 25 mol %. The molar weight of the PLA (A) was thus varied from 10,000 to 20,000 and 40,000 Da.

In total, 9 polymers were successfully synthesized. In these polymers, different multiblock structures with the majority of the PCL in the soft block and the majority of the PLA in the hard block were found to having varying properties. The polymer composition and structural properties are presented in Table 4.

The Young's modulus of the synthesized polymers ranged from 1 to 220 MPa, and elongation at break varied from 7% to 1800%. Shape recovery was different for different polymer structures, with the lowest measured shape recovery at 32% and the best at 87% after 100% elongation.

Table 4. Polymer composition and structural properties

N	PLA targeted M_n, kDa	% LA in PCL, mole	% CL in PLA, mole	*PCL/PLA crystallinity, %	*PCL/PLA segment lengths, monomer unit	Shape recovery, %
1	10	10	0	39.1/27.9	12.8/18.0	57
2	20	10	10	25.6/47.1	11.2/20.1	61
3	40	10	25	25.5/31.6	9.9/16.1	70
4	10	25	10	20.8	5.9/6.9	83
5	20	25	25	1.3	6.7/8.9	87
6	40	25	0	50.4	7.4/35.6	32
7	10	50	25	34.8	3.6/8.3	80
8	20	50	0	47.9	4.3/20.6	73
9	40	50	10	45.6	3.7/23.5	0

*Crystallinity of PCL and PLA was calculated based on DSC data, PCL and PLA segment lengths and 13C NMR analysis.

For example, Polymer 1, with the structure 10-40-10 kDa (PLA-b-PCL-b-PLA) with 10% L-lactide in the soft block and homopolymer PLA in the hard block, demonstrated very high deformation after stretching. This polymer also had high elongation at break (more than 1200%) but poor shape recovery. Polymer 5, with 25% L-lactide in the PCL and 25% ε-caprolactone in the PLA, had good shape recovery and a elastic behavior. Furthermore, Polymer 9, with 50% L-lactide in the PCL and 10% ε-caprolactone in the PLA, had the highest modulus (237 MPa), very low elongation at break (42%) and no shape recovery.

The influence of structural parameters of the synthesized copolymers on shape recovery was surprising. Additional copolymers have thus been synthesized to further elucidate the influence of structure, particularity on shape recovery. If we compare the influence of molar mass of a PLA block containing 25% ε-caprolactone on shape recovery, we could not find any significant difference (Figure 7). Although the hysteresis and point of recovery on the x-axis were similar, one difference existed: Young's modulus and tensile stress increased with the increasing of molar mass of the PLA block.

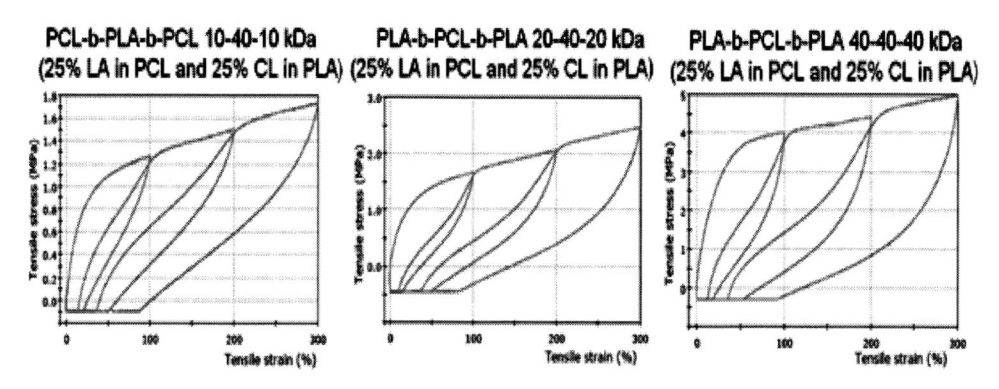

Figure 7. Influence of molar mass of a PLA block containing 25% ε-caprolactone and identical middle blocks on triblock shape recovery.

The second interesting relationship between polymer structure and shape recovery was the influence of the amount of ε-caprolactone added to the PLA block had on elasticity. We could alternatively say that PLA crystallinity affected shape recovery because PLA crystallinity decreased with increasing of ε-caprolactone amount. Figure 8 presents the cyclic tests of three PLA-PCL-PLA copolymers with the same targeted molar mass of 20-40-20 kDa and the same quantity of L-lactide in the middle block (25%) but with different quantities of ε-caprolactone in the PLA (0, 10 and 25%).

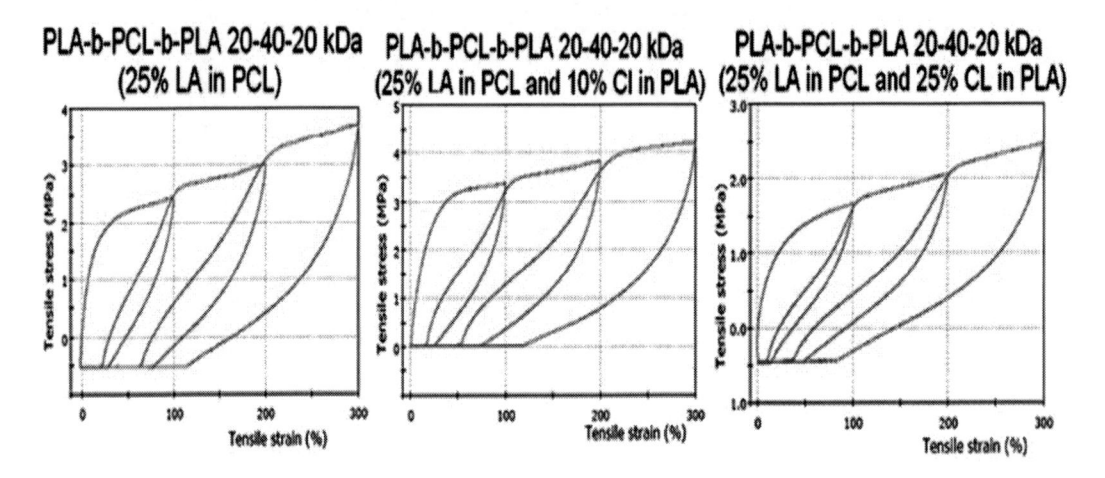

Figure 8. Influence of hard block composition on the shape recovery of triblock PLA-PCL-PLA copolymers (the PLA block is a homopolymer in the left graph).

It can be seen that the addition of 10% ε-caprolactone (middle graph) did not change the shape hysteresis, possibly because the PLA retained a certain degree of crystallinity. However, shape recovery was improved when 25% ε-caprolactone was added to the PLA (Figure 8, right). In this case, the crystallinity of the PLA was only 1.3% (Polymer 5 in Table 4).

The dependence of shape recovery (R) on PLA molar weight (M_{PLA}), quantity of L-lactide in the PCL soft block (%LA) and the quantity of ε-caprolactone in the PLA hard blocks (%CL) can be described by pair multiplication equations with correlation coefficients equal to 0.93 (Equation 8).

$$R=60.70-1.21\times M_{PLA}+0.76\times\%LA+2.86\times\%CL-6.44\times10^{-4}\times M_{PLA}\times\%LA-5.58\times10^{-3}\times M_{PLA}\times$$
$$\%CL-6.04\times10^{-2}\times\%LA\times\%CL \tag{8}$$

where M_{PLA} is the molar mass of the PLA block, in kDa; %LA and %CL are the amount of L-lactide added to the synthesis of the PCL block and the quantity of ε-caprolactone added to the synthesis of the PLA block, respectively, in mol %.

We can see from the equation 8 that the molar mass of the PLA block negatively influenced shape recovery because of negative coefficients in front of M_{PLA}. According to the equation, the shape recovery of polymers decreases with the growth of the molar mass of the PLA block.

Using this equation, we sought to optimize the structure of the polymer for maximum shape recovery. The optimized polymer was synthesized with the following targeted structure: PLA-PCL-PLA 10-40-10 kDa and 46% L-lactide in the PCL block and 13% ε-caprolactone in the PLA block. This polymer revealed good shape recovery with the degree of recovery equal to 90% after 100% elongation.

4.3. Shape Recovery of Reverse Multiblocks with Middle PLA Block

The organization of irregular structures in linear amorphous macromolecules can significantly improve shape recovery because irregularity, like crosslinking, does not allow macromolecules to flow during stretching. The quantity of irregular structures can be increased through transesterification reactions.

There is a defined order for monomer addition in PCL-PLA block synthesis when stannous octoate is used as the catalyst. Two competing reactions during coordinated anionic ring opening polymerization (CAROP) occur: the opening of ester bonds in molecules of a cyclic monomer and the breakage of ester bonds in the macromolecules of a polymer. These reactions depend on the type of catalyst, the existing polymer, which plays the role of macroinitiator, and the type of monomer added to the second step of synthesis.

A series of polymers were thus prepared to study the influence of transesterification on the mechanical properties of polymers. The method of synthesis sought to create the best condition for transesterification and therefore build as much irregularity into the structure of the macromolecules as possible. We synthesized the PLA block by first adding small quantities of ε-caprolactone to disrupt the PLA crystallinity. Next, the PCL block was synthesized with the addition of L-lactide. The experiment was designed to make the modeling and prediction of polymer properties possible. Nine polymers were successfully made, and their methods of preparation and structures are listed in Table 5.

Table 5. Design of PCL-b-PLA-b-PCL polymer synthesis

N	Molar weight of PLA, kDa (first step)	Molar weight of PCL (each), kDa (second step)	% of CL added to the first step, mole	% of LA added to the second step, mole
1	20	20	10	0
2	40	20	30	25
3	80	20	50	50
4	20	40	30	50
5	40	40	50	0
6	80	40	10	25
7	20	80	50	25
8	40	80	10	50
9	80	80	30	0

The corresponding mechanical properties of the synthesized polymers are presented in Table 6.

It can be seen that Polymer N2 had good shape recovery, which was consistent across different elongations (up to 300%). Cyclic tests of polymer N2 and industrial polymer PLC 70/30 were almost identical, but polymer N2 recovered faster, as evidenced by the smaller size of the hysteresis loops (Figure 9).

Table 6. Mechanical properties of PCL-PLA-PCL copolymers

N	Modulus, MPa	Elongation at break, %	Shape recovery by DMA, %	Shape recovery by cyclic test after elongation, %		
				100%	200%	300%
1	15.21	1510	75	80	73	63
2	3.42	1360	88	91	90	82
3	5.45	150	78	89	0	0
4	8.01	480	80	85	79	62
5	26.97	1170	59	65	50	42
6	5.31	10	40	90	0	0
7	4.58	100	15	35	0	0
8	25.26	15	42	60	45	0
9	45.84	906	53	65	50	42

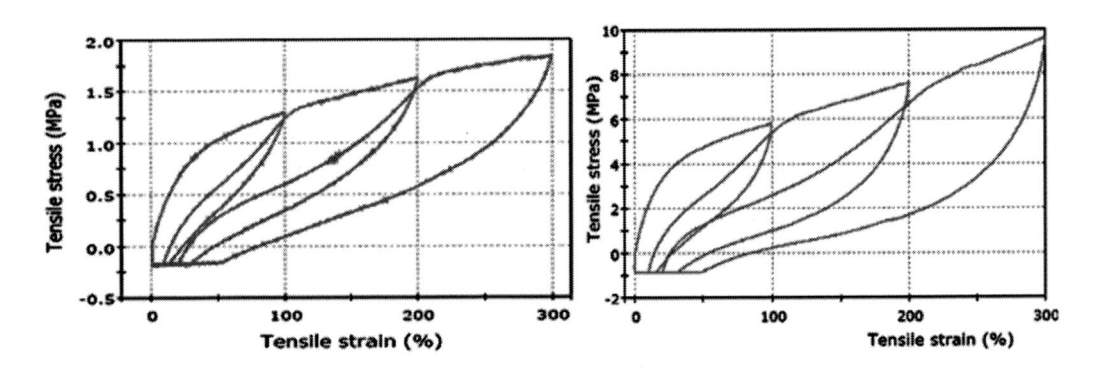

Figure 9. Cyclic tests of polymer N2 (PCL-PLA-PCL, 20-40-20 kDa, 25% LA in the PCL and 30% CL in the PLA) on the left and industrial copolymer PLC 70/30 on the right.

The simulation of the relationship between polymer structure and shape recovery can be described by the following parabolic equation with a coefficient of correlation equal to 0.99.

$$R=69.0347+0.2875\times M_{PLA}-0.075\times M_{PCL}+1.19167\%\times CL+0.03\%\times LA-4.8611\times10^{-4}\times M_{PLA}^{2}-4.8611\times10^{-3}\times M_{PCL}^{2}-0.0256\times\%CL^{2}+0.0028\times\%LA^{2} \tag{9}$$

where M_{PLA} is the molar mass of the PLA block, in kDa; M_{PCL} is the molar mass of the PCL block, in kDa; %LA and %CL are the amount of L-lactide added to the synthesis of the PCL block and the quantity of ε-caprolactone added to the synthesis of the PLA block, respectively, in mol %.

Analyzing this equation, we could conclude that increasing the PCL block molar mass negatively impacted shape recovery because of the negative coefficients in front of this variable. Looking for an extreme recovery value and taking the derivation of each variable of Equation 9, we noted that increasing the molar mass of the middle PLA block improved shape recovery. In addition, the optimized quantity of ε-caprolactone added to PLA was 23%. The optimization of the PCL block molar mass and the quantity of L-lactide added could not be performed, however, as theoretically the PCL molar mass should be reduced to a minimum and the quantity of L-lactide added is insignificant according to the coefficients of the above equation.

5. SHAPE RECOVERY OF PLA-B-(PCL-CO-PLA)-B-PLA TRIBLOCK COPOLYMERS

The triblock copolymer of the type ABA has the potential to be a good elastomer when one of its blocks is amorphous and supple and another is hard and crystalline. We designed a series of triblock copolymers with high molar mass PCL-co-PLA middle blocks (B) with different monomer ratios and short (A) blocks made of PLA. During the experiment, three parameters were varied: the molar mass of the middle block, the quantity of L-lactide in the middle block and the molar mass of the hard block (PLA). The targeted structural parameters of the synthesized copolymers and their physical characteristics are presented in Table 7.

**Table 7. Targeted structural and mechanical properties
of PLA-b-(PCL-co-PLA)-b-PLA copolymers**

N	Targeted Mn of PCL, kDa	% L-lactide in PCL	Targeted Mn of PLA, kDa	PCL/PLA crystallinity, %	PCL/PLA block length, monomer units
1	30	25	5	0/26.8	7.3/9.0
2	60	25	15	0/25.4	6.9/11.2
3	100	25	30	0/32.4	7.4/10.3
4	30	50	15	0/31.5	3.4/16.1
5	60	50	30	0/33.6	3.4/15.3
6	100	50	5	0/3.2	4.5/6.3
7	30	75	30	0/42.0	2.5/35.8
8	60	75	5	0/3.2	1.9/9.4
9	100	75	15	0/21.9	2.3/12.4

The range in mechanical properties of the synthesized triblock copolymers depended on the polymer structure and ranged from 10 to 1500% for elongation at break, 1.5-15 MPa for Young's modulus and 0-95 % shape recovery (Table 8).

Table 8. Mechanical properties of PLA-b-(PCL-co-PLA)-b-PLA copolymers

N	Mn PCL, kDa	%L-lactide in PCL	Mn PLA, kDa	Elongation at break, %	Young's modulus, MPa	Shape recovery after elongation to		
						100%	200%	300%
1	30	25	5	123	1.56	10	0	0
2	60	25	15	979	1.38	89	72	63
3	100	25	30	1613	4.02	70	64	58
4	30	50	15	689	7.95	79	60	42
5	60	50	30	1010	5.84	73	55	40
6	100	50	5	606	2.38	92	93	78
7	30	75	30	9	37.33	0	0	0
8	60	75	5	564	10.97	80	75	53
9	100	75	15	1017	14.98	75	69	47

Polymer N6 displayed the highest degree of shape recovery, the cyclic test of which is presented in Figure 10.

It can be seen from Figure 10 that the polymer was quite weak and the modulus value was low; however, the degree of recovery was high enough to consider it a good elastomer, especially at small (up to 100% elongation) deformations.

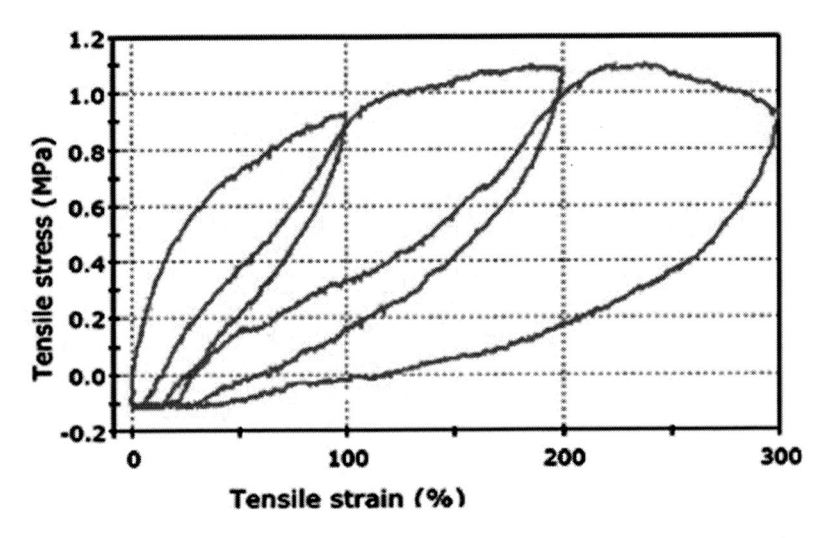

Figure 10. Cyclic test of Polymer N6 from Table 8.

We next sought to describe the shape recovery behavior of the synthesized triblock copolymers with the soft middle block and hard outer blocks. The following mathematical equation (10), with the coefficient of correlation R = 0.98 describing the relationship between shape recovery (R) and polymer structure, was constructed on the basis of the obtained shape recovery data:

$$R= -65.057-0.374 \times M_{PCL} +0.139 \times \%LA+21.065 \times M_{PLA} +4.149 \times 10^{-2} \times M_{PCL} \times \%LA-0.128$$
$$\times M_{PCL} \times M_{PLA} -0.241 \times \%LA \times M_{PLA} \tag{10}$$

where M_{PCL} is the molar mass of the middle soft PCL block, in kDa; M_{PLA} is the molar mass of the PLA hard block, in kDa; and %LA is the quantity of L-lactide in the soft block, in mol %.

To make the analysis of Equation 10 easier, we fixed one variable at a time. For example, if we set M_{PCL}=100 kDa, we received the following equation (11):

$$R = -102.457+8.265 \times M_{PLA} +4.288 \times \%LA-0.241 \times \%LA \times M_{PLA} \tag{11}$$

Trying to find the maximum recovery value by taking derivations of variables, we discovered that the theoretical maximum recovery value should be reached under the following conditions: M_{PCL} = 100 kDa (fixed), M_{PLA} = 17 kDa, amount of L-lactide in the PCL = 34%. According to these values, we could conclude that L-lactide played an important role in the copolymer structure by influencing shape recovery. In addition, the PLA block should remain small to avoid excessive brittleness.

The ability of a polymer to recover its shape becomes better when the molar mass of PLA block goes to the minimum. However, the PLA block should not be too small, as a polymer would become amorphous and weak. The quantity of L-lactide in the middle block is also very important. A small quantity of L-lactide in the middle soft block does not destroy PCL crystallinity completely, and the polymer is not elastic. Incomplete disruption of crystallinity and poor elasticity then lead to poor shape recovery. Larger amounts of L-lactide in the middle block (>40%), however, create excessively crystal structure in the middle block also make shape recovery worse.

To more clearly see the relationship between shape recovery and the structural parameters and other mechanical characteristics of a polymer, we calculated the coefficients of correlation. The results of this calculation are presented in Table 9.

Table 9. Correlation (R) between shape recovery and polymer properties

PLA block length	PCL block length	PLA crystallinity, %	% LA in PCL	Mn of PCL, Da	Mn of PLA, kDa	Elongation at break, %	Modulus, MPa
-0.63	-0.06	-0.54	-0.05	0.59	-0.21	0.67	-0.57

It can be seen that strong negative correlations between shape recovery and PLA block length and PLA crystallinity exist. Conversely, the molar mass of the soft block and shape recovery are correlated positively: a larger molar mass of the middle block provides better shape recovery. Logically, shape recovery correlates with elongation at break, and there is negative correlation between shape recovery and Young's modulus. Similar tendencies can be seen in the work of Lu et al. [21]

The influence of L-lactide quantity in the middle soft block on shape recovery can be seen in three examples of analogous polymers with the same structure but with different amounts of L-lactide in the middle block (Figure 11).

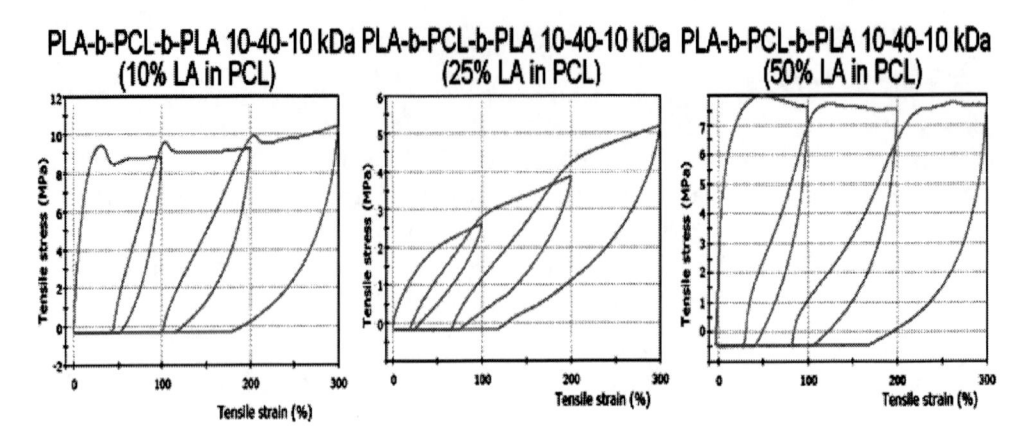

Figure 11. Influence of L-lactide amount in the middle block of PLA-b-(PCL-co-PLA)-b-PLA polymers on shape recovery.

It is evident that the L-lactide quantity in the middle block should not be too small or too large. At 10% L-lactide (Figure 11, left) in the PCL middle block, PCL remains crystalline and does not recover well. At 25% L-lactide, the middle block becomes softer, and tensile stress is two times lower than for the polymer with 10% L-lactide. The shape recovery of the polymer is also much better in this case. At large amounts of L-lactide (50%, Figure 11, left), the polymer acquires high PLA crystallinity, and shape recovery is less than 50%. Considering these data, it is possible to say that a golden middle value of copolymer in the middle block is needed for good shape recovery.

The influence of PLA hard block molar mass on shape recovery is displayed in Figure 12.

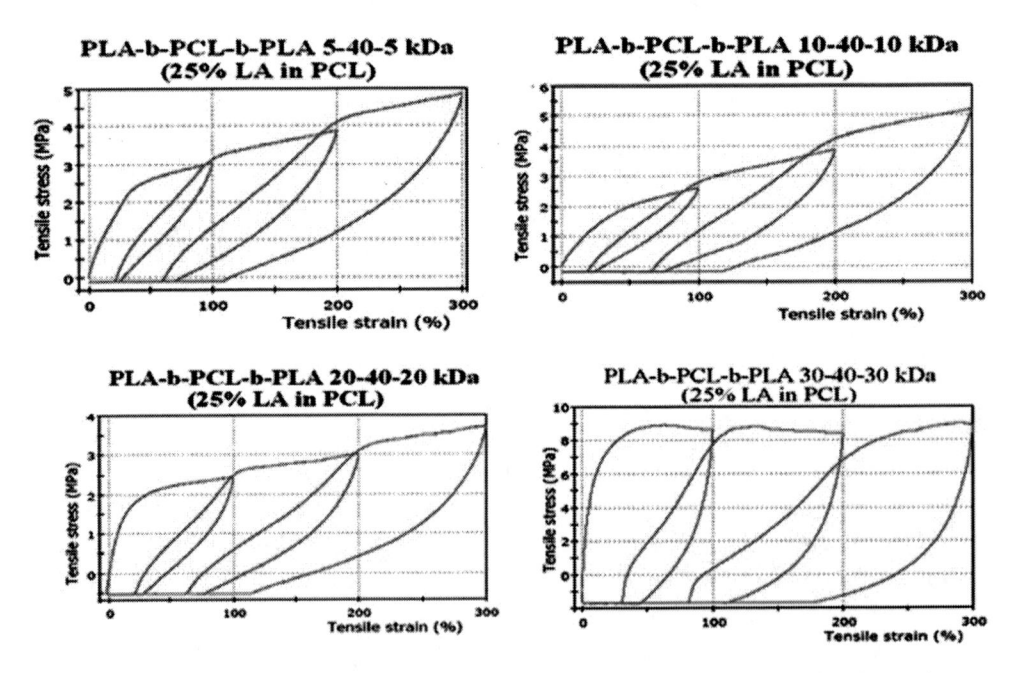

Figure 12. Influence of the molar mass of PLA hard block in PLA-PCL-PLA ABA block copolymer on shape recovery.

It can be seen from Figure 12 that there is no major difference in shape recovery between the samples with PLA hard block molar masses of 5, 10 and 20 kDa. The same conclusion could be drawn from the coefficients of correlation between PLA molar mass and shape recovery in Table 9 (-0.21). The weakly negative correlation means that shape recovery slightly decreases with increasing PLA molar mass. Only when the total molar mass of the hard PLA block starts to exceed the molar mass of the PCL soft block (when the polymer contains more crystalline phase than amorphous) does the polymer shape recovery deteriorate quickly. In this case, the polymer becomes hard; the tensile stress for the polymer with PLA molar mass of 30 kDa is almost three times higher than for the analogous polymer with a molar mass for PLA of 20 kDa (last two graphs in Figure 12).

We noted that shape recovery optimization in a polymer structure depended strongly on the environment temperature. Samples presented in Figure 12 were tested at room temperature. Higher temperatures request higher crystallinity from copolymers and higher quantities of L-lactide incorporation in the middle block. For example, the optimization of a ABA triblock polymer structure such as PLA-b-(PCL-co-PLA)-b-PLA revealed the following best polymer structure for use at 37°C: PLA-PCL-PLA, 16-80-16 kDa and 46% of L-lactide in the middle block. This polymer has a high value of elongation at break at 37°C (850%) and retains favorable mechanical properties even at 50°C (elongation at break was 275%). It should be noted, however, that the elongation at break for this polymer measured at 25°C was 1850%. In addition, the glass transition temperature for this polymer was 8.5°C, and the polymer possessed 21.5% PLA crystallinity. Shape recovery behavior in tensile and cyclic tests of the triblock copolymer PLA-PCL-PLA, 16-80-16 kDa and 46% of L-lactide in PCL at 37^0 C is presented in Figure 13.

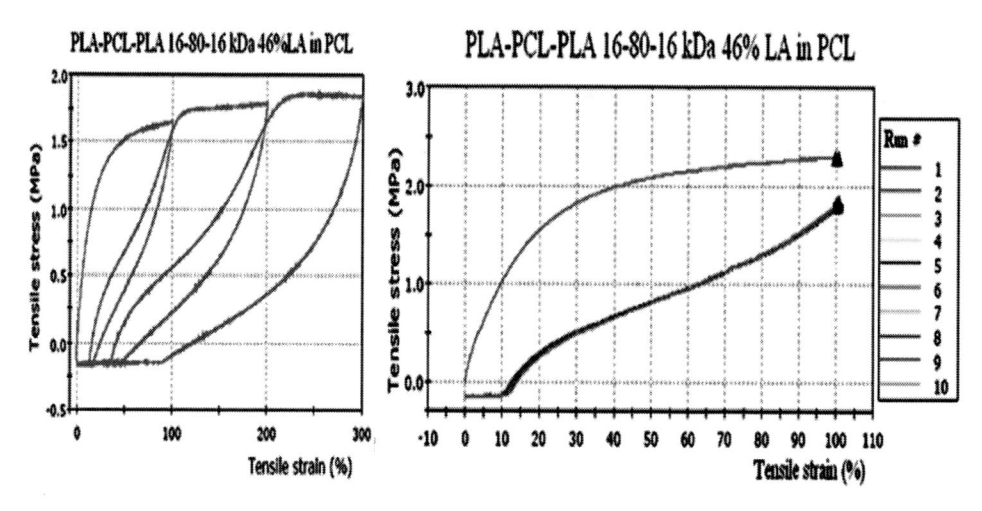

Figure 13. Cyclic and multitensile tests of triblock copolymer PLA-PCL-PLA, 16-80-16 kDa and 46% of L-lactide in PCL at 37°C.

Figure 13 illustrates the shape recovery behavior of polymer. From the cyclic test, we can see, for example, the starting point for the line of the second cycle. If this point is very close to zero, we can say that the polymer has good shape recovery. The shape recovery becomes worse when the starting point of the second (or third) cycle is farther away from zero.

During the tensile test, one specimen stretched to 100% and returned to the initial state 10 consecutive times. This sample clearly recovered almost 90% of its shape even after repeated stretching. The difference between the first stretching and subsequent ones is known fact of thermoelastomers behavior. The difference between the first run and other elongations cause the reorganization of macromolecules, crystallization or weak destruction in polymer structure from the stress of the first test. [22]

6. SHAPE RECOVERY OF PLA-B-(PCL-CO-PDLA)-B-PLA TRIBLOCK COPOLYMERS

We synthesized triblock copolymers using DL-lactide targeting the following structure: PDLA-b-(PCL-co-PDLA)-b-PDLA. Unfortunately, all of the resulting copolymers were very soft. Some of them possessed very high values of elongation at break, but their shape recovery was poor. These polymers were almost amorphous, so the poor performance was likely due to a lack of crystalline points of physical crosslinking.

Thus, the next tested structure was PLA-b-(PCL-co-PDLA)-b-PLA. We suspected that basing the middle block on PCL and PDLA would create a more amorphous structure when compared to the PCL-co-PLA combination. However, as stated above, some crystallinity is needed to obtain good shape recovery. Therefore, a hard block was synthesized from crystalline PLA. A series of triblock copolymers was subsequently synthesized using design of experiment. Three factors were changed for investigation: molar masses of the middle and hard blocks and the monomer molar ratio in the middle block. Targeted copolymer structures and mechanical characteristics of the copolymers are presented in Table 10.

Table 10. Targeted structural characteristics and mechanical properties of PLA-(PCL-PDLA)-PLA triblock copolymers

Targeted polymer molar mass (for blocks), kDa	Ratio CL:DLA in the middle block	PCL/PLA segment length, monomer units	Elongation at break, %	Young's modulus, MPa
5-10-5	25:75	1.91/19.62	590	18.05
10-20-10	25:75	2.13/16.84	984	9.25
20-40-20	25:75	2.96/18.57	803	7.91
10-10-10	50:50	2.66/25.46	27	26.88
20-20-20	50:50	2.88/27.32	455	103.07
5-40-5	50:50	3.92/6.62	981	0.31
20-10-20	75:25	5.93/31.14	2.8	257.85
5-20-5	75:25	4.82/9.52	230	2.71
10-40-10	75:25	7.71/11.25	863	5.75

Shape recovery of these polymers stretched to 100, 200 and 300% is summarized in Table 11.

Table 11. Shape recovery of PLA-b-(PCL-co-PDLA)-b-PLA triblock copolymers

Polymer	Ratio CL:DLA in the middle block	Shape recovery property after elongation to		
		100%	200%	300%
5-10-5	25:75	65	62	35
10-20-10	25:75	75	72	52
20-40-20	25:75	70	65	32
10-10-10	50:50	-	-	-
20-20-20	50:50	50	35	20
5-40-5	50:50	93	89	87
20-10-20	75:25	-	-	-
5-20-5	75:25	92	83	-
10-40-10	75:25	88	74	66

The polymer with the highest shape recovery was PLA-b-(PCL-co-PDLA)-b-PLA, 5-40-5 kDa with 50% DL-lactide in the middle block. This polymer had traces of crystallinity which is displayed as a small deviation in curve area at 100^0C (Figure 14). This phenomenon could have resulted from small crystals with irregular structures that melted at a temperature much different than the normal melting temperature of PLA. This effect could not be seen for PLA with high molar masses (Figure 14, right).

Lower crystallinity of polymer PLA-b-(PCL-co-PDLA)-b-PLA, 5-40-5 kDa with 50% DL-lactide in the middle block brought a high degree of softness. Additionally, Young's modulus was very low (0.31 MPa), indicating that the polymer was weak.

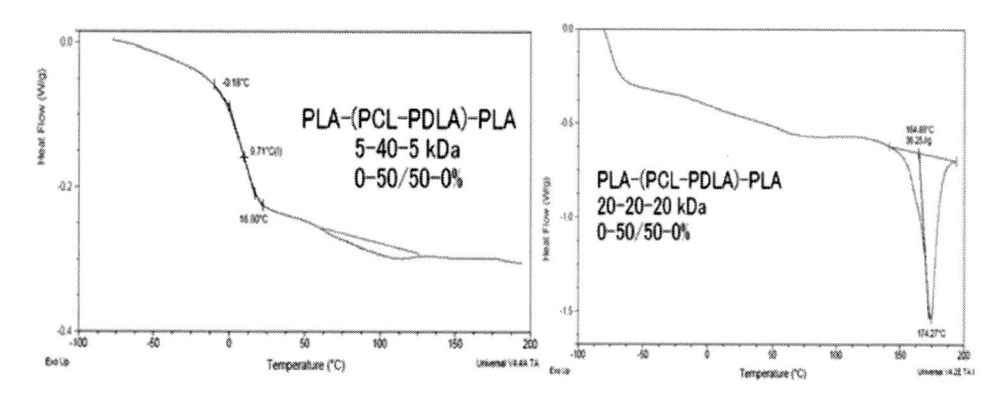

Figure 14. DSC analysis of triblock copolymers: PLA-b-(PCL-co-PDLA)-b-PLA, 5-40-5 kDa with 50% DL-lactide in the middle block (left); PLA-b-(PCL-co-PDLA)-b-PLA, 20-20-20 kDa with 50% DL-lactide in the middle block (right).

When we increased the molar mass of the PLA block to 10 kDa, the mechanical properties of the polymer were significantly improved: elongation at break was 1150% and Young's modulus was 3.29 MPa. Furthermore, the degree of shape recovery was almost 97% after elongation to 100% (Figure 15).

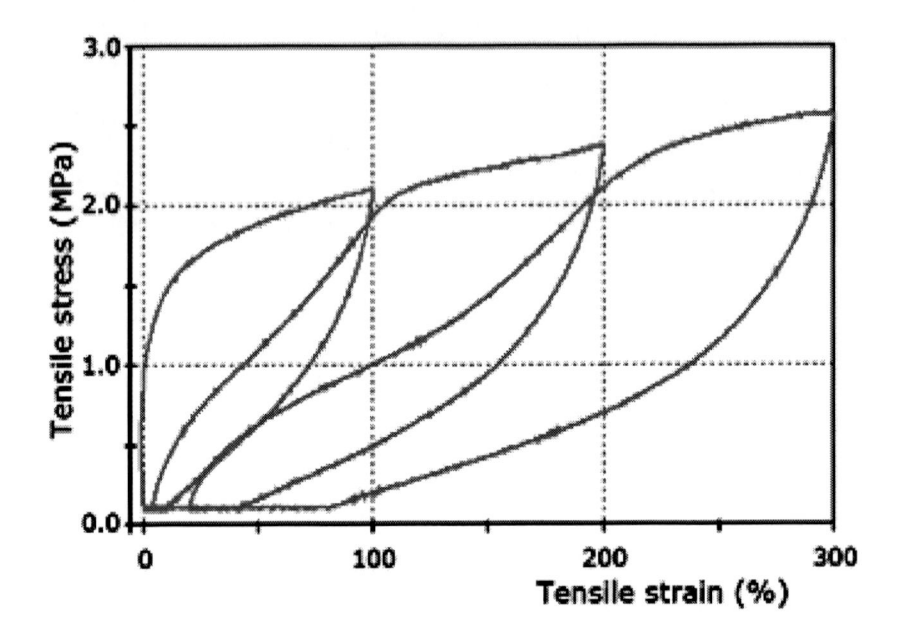

Figure 15. Cyclic test of triblock copolymer PLA-b-(PCL-co-PDLA)-b-PLA, 10-40-10 kDa with 50% DL-lactide in the middle block.

Based on the recorded data, a parabolic simulated model was built (R=0.987) to describe the relationship between shape recovery and polymer structure (Equation 12).

$$R=63.2222+8.90833 \times M_{PCL}-2.30667 \times \%CL-8.43333 \times M_{PLA}-0.139167 \times M_{PCL}^2 +0.0210667 \times \%CL^2+0.22222 \times M_{PLA}^2 \tag{12}$$

where M_{PCL} is the molar mass of the middle block, in kDa; %CL is the amount of ε-caprolactone in the middle block, in mol %; and M_{PLA} is the molar mass of the hard PLA block, in kDa.

Using specific derivations, we found the optimized values of each variable to maximize shape recovery. According to the calculation, the best shape recovery could be reached with the polymer PLA-b-(PCL-co-PDLA)-b-PLA with the following structural parameters: molar mass of the middle block of 32 kDa, quantity of ε-caprolactone in the middle block of 55% and a molar mass for the PLA hard block of 19 kDa. Fixing the molar mass of PLA at 19 kDa, we obtained an equation containing only two variables. The graphical representation of this equation is given in Figure 16.

The simulated graph illustrates that the best shape recovery could be reached when the molar mass of the middle block was in the range of 20 to 40 kDa.

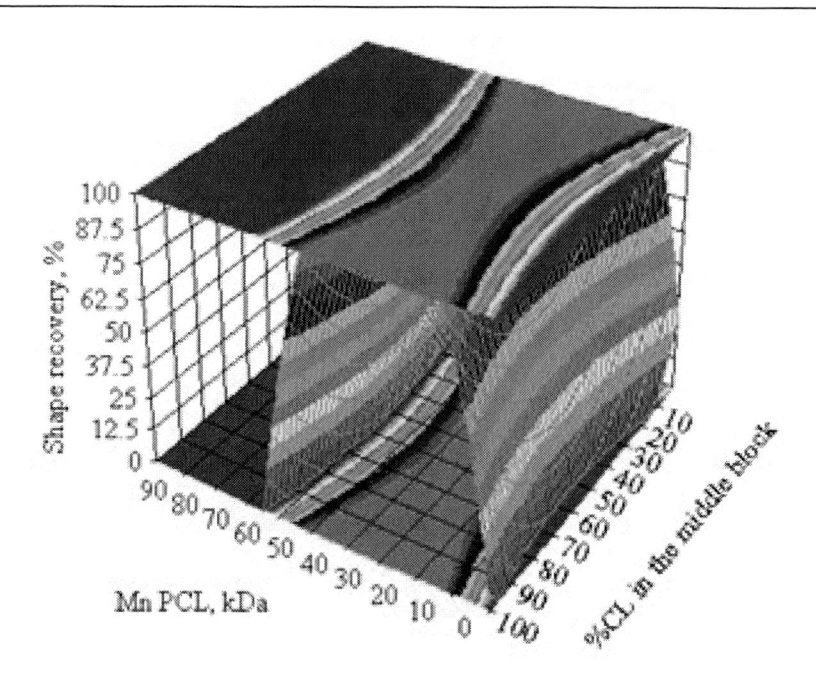

Figure 16. Simulated shape recovery of polymer PLA-b-(PCL-co-PDLA)-b-PLA with a fixed molar mass of PLA = 19 kDa.

The relationship between shape recovery and structural and mechanical characteristics of the calculated polymer is presented in Table 12.

Table 12. Correlation (R) between shape recovery and structural and mechanical properties of the calculated polymer

PLA block length	PCL block length	PLA crystallinity, %	% DLA in PCL	Mn of PCL, kDa	Mn of PLA, kDa	Elongation at break, %	Modulus, MPa
-0.89	0.09	-0.67	0.12	0.68	-0.49	0.77	-0.70

From Table 12, it can be seen that PLA segment length and PLA crystallinity have the largest influence on shape recovery of the triblock copolymer with the structure PLA-b-(PCL-co-PDLA)-b-PLA. Both relationships have negative coefficients of correlation, meaning that reducing the PLA block length and balancing the amorphous and crystalline properties of the polymer allows for good shape recovery. Interestingly, according to table 12, the shape recovery of this type of triblock copolymer did not depend on PCL block length. This could be because the crystals of PLA play a major role in shape recovery and PCL block length and crystallinity influence on shape recovery are is incomparably smaller than impact of the PLA. Another possible explanation is that PCL crystals do not form physical crosslinking points, and all recovery is based completely on the PLA crystal. Therefore, we could not obtain good shape recovery when using PLDA for hard block synthesis, which possesses lower crystallinity than PLA.

CONCLUSION

Elastomers usually possess points of hardness connected by the amorphous phase. The separation of phases acts like a mechanical corset on the material, returning macromolecules to their initial state after removal of the stress and providing good shape recovery. There is an influence of micro- and macrostructure of polymer on shape recovery property.

Testing shape recovery of different types of PCL-PLA copolymers, we discovered several relationships between this property and the polymer structure. The crystallinity of the polymer strongly influenced shape recovery in all thermoplastic elastomers. Additionally, a polymer should not be highly crystalline; otherwise, it will be brittle. Polymers with high crystallinity may occasionally have high values of elongation at break, but such polymers will not have any shape recovery ability. On the other hand, completely amorphous polymers also do not possess good shape recovery. Soft materials without any trace of crystalline structure flow after an applied force, easily changing shape and unable to recover their original shape. In some cases, the absence of crystallinity can be compensated for by a high molar mass of polymer especially when macromolecules have an irregular structure that can act like physical crosslinking, restraining the displacement of macromolecules during stress. Long macromolecules have more irregular structures and can therefore return to their initial state after removing the applied stress. In addition, we observed that molar mass strongly influenced the mechanical properties of a polymer above values of 50,000 Da.

Crystallinity is closely connected with the segment length of a copolymer, which is built from two or more comonomers. The correlation is simple, longer segment lengths were found to provide higher crystallinity and decrease the degree of recovery. Therefore, if we considered biodegradable polymers, copolymers possessed higher degrees of recovery than homopolymers.

When a block structure is designed to have good shape recovery, there are a host of factors influencing this mechanical property. As a rule, total crystallinity of a polymer influences shape recovery. In the case of block structures, crystallinity comes from two or more blocks. To achieve a certain level of mechanical properties, crystallinity can be increased differently. For example, increasing the ratio of crystalline polymer in the soft block or increasing the molar mass of the hard block both improves a polymer's mechanical properties. Based on this rationale, an optimized value for targeted mechanical properties can be found by balancing the crystallinity of blocks in the polymer. The best shape recovery was revealed for the copolymer possessing 10-20% polylactide crystallinity.

Macrostructure of elastomers also significantly influences chape recovery. We have successfully shown that triblock copolymers of ε-CL, and L-or (DL)-lactide exhibit true thermoplastic elastomeric behavior when the middle segment is completely amorphous, when the overall polymer molar mass is reasonably high and when the total molar mass of the hard blocks does not exceed the mass of the soft block. The best shape recovery was achieved when the ratio of molar mass of soft: hard blocks were in the range of 4:1 to 2:1.

REFERENCES

[1] Van der Mee, L.; Helmich, F.; de Bruijn, R.; Vekemans, J.; Palmans, A. R. A.; Meijer, E. W. Macromolecules 2006, 39(15), 5021-5027.

[2] Amsden, B. Soft Matter 2007, 3(11), 1335-1348.

[3] Hassan, M. K.; Kennath, A.; Storey, R. F.; Wigging, J. S. J. Polym. Sci.: Part A: Polym. Chem. 2006, 44, 2990-3000.

[4] Lim, K. Y.; Kim, B. C.; Kee, J. Y. J. Appl. Polym. Sci. 2003, 88, 131-138.

[5] Maeda, Y.; Maeda, T.; Yamaguchi, K.; Kubota, S.; Nakayama, A.; Kaasaki, N.; Yamamoto, N.; Aiba, S. J. Polym. Sci.: Part A: Polym. Chem. 2000, 38, 4478-4489.

[6] Lakshmi, S.; Laurencin, N. C. T. Prog. Polym. Sci. 2007, 32, 762-798.

[7] Albertsson, A. C.; Edlund, U.; Stridsberg, K. Macromol. Symp. 2000, 157, 39-46.

[8] Sivalingam, G.; Madras, G. Polym. Degrad. Stab. 2003, 80, 11-16.

[9] Fukuzaki, H.; Yoshidat, M.; Asano, M.; Kumakura, M. Polymer 1990, 31, 2006-2014.

[10] Callister, W. D. Advanced Polymeric Materials. Materials Science and Engineering: An Introduction. John Wiley and Sons, Inc. USA. 2007, 327.

[11] In't Veld, P. J. A.; Ester, M.; Van De Witte, P. V.; Hamhuis, J.; Dijkstra, P. J.; Fiijen, J. J. Polym. Sci.: Part A: Polym. Chem. 1997, 35, 219-226.

[12] Odelius, K.; Albertsson, A. C. J. Polym. Sci.: Part A: Polym. Chem. 2008, 46, 1249.

[13] Huang, M. H.; Li, S.; Vert, M. Polymer 2004, 45, 8675-8681.

[14] Baimark, Y.; Molloy, R. Polym. Adv. Technol. 2005, 16 (4), 332-337.

[15] Meng, Q.; Hu, J.; Ho, K. C.; Ji, F.; Chen, S. J. Polym. Environ. 2009, 17, 212-224.

[16] Behl, M.; Lendlein, A. Materials Today 2007, 10(4), 20-28.

[17] Xue, L.; Dai, S.; Li, Z. Biomaterials 2010, 31, 8132-8140.

[18] Lu, X. L.; Sun, Z.; Cai, W. Phys. Scrypta 2007, T129, 231-235.

[19] Lipik, V. T.; Fong, J. K.; Chattopadhyay, S.; Widjaja, L. K.; Liow, S. S.; Venkatraman, S. S.; Abadie, M. J. M. Acta Biomaterialia 2010, 6, 4261-4270.

[20] Min, C.; Cui, W.; Bei, J.; Wang, S. Polym. Adv. Technol. 2005, 16(8), 608-615.

[21] Lu, X. L.; Cai, W.; Gao, Z. Y. J. Appl. Polym. Sci. 2008, 108, 1109-1115.

[22] Nagata, M.; Yamamoto, Y. J. Polym. Sci.: Part A: Polym. Chem. 2009, 47, 2422-2433.

In: Polymer Synthesis
Editor: E. Kowsari

Chapter 11

UTILIZATION OF CHEAP CARBON SOURCES FOR MICROBIAL PRODUCTION OF POLYHYDROXYALKANOATES

Anupama Shrivastav, Sanjiv K. Mishra and Sandhya Mishra[*]

Marine Biotechnology and Ecology Discipline,
Central Salt and Marine Chemicals Research Institute
Council for Scientific and Industrial Research (CSIR),
G. B. Marg, Bhavnagar, India

ABSTRACT

Recent growing use of plastics has led to the design and development of new degradable thermoplastic materials that are more "friendly" to the environment. Among the various biodegradable plastics available, there is a growing interest in polyhydroxy-alkanoates (PHAs) or green plastics, due to their properties which resemble those of conventional petrochemical based polymers. PHAs are polyesters of various hydroxyalkanoate monomers accumulating as energy/carbon storage materials by granular inclusions in the cytoplasm of various bacterial cells under unfavorable growth condition along with the presence of excess carbon source. Their production from renewable resources and their complete biodegradibility give PHAs promising advantages from an environment point of view. All PHAs are completely degradable to carbon dioxide and water. PHAs have versatile applications in packaging films, disposable items etc. PHAs have been recognized as a good tool for biodegradable polymers. The high production cost of PHA due to the substrate cost, constricts their industrial applications. Several processes for production of PHA from inexpensive carbon sources and waste products have been investigated to utilize bountiful organic compounds present in the waste. If waste products are used as substrate for the production of PHA, dual advantage of reducing waste disposal cost and production of value-added products could be actualized. Although, PHAs have been recognized as a good candidate for biodegradable polymers, their high production cost limits their

---

[*] Corresponding author: Dr. Sandhya Mishra. Phone: +91-278-2563805, 2567760, 616 (ext.). Fax No. +91-278-2567562. E-mail: smishra@csmcri.org , smishracsmcri@rediffmail.com.

industrial application. For the economical production of PHAs, various bacterial strains, either wild-type or recombinant strains along with new fermentation strategies have been developed through metabolic engineering for the production of PHAs with high concentration and productivity.

The most significant factor for the increased cost of production in PHA is mainly through the substrate as a carbon and energy source. Several processes for PHAs production from cheap carbon sources and waste products have also been investigated in order to utilize abundant organic compounds in waste. If waste products are used as a substrate for the production of PHA, dual advantage of reducing waste disposal cost and production of value-added products can be realized. This review paper will be focused on the production of PHA by microorganisms from cheap and inexpensive carbon sources, biosynthetic pathways, as well as factors affecting the production and its composition will also be discussed.

Keywords: biopolymer; Polyhydroxyalkanoate; cheap; carbon sources; waste streams; byproducts; biodegradable

1. INTRODUCTION

Problems associated with global environment and solid waste management have generated interest in the development of novel plastics. Among the various biodegradable plastics available, there is growing interest in the group of polymers known as polyhydroxyalkanoates (PHAs) which are polyesters of various hydroxyalkanoate monomers accumulating as energy/carbon storage materials by granular inclusions in the cytoplasm of various bacterial cells under unfavorable growth condition along with the presence of excess carbon source. Of all the PHAs, poly (3-hydroxybutyrate) (PHB) has attracted considerable interest as a candidate for biodegradable and biocompatible plastics. Beijerinck first observed PHAs as refractile bodies inside bacterial cells in 1888. However, PHA composition was established by Lemoigne only in 1926 [1]. Bacterial polyhydroxyalkanoates (PHA) have attracted much attention as environmentally degradable thermoplastics. They are being viewed as potentially useful for replacing many synthetic plastics in a wide range of agriculture, marine and medical applications. While, the conventional synthetic plastics contribute to serious pollution problems, PHAs are of biological origin, they could be completely broken down to water and carbon dioxide by microorganisms found in wide range of environments, such as soil, water and sewage [2, 3, 4]. In June 2005; a US company (Metabolix, Inc.) was bestowed with the US Presidential Green Chemistry Challenge Award for development and commercialization of cost-effective method for manufacturing PHAs. In 1976, ICI of England explored if PHB could be satisfactorily produced by microbial fermentation. In 1993, Zeneca Bioproducts took over ICI's activities and in 1996 Monsanto bought the bioplastics production business from Zeneca. Monsanto closed down its activities in 1998 but many new players became active in this field since then. Some prominent ones are Metabolix, Proctor and Gamble, DuPont, General Motors, and Toyota etc. BIOPOL is a marketed bioplastic product.

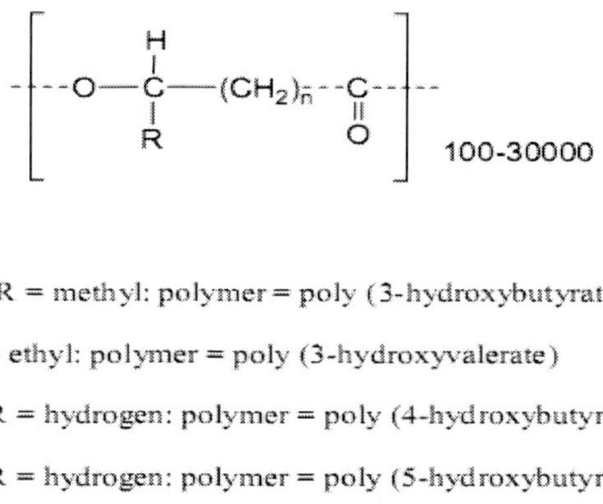

$$100\text{-}30000$$

n = 1 R = methyl: polymer = poly (3-hydroxybutyrate)

R = ethyl: polymer = poly (3-hydroxyvalerate)

n = 2 R = hydrogen: polymer = poly (4-hydroxybutyrate)

n = 3 R = hydrogen: polymer = poly (5-hydroxybutyrate)

Figure 1. General structure of PHA.

3-HB unit   3-HV unit

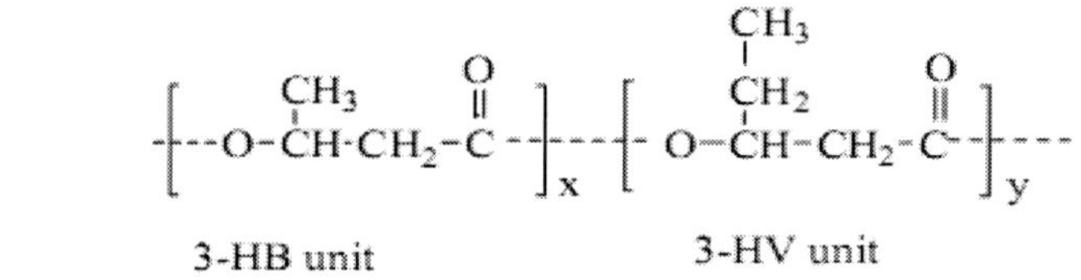

Figure 2. Structure of Copolymer 3HB-3HV.

PHAs are microbial polyesters, synthesized by numerous bacteria, accumulated as a carbon/energy and/or reducing power storage material usually under the condition of limiting nutritional elements such as N, P, S, O or Mg in the presence of excess carbon source [5]. More than 300 different microorganisms are known to synthesize and accumulate PHAs intracellularly [6]. The homopolymer poly (3-hydroxybutyrate) (PHB) was the first of the PHA discovered and further identified by Maurice Lemoigne in 1926 [7]. Approximately, 150 different hydroxyalkanoic acids are known as constituents of these bacterial storage polyesters [8]. General structure of PHA is shown in figure 1 and figure 2. Bacterial PHAs could be bifurcated into two groups depending on the number of carbon atoms in the monomeric units [9]. One group of PHAs is short-chain-length PHA (SCL-PHA) consisting of 3 to 5 carbon atoms, while the other group being medium-chain-length PHA (MCL-PHA) consisting of 6 to 14 carbon atoms. Biologically produced PHAs are composed only of chirally pure (R)-configuration monomers.

2. PHA BIOSYNTHESIS

General PHA biosynthetic pathways represented by *Cupriavidus necator* (formerly known as *Ralstonia eutropha*) which has been extensively investigated and present in wide range of bacteria. A β- ketothiolase catalyzes the formation of a carbon-carbon bond of two acetyl-CoA moieties. NADPH-dependent acetoacetyl-CoA reductase catalyzes the reduction of acetoacetyl-CoA formed in the first reaction to 3-hydroxybutyryl-CoA. The third reaction of this pathway is catalyzed by the PHA synthase, which linked the 3-hydroxybutyryl moiety to an existing polymer molecule by an ester bond. Two moles of acetyl-CoA are used to form a HB unit of the copolymer, while a HV unit is formed by the reaction of acetyl CoA and propionyl-CoA [10]. The degradation of PHA by *C. necator* could occur simultaneously with its biosynthesis under nitrogen limitation known as "cyclic nature of PHA metabolism" [11]. PHA synthesis with an enoyl-CoA hydratase, *Rhodospirillum rubrum* PHA biosynthetic pathway which includes two enoyl-CoA hydratases in the second step of catalyzing the conversion of L-3-hydroxybutyryl-CoA to D-3-hydroxybutyryl-CoA via crotonyl-CoA [12].

PHA biosynthesis from fatty acids is represented by *Pseudomonas oleovorans,*. The current classification of genus *Pseudomonas* is based on ribosomal RNA (rRNA)/DNA homology. rRNA homology group I is the largest one including species common in soil such as *P. aeruginosa, P. putida* and *P. fluorescens.* Characteristic of many of these is the production of fluorescent pigments, while species belonging to rRNA homology group II might be animal or plant pathogens. PHAs formed by *Pseudomonas* sp. of the rRNA homology group I was directly related to the structure of the alkane, alkene, or fatty acid carbon source. The composition of the polymer depended on the length of the carbon backbone of the substrate used. This biosynthetic pathway is found in *P. oleovorans* and most *Pseudomonas* sp. from the rRNA homology group I [13]. These organisms produce medium-chain-length (MCL) PHAs (from C6-C9) from MCL-alkanes, alcohols, or alkanoates. This PHA biosynthesis involves the cyclic-β-oxidation and thiolytic cleavage of fatty acids, i.e., 3-hydroxyacyl-CoA, and intermediates of the β -oxidation pathways, are used for PHA biosynthesis.

PHA biosynthesis from carbohydrate: This is mostly found in *Pseudomonas aeruginosa* of the rRNA homology group I *Pseudomonads,* with the exception of *P. oleovorans.* MCL-PHAs using this pathway are produced from carbohydrates like gluconate, fructose, acetate, glycerol and lactate. Copolymers consisting of medium chain length 3-hydroxyalkanoates are synthesized from acetyl-CoA. Pseudomonads that possess this pathway are *P. aeruginosa, P; aureofaciens, P. citronellolis, P. mendocina, P. putida, P. chlororaphis* and *P. marginalis* [14] and this pathway is called the *P. aeruginosa* PHA biosynthetic pathway [15].

3. FACTORS AFFECTING THE SYNTHESIS OF PHA AND ITS COMPOSITION

PHA production by wild type strains and recombinants is usually performed in two stage cultures, which consist of a cell-growth phase and a PHA-production phase. In cell growth phase, nutritionally enriched medium used to obtain high cell mass during early cultivation. In second phase, the cell growth is restricted by depletion of some nutrients such as nitrogen, phosphorus, oxygen, or magnesium. This depletion acted as a trigger for the metabolic shift to PHA biosynthesis.

3.1. Type of Substrate, its Concentration and the Growth Conditions

Recently, Quillaguaman reported that *Halomonas boliviensis* LC1 accumulate PHB under controlled conditions in fermentor using 1.5% (w/v) yeast extract as N source, and intermittent addition of sucrose to provide excess carbon source, polymer accumulation was of 44wt% and CDW 12 g.L^{-1} after 24h of cultivation [16]. Steinbuchel and Pieper had studied the production of P(3HB-co-3HV) copolymer by *C. necator* strain R3 under nitrogen limitation. PHA contents obtained were 47%, 35.7%, 29.5%, 21.5% and 43.2% when fructose, gluconate, acetate, succinate and lactate were used as carbon source, respectively [17]. Bourque investigated production of PHA by 118 methylotrophic microorganisms grown on methanol. *Methylobacterium extorquens* accumulated a high PHA content 60-70% with 20% PHV when grown on the mixture of methanol and valerate [18]. Lee and Yu operated a two-stage bioprocess for PHA production. The first stage was anaerobic digester [19]. The mixture of volatile fatty acids produced in first stage was used by *C. necator* for PHA production in a subsequent stage. *C. necator* was grown under aerobic and nitrogen-limiting conditions. PHA production of about 34% of cell mass was obtained using digested sludge supernatant. Acetic acid was the most effective fatty acid used by *C. necator* followed by propionic acid, butyric acid, and valeric acid. Suzuki reported the maximum PHB production of 66% by *Pseudomonas sp. K* using methanol as a sole carbon and energy source and nitrogen deficiency was found to be the most effective way to stimulate the accumulation of PHB [20]. Daniel reported maximum PHB content of 55% by *Pseudomonas 135* when grown in ammonium-limited fed batch culture using methanol as a sole carbon and energy source. Some wild type cyanobacteria are also capable of accumulating P(3HB) in the cells from carbon dioxide [21]. Laxuman and Nirupma had reported that *Nostoc muscorum* produced upto 35% of PHB when grown in Mixotrophic conditions with 0.4% (w/v) glucose and acetate [22]. The PHB accumulation increased to 40-43% (w/w) dry cells with gas exchange limitations. Shrivastav et al. have studied marine cyanobacteria Spirulina subsalsa from Veraval coast, Gujarat, India, producing PHA under increased sodium chloride (NaCl) concentration with 7.5% PHA content [23].

The copolymer composition produced by *C. necator* H16 grown in medium containing sodium acetate and sodium propionate was studied by Doi [24]. The PHV content of copolymer increased as propionate concentration in medium increased. P(3HB-co-3HV) copolymer was obtained when propionate was used as a sole carbon source. PHA contents produced by *C. necator* H16 also varied as the type of substrate and its concentration varied. Kellerhals stated that some substrates like octanoic acid, alkanoic acid and halogenated derivatives were toxic when present in excess and, therefore, required control of the carbon source concentration in the media [25]. They have developed a closed loop control system based on online gas chromatography to maintain continuously fed substrates at desired levels to control the concentration of the substrates.

The composition of PHA produced was related to the substrate used for growth and the concentration of a substrate supplied also affected the amount of polymer produced.

3.2. Organism

Valappil have acquired new PHA producing *Bacillus* spp. named *B.cereus SPV*. This strain was capable of using wide range of carbon sources including glucose, fructose, sucrose, various fatty acids and gluconate for production of PHA [26]. The PHA once produced remained at constant maximal concentration, without any degradation. Page had stated that *Azotobacter vinelandii* was not considered for commercial production because it produced PHA with low yield and forms capsules. While, the strain UWD of this organism assumed interest because of its capsule-negative mutant and also produced PHA content of approximately 70-80%. *Alcaligenes latus* can store PHA up to 80% under normal growth condition. Therefore, one-step PHA production process could be used with this organism[27]. Yamane studied the production of PHA by *A. latus* using sucrose as feed substrate. High cell concentration (142 g/L) was obtained in a short culture time (18 hours incubation time of culture) and PHB content at the end of culture time was 50% [28]. They attained a cell concentration of 125 g/L with 29 hours of culture time for an initial cell concentration of 1.14 g/L, 20 hours for 4.4 g/L and 16.5 hours for 13.7 g/L. They concluded that inoculum size reduces the culture time. Renner described the production of copolymer P(3HB-co-HV) and P(3HB-co-4HB) by 13 bacteria from the rRNA super family III of bacteria, since most of its members are able to accumulate P(3HB) they reported that different bacteria could produce PHAs with different PHV/PHB compositions when grown on the same substrate [29]. Anderson had studied the production of copolymer by the genera *Rhodococcus*, *Nocardia*, and *Corynebacterium* [30]. They found that these genera had differences in accumulation and composition of PHA when grown on the same single carbon source.

At present, the scale of industrial production of biodegradable plastics from PHA is relatively small, and it takes place largely via bacterial fermentation process. [31] For PHA production estimates of the contribution of the substrate cost to total production costs were 28–50 %.[32,33]

From an economical point of view, the cost of substrate (mainly carbon source) contributed substantially to overall production cost of P(3HB). Industrial waste products or residual compounds from biotechnological processes might be potential carbon sources for production of PHAs [34]. Such substrates assumed particular interest, since disposal of these compounds added to the extra costs due to waste management.

In general, disposal of the waste stream caused an enormous environmental problem due to its high biological/chemical oxygen demand. Furthermore, the treatment of waste stream to purified effluent needs much effort and was very difficult, because the waste stream often contained various organic compounds. PHA production from waste could provide double benefits because polluting waste was converted into environmental friendly biodegradable polymer and the use of waste stream for PHA production as a substrate will prove to be relatively inexpensive as compared to other conventional carbon substrates.

4. PHA PRODUCTION FROM WASTE PRODUCTS/STREAM

Despite its basic attractiveness as a substitute for petroleum-derived polymers, the major problem faced by commercial production and application of PHA in consumer products has been the high cost of bacterial fermentation, making bacterial PHA 5–10 times more expensive than the petroleum-derived polymers, such as polypropylene and polyethylene, which costs approximately US$ 1.6 kg^{-1}. Low-cost raw materials can be used to reduce significantly the production cost of polyhydroxyalkanoates (PHA). It has been investigated whether the abundant organic compounds in wastewater could be converted to useful products. [35] This makes it possible to produce PHA more economically, and at the same time to treat wastewater without any additional disposal cost. Several bacterial strains have been reported which could produce PHA from waste byproducts/streams. [36]

Hu reported that PHA could be produced in an activated sludge, P(3HB) and poly (3-hydroxy-butyrate-co-3-hydroxyvalerate) [P (3HB-co-3HV)] were produced from butyrate and/or valerate [37]. When butyrate was used as a sole carbon source, P(3HB) was produced comprising up to 37% of the dry cell weight. When valerate was added to the medium, P(3HB-co-3HV) copolymer was produced. The 3-hydroxy-valerate mole fraction in P(3HB-co-3HV) reached a maximum of 54% when valerate was used as sole carbon source. Fernandez used agro industrial oily wastes like residual waste frying oil and other oily wastes as substrates for PHA production by *Pseudomonas aeruginosa 42A2* strain, which accumulated PHA up to 54.6% [38]. The PHA production and monomer composition by *Pseudomonas aeruginosa 42A2* strain were affected by K_{La} and temperature. Steinbuchel had studied the production of PHA using residual oil from biotechnological rhamnose production process as a carbon source for growth of *C. necator* H16 and *P. oleovorans*, which accumulated PHA amounting to 41.3% and 38.9%, respectively, of the cell dry mass when these strains were cultivated in mineral salt medium with the waste oil as the sole carbon source. The accumulated PHA isolated from *C. necator* consisted of only 3-hydroxybutyric acid, whereas the PHA isolated from *P. oleovorans* consisted of 3-hy-droxyhexanoic acid, 3-hydroxyoctanoic acid, 3-hydroxy-decanoic acid, and 3-hydroxydodecanoic acid. Approximately, 20–25% of the components of the residual oil were converted into PHA. *A. vinelandii UWD* was able to produce PHA without nutrient limitation. This strain was employed for the production of PHA from swine waste liquor, which is rich in organic and inorganic compounds, as reported by Cho [39]. *A. vinelandii UWD* grew to 2.0 g dry cell weight/l with concomitant production of P(3HB-co-3HV) up to 34% of dry cell weight, in two-fold diluted swine waste liquor. Dry cell weight and P (3HB-co-3HV) content reached 9.4 g.L^{-1} and 58.3%, respectively when 30 g.L^{-1} of glucose was supplemented. Taniguchi reported that waste plant oils as well as waste tallow were assimilated and successfully converted to PHA with relatively high yield by the bacterial fermentation. The waste plant oils usually gave poly (3-hydroxybutyrate) (PHB) while waste tallow gave poly (3-hydroxybutyrate-co-3-hydroxyvalerate) (PHBV). The ratio of 3-hydroxyvalerate (3HV) unit in the copolyester was controlled by the addition of sodium propionate to the cultivation medium, containing waste plant oils as carbon sources. These authors also stated that the percentage of PHA accumulated was quite high, up to 80% of the cell dry weight. To develop more efficient way of PHA production from waste products, attempts were made by Son to isolate microorganisms that can accumulate PHA in wastewater [40]. *Actinobacillus sp.* was isolated from the soil sample near alcoholic distillery factory. Alcoholic distillery waste water contains sugar and nitrogen compounds, and thus, has been considered as a potential feedstock for the fermentation industry. When *Actinobacillus sp.* EL-9 was cultivated on alcoholic distillery wastewater, cells accumulated P(3HB) up to 42% of dry cell weight. PHB production in *Actinobacillus sp.* EL-9 was not dependent on nutrient limitation. The PHB accumulation of *Actinobacillus sp.* EL-9 was growth associated where, cell growth and PHB accumulation were carried out simultaneously. The soya wastes from soya milk dairy and malt wastes from a beer brewery plant were used as the carbon sources by Yu for the production of polyhydroxyalkanoates (PHA) by selected strain of microorganism, *Alcaligenes latus DSM 1124*. The final dried cell mass and specific polymer production of *A. latus* DSM 1124 were 32g/L and 70% polymer/cells (g/g), 18.42 g.L^{-1} and 32.57% polymer/cell (g/g), and 28 g.L^{-1} and 36% polymer/cells (g/g), from malt waste, soya waste, and from sucrose, respectively[41]. These results suggested that several types of food wastes could be used as the carbon source for the production of PHA. Wong isolated a Gram-positive coccus-shaped bacterium capable of synthesizing relatively higher molecular weight poly-hydroxybutyrate from sesame oil, which was identified as *Staphylococcus epidermidis* [42]. Cell growth up to a dry mass of 2.5 g.L^{-1} and PHB accumulation up to 15.02% of cell dry wt was obtained. Various industrial food wastes including sesame oil, ice cream, malt, and soya wastes were investigated as nutrients for *S. epidermidis* to reduce the cost of the carbon source. By using malt wastes as nutrient for cell growth, PHB accumulation of *S. epidermidis* was much better than using other wastes as nutrient source. The final dried cell mass and PHB production using malt wastes were 1.76 g.L^{-1} and 6.93% polymer/cells (gms/gm), and 3.5 g.L^{-1} and 3.31% polymer/cells (gms/gm) in shake flask culture and in fermentor culture. The feasibility of using olive oil mill effluents as a substrate in biodegradable polymer production was studied by Dionisi. where, olive oil mill effluents were anaerobically fermented at various concentrations along with different pretreatments and without pretreatments to obtain volatile fatty acids, which are the most highly used substrate for polyhydroxyalkanotes production [43]. Olive oil mill effluents were also tested for PHA production by using mixed culture from an aerobic Sequencing Batch Reactor (SBR) where, olive oil mill effluents were centrifuged and was tested with or without fermentation. Best results with regard to PHA production were obtained in fermented olive oil mill effluents because of the higher Volatile fatty acids (VFA) concentration. Rafael studied the use of alpechin [44]. The olive-mill waste waters produced from olive oil extraction are known as "alpechín". This is regarded as a severe environmental problem, because of its high organic content and large simple phenolic compounds, that are antimicrobial and phytotoxic. *Pseudomonas putida KT2442* was capable of growing in olive oil waste water, which is toxic for many other microorganisms. The transformation of *Pseudomonas putida KT2442* with the plasmid pSK2665, harboring *Alcaligenes eutrophus* genes, increased the PHA production against the parental strain after 72 hours. Alpechín was a cheap and complex carbon source and *Pseudomonas putida KT2442* harboring the plasmid pSK2665 was found to grow at high concentration of this residue and accumulate PHA granules. PHA production was attempted by Rhu with Sequencing Batch Reactors (SBRs) from food waste [45]. Seed microbes were collected from a sewage treatment plant with a biological nutrient removal process, and acclimatized with synthetic substrate prior to the application of the fermented food waste. The maximum content of 51% PHA was obtained with an anaerobic/aerobic cycle with P limitation, and yield was estimated to be about 25 kg of PHA/dry ton of food waste. Economical analysis done by the same authors suggested that PHA produced from the food waste could be an alternative material to produce the biodegradable plastic.

A different approach taken by Lee was metabolic engineering of non-PHA producing bacteria, recombinant *E. coli* strain harboring the *C. necator* PHA biosynthesis genes, for the production of PHA from whey [46]. Whey is a major dairy byproduct from the production of cheese, representing 80-90% of the volume of milk transformed. It contains approximately 4.5% (w/v) lactose, 0.8% (w/v) protein, 1% (w/v) salts and 0.1-0.8% (w/v) lactic acid. Many *E. coli* strains can utilize lactose for their growth. Its high biological oxygen demand made it very difficult to dispose off. Using flask cultures of metabolically engineered *E. coli* strain using whey as a nutrient, the P(3HB) concentration and P(3HB) content obtained were as high as 5.2 g.L^{-1} and 81% of dry cell weight, respectively. In fed-batch cultures of this recombinant *E. coli* strain using concentrated whey solution as the feeding solution, dry cell weight and P(3HB) concentration of 87 and 69 g.L^{-1}, respectively, were obtained in 49 h of cultivation [47]. These results demonstrated for the first time that P(3HB) could be efficiently produced from whey. Povolo and Casella studied the bacterial conversion of lactose and whey permeate to PHA using *Sinorhizobium meliloti* 41 and *Hydrogenophaga pseudoflava* DSM 1034 resulting in 0.48 and 0.38 g cell dry weight (CDW) L^{-1}, and produced P(3HB) at 3.5 and 4.4%of CDW, respectively [48]. Marangoni studied the influence of two strategies- pH regulation and periodic feeding of propionic acid on the incorporation of 3-hydroxy valerate monomer in the copolymer poly (3-hydroxybutyrate-co-3-hydroxyvalerate) (P(3HB-co-3HV)) by *C. necator* grown on whey supplemented with invert sugar, which resulted in the productivity of only 0.17 g/l/h, with incorporation of 37 % of 3-hydroxyvalerate comonomer in the polymer, but the lactose was not utilized by the strain [49]. A recombinant *E. coli* strain has been reported by Liu to carry out the production of PHB on molasses as carbon source [50]. The final dry cell weight, PHB content and PHB productivity were 39.5 g/l, 80% (w/w) and 1 g/(l h), respectively. Yellore and Desai discussed the potential of *Methylobacterium* sp. ZP24 in PHB production from cheese whey with the polymer yield of 1.1 g.L^{-1}, yield increased upto 3 g.L^{-1} with supplementation of the whey medium with ammonium sulphate [51]. Dhanasekar and Viruthagiri studied the effect of substrate concentrations and inoculum concentrations on cell mass and PHB production by mathematically modelling and experimental examining the batch production of P(3HB) by *Azotobacter vinelandii,* using sucrose or cheese whey as a substrate with yield of 1.8 g.L^{-1} of P(3HB) in 48h with an initial biomass of 0.3 g.L^{-1} [52]. Their results had shown that polymer biosynthesis was highly dependent on the initial biomass concentration. Salmiati produced PHA from organic wastes using mixed bacterial cultures by anaerobic-aerobic fermentation systems using Palm oil mill effluent (POME) as an organic source [53], which was cultivated in a two-step-process of acidogenesis and acid polymerization. PHA production was carried out using mixed culture in aerobic bioreactor. Production of PHAs was high using a high volume of substrates because of the higher VFA concentration. The maximum PHA content was observed at 40% of the cell dried weight.

Crude glycerol derived through biodiesel production due to its impurities and huge volume poses the problem in handling and further affecting the economics involved in the whole process. Since, crude glycerol is continuously generated from the biodiesel process it is now indispensable to develop some means of its utilization as such without any treatment and converting to a value added byproduct thereby, making the whole Biodiesel process economically viable.

Few attempts have been done in the last few years for the utilization of glycerol as a substrate for PHA production. Aldor et al. reported a metabolically engineered pathway for the production of poly(3-hydroxybutyrate-co-3-hydroxyvalerate) (PHBV) in *S. enteric* and it accumulated PHBV with significant HV incorporation when grown aerobically with glycerol as the sole carbon source [54]. Andreessen et al. developed a novel, non-naturally existing pathway for the biotechnological conversion of glycerol into poly(3HP) in *Escherichia coli* [55]. Ibrahim and Steinbuchel isolated and identified a new PHA accumulating strain of genus *Zobellella* which was able to utilize glycerol for growth and PHB accumulation [56].

Ashby et al. used *Pseudomonas oleovorans* and *Pseudomonas corrugata* for PHA production from co-product stream of soya based biodiesel production (CSBP) stream containing glycerol, fatty acid soaps and residual fatty acid methyl esters at 1% to 5% concentration in a 2-stage fermentation process. The alkaline co-product stream (pH 13) was neutralized with 1 N HCl to pH 7 before using as substrate. The polymer cell productivity was only 42% of cell dry weight (CDW) with *Pseudomonas corrugata* while polymer yield with respect to glycerol was < 5% even under optimized conditions [57]. Recently Shrivastav et al. have used Jatropha biodiesel byproduct as carbon source for production of Polyhydroxyalkanoate using two promising bacterial isolates SM-P-1S, *Bacillus sonorensis* and SM-P-3M, *Halomonas hydrothermalis* isolated from soil and marine environment with 71.8% and 75% PHA/CDW respectively [58, 59]. Ciesielski et al. have characterized the mixed population consisted of microorganisms affiliated with four bacterial lineages- a, c-Proteobacteria, Actinobacteria, and Bacteroides responsible for the conversion of crude glycerol into PHAs by cultivation dependent and independent methods [60]. Zhu et al. used *Burkholderia cepacia ATCC 17759* for synthesizing poly-3-hydroxybutyrate (PHB) from glycerol concentrations ranging from 3 to 9% (v/v) and reported that increasing the glycerol concentration results in a gradual reduction of biomass, PHA yield, and molecular mass (Mn and Mw) of PHB [61]. Ibrahim and Steinbuchel isolated and identified a new PHB accumulating strain *Zobellela denitrificans* MW1 which is able to utilize glycerol for growth and PHB accumulation to high content in the presence of NaCl and in further study used the same strain for large scale production of PHB by use of glycerol and obtained PHB content of 66.9 ± 7.6% of cell dry weight [62]. Nikel et al. have analysed PHB synthesis under microaerobic conditions in a recombinant *Escherichia coli* arcA mutant using glycerol as main source and obtained PHB content up to 51% [63]. Crude Glycerol, a by-product from the production of biofuels from oilseeds, emerged as possible substrate in PHB production. In an economical point of view, the cost of substrate, mainly carbon source that contributes most significantly to the overall production cost of PHAs can be decreased if waste product/stream is used as a substrate.

5. PHA PRODUCTION FROM CHEAP CARBON SOURCE

There have been many reports on the production of PHAs from cheap carbon sources by wild-type PHA producers. Xylose is a significant component of hemicellulose of hardwoods and crop residues, which has been studied extensively in the fuel alcohol research area. Ramsay reported that *Pseudomonas cepacia* ATCC 17759, when grown on xylose produced $2.6gL^{-1}$ biomass with 60% (w/w) PHB in shake flasks, on ammonium-limited mineral salts culture medium with 10 g.L^{-1} xylose [64]. Substrate cost (in terms of hydrolyzed hemicellulose) for PHB production was comparable to that of cane molasses and half that of bulk glucose. Lee, developed a recombinant *E. coli* strain harboring the *C. necator* PHA biosynthesis genes which can produce P(3HB) from xylose [65]. In a chemically defined medium supplemented with xylose, P(3HB) concentration and P(3HB) content obtained were 1.7 g.L^{-1} and 35.8%, respectively, in flask culture. Supplementation of a small amount of complex nitrogen source such as soybean hydrolysate improved P(3HB) production to a level of 74% of cell dry weight. Since, recombinant *E. coli*, could be easily grown to high cell density by utilizing xylose, recombinant *E. coli* would be useful for the production of P(3HB) from xylose or hemicellulose hydrolysate [10]. Omar reported the ability of *Bacillus megaterium* to grow on various carbon sources such as date syrup, beet molasses, fructose, lactose, sucrose, glucose in mineral salts medium [66]. Best results in relation to growth and PHB production were obtained in the cheaper carbon sources like date syrup and beet molasses. Beet molasses has proven to be an excellent feedstock for polyhydroxyalkanoate (PHA) production by *Azotobacter vinelandii* UWD. The substrate-cost for PHA production from beet molasses in fed-batch culture was one-third of that using glucose. Copolymers containing β-hydroxyvalerate were readily formed in beet molasses medium when valerate was used as a precursor. The origin of the hydroxyvalerate monomer was most likely a β-ketoacyl-CoA intermediate in the β-oxidation of the odd-length *n*-alkanoates. Production of P(3HB) by *Rhizobium meliloti* and by *Bacillus cereus* M5 strain was demonstrated using sugar beet molasses as a carbon source [67,68]. P(3HB) contents of 56 and 74 % of cell dry weight for *R. meliloti* and *B. cereus*, respectively, were reported. Solaiman screened several mcl-PHA-producing *Pseudomonas* species and identified *P. corrugata* as capable of growing and producing biopolymer when cultivated on soy molasses medium [69]. The biopolymer yields from *P. corrugata* cells grown in shake-flask cultures with a chemically defined medium supplemented with 5 % (w/v) soy molasses were 0.6 gL^{-1} mcl PHA.

Yezza isolated a new bacterial strain, from groundwater contaminated with explosives, affiliated to the genus *Methylobacterium* [70]. The bacterial isolate designated as strain GW2 was found capable of producing the homopolymer poly-3-hydroxybutyrate (PHB) from various carbon sources such as methanol, ethanol, and succinate. Methanol acted as the best substrate for the production of PHB reaching 40 % w/w dry biomass. Optimal growth occurred at 0.5 % (v/v) methanol concentration and growth was strongly inhibited at concentration above 2 % (v/v). *Methylobacterium sp.* strain GW2 was also able to accumulate the copolyester poly-3-hydroxybutyrate-poly-3-hydroxyvalerate (PHB/HV) when valeric acid was supplied as an auxiliary carbon source to methanol. After 66 h, a copolymer content of 30 % (w/w) was achieved with a PHB to PHV ratio of 1:2. Shunsaku demonstrated that two types of methylotrophic bacterial strains, *P. denitrificans* and *M. exotorquens*, synthesized the copolyester poly (3-hydroxybutyrate-co-3-hydroxyvalerate) when methanol and n-amyl alcohol were added together to N-limited medium [71]. The composition of the copolyester differed considerably between the 2 strains: the copolyester from *P. denitrificans* was comparatively rich in 3-hydroxyvalerate (3HV). The 3HV content of the copolyester synthesized by this strain increased with increasing concentrations of n-amyl alcohol. Its maximum content was 91.5 mol%. In *M. extorquens*, the maximum 3HV content was confined to only 38.2 mol%. Loo studied the suitability of palm kernel oil, crude palm oil and palm acid oil as substrates for scl-PHA synthesis by *C. necator*, PHB⁻4 containing the PHA synthase gene of *Aeromonas caviae* [72]. Copolymer P(3HB-co-3HHx) was synthesized at yields ranging from 1.5-3.7 g.L^{-1} and the content of 3HHx in the polymer was 5%. Alias and Tan isolated a Gram-negative bacterium FLP1, from palm oil mill effluent by using a culture enrichment technique and tentatively identified it as closely related to *Burkholderia cepacia*, when this organism was grown on crude palm oil and palm kernel oil it produces P(3HB) and supplementation of odd number fatty acids resulted in production of poly(3-hydroxybutyrate-co-3-hydroxyvalerate) [73]. Yezza employed maple sap, an abundant natural product rich in sucrose as carbon source to *Alcaligenes latus* for the production of PHB and obtained 77.6±1.5 % PHB in shake flask and PHB content of 77.0±2.6% in Batch fermentation [74]. Nehal investigated that *Comamonas testosteroni* accumulated PHAs up to 78.5–87.5% of the cellular dry material during cultivation on various vegetable oils [75]. The efficiency of the culture to convert oil to PHAs ranged from 53.1% to 58.3% for different vegetable oils.

Table.1. Different waste and cheap carbon sources investigated for PHA production

Microorganism	Carbon source	PHA content (% Cell dry weight)	References
Pseudomonas aeruginosa 42A2	Agro industrial oily wastes	54.6	Fernandez et al. (2005)
Cupriavidus necator H16 *P. oleovorans*	Residual oil from rhamnose production	41.3 38.9	Steinbuchel et al. (2000)
Azotobacter vinelandii UWD	Swine waste liquor	58.3	Cho et al. (1997)
Actinobacillus sp. EL-9	Alcoholic distillery waste water	42	Son et al. (1996)
Alcaligenes latus DSM 1124	Soya wastes Malt wastes	70 32.57	Yu et al. (1999)
Staphylococcus epidermidis	Sesame oil	15.02	Wong et al. (2000)
Recombinant *E. coli*	Whey	81	Lee et al. (1997)
Sinorhizobium meliloti 41 *Hydrogenophaga pseudoflava DSM 1034*	Whey	3.5 4.4	Povolo and Casella (2003)
C. necator	Whey	20.6	Marangoni et al. (2002)
Recombinant *E. coli*	Molasses	80	Liu et al. (1998)
Methylobacterium sp. ZP24	Cheese whey	59	Yellore and Desai (1998)
Mixed bacterial cultures	Palm oil mill effluent	40	Salmiati et al. (2007)
Pseudomonas oleovorans *Pseudomonas corrugata*	Soya Biodiesel co-product stream	42	Ashby et al. (2004)
Bacillus sonorensis *Halomonas hydrothermalis*	Jatropha biodiesel byproduct	71.8 75	Shrivastav et al. (2010)
Burkholderia cepacia ATCC 17759	GlycerolGlycerol	3-9	Zhu et al. (2010)
Zobellela denitrificans	Glycerol	66.9 ± 7.6	Ibrahim and Steinbuchel (2010)
Escherichia coli arcA mutant	Xylose	51	Nikel et al. (2008)
Pseudomonas cepacia ATCC 17759	Sugar beet molasses	60	Ramsay et al. (1995)
Rhizobium meliloti		56	Mercan and Beyatli, (2005)
Bacillus cereus	Methanol	74	Yilmaz and Beyatli (2005)
Methylobacterium GW2	Maple sap	40	Yezza et al. (2006)
A. latus	Vegetable oils	77	Yezza et al. (2007)
Comamonas testosteroni		78.5–87.5	Nehal et al. (2005)

The composition of the PHAs formed was not substrate dependent as PHAs obtained from *C. testosteroni* during growth on variety of vegetable oils showed similar compositions; 3-hydroxyoctanoic acid and/or 3-hydroxydecanoic acid being always predominant. Jin Wang and Han-Qing Yu investigated the production of polyhydroxybutyrate (PHB) and extracellular polymeric substances (EPS) by *C. necator* ATCC 17699 at various glucose and $(NH_4)_2SO_4$ concentrations in batch cultures [76]. The PHB yield on glucose reached a maximum value of 0.34 g/g at glucose concentrations of 38.2 $g.L^{-1}$ and $(NH_4)_2SO_4$ of 3.2 $g.L^{-1}$. and the biosynthesis of EPS by *C. necator* was closely coupled with cell growth, while PHB was synthesized only under nitrogen-deficient and cell-growth-limited conditions. By changing the carbon source and bacterial strain used in the fermentation process, it is possible to produce related biopolymers having properties ranging from stiff and brittle plastics to rubbery plastics. Table1 shows a comparative data of the various waste products and cheap carbon sources been investigated for PHA production.

CONCLUSION

The utilization of waste product and cheap carbon sources will provide green plastics through non-conventional means. Conventional plastics/polymers being not readily biodegradable and produced from petroleum is hazardous to environment, polymers produced through biological means (microbes, plants etc.) is biodegradable conducive for conservation of the environment. The economic feasibility of PHAs would depend on few important factors like growth rate of microbe for generation of biomass, substrate cost, and recovery process including the solvent involved. Bacterial strains which could utilize the cheap carbon source were metabolically engineered for the production of P(3HB) with high yield and for efficient utilization of the inexpensive carbon source. However, polymer concentration obtained was much lower than those obtained by using purified carbon substrates. Therefore, there is a need for development of better strains and more efficient fermentation strategies for production of polymers from cheap carbon sources. Extensive and intensive research has been done on recombinant strains of *E.coli, Ralstonia* spp. for utilization of cheap carbon sources/waste. The raw material utilized dictated the price of PHA to a large extent and could influence its production price to such a level which is closer to other biopolymers like polylactic acid (PLA). The isolation and development of bacterial strains that could utilize cheap carbon substrates should also be pursued vigorously. Removal of fermentable organic substrates by converting them into PHAs can reduce waste treatment costs and generate revenue through the production of value-added products, a biodegradable polymer (PHA).

ACKNOWLEDGMENT

We gratefully acknowledge Dr. P. K. Ghosh (Director, CSMCRI) for valuable suggestions while preparing this manuscript. We acknowledge Mr. M. R. Gandhi and the members of PDEC division for providing the raw material being utilized in making the PHA. We also acknowledge Dr. P. Paul and the members of the Analytical science division of CSMCRI for help in characterization of PHA. SM acknowledges Daimler Chrysler and CSIR for financial support. AS and SKM wishes to acknowledge CSIR for Senior Research Fellowship.

REFERENCES

[1]   Swift, G. Acc. Chem. Res. 1993, 26, 105.
[2]   Shrivastav, A; Mishra, SK; Pancha, I; Jain, D; Bhattacharya, S; Patel, S; Mishra, S. World. J. Microb. Biotech. 2010; DOI: 10.1007/s11274-010-0605-2.
[3]   Brandl, H; Gross, RA; Lenz, RW; Fuller, RC. Adv. Biochem. Eng. Biotechnol. 1990, 41, 77.
[4]   Byrom, D. Trends. Biotechnol. 1987, 5, 246.
[5]   Madison, LL; Huisman, GW. Micro. Mol. Biol. Rev. 1999, 63, 21.
[6]   Steinbuchel, A; Fuchtenbusch, B. Trends. Biotechnol. 1988, 16, 419.
[7]   Lemoigne, M. Bull. Soc. Chimique. Belgique. 1926, 8, 770.
[8]   Steinbuchel, A; Valentin, H. FEMS. Microbiol. Letters. 1995, 128, 219-228.
[9]   Lee, SY; Chang, HN. Adv. Biochem. Eng/Biotechnol. 1995, 52, 27.
[10]   Poirier, Y; Nawrath, C; Somerville, C. Biotechnol. 1995, 13, 142–150.
[11]   Doi, Y. Microbial Polyester, VCH, New York,1990.
[12]   Brandl, H; Knee, EJ.Jr; Fuller, RC; Gross, RA; Lenz, RW. Int. J. Biol. Macromol. 1989,11, 49.
[13]   Lageween, RG; Huisman, GW; Preustig, H; Ketelaar, P; Eggink, G; Witholt, B. Appl. Env. Microbiol. 1988, 54, 2924.
[14]   Timm, A; Steinbüchel, A. Appl. Env. Microbiol. 1990, 56, 3360.
[15]   Choi, J; Lee, SY. Appl Microbiol Biotechnol. 1999, 51, 13.
[16]   Quillaguaman, J; Munoz, M; Mattiasson, B; Kaul, RH. Appl. Microbiol. Biotechnol. 2007, 74, 981.
[17]   Steinbuchel, A; Pieper, U. Appl. Microbiol. Biotechnol. 1992, 37, 1.
[18]   Bourque; Ouellette, B;, Andre, G; Groleau, D. Appl. Microbiol. Biotechnol. 1992, 37, 7.
[19]   Lee, S; Yu, J. Resour. Conser. Recy. 1997, 19, 151.
[20]   Suzuki, T; Yamane, T; Shimizu, S. Appl. Microbiol. Biotechnol. 1986, 23, 322.
[21]   Daniel, M; Choi. JH; Kim, JH; Lebeault, JM. Appl. Microbiol. Biotechnol. 1992, 37, 702.
[22]   Sharma, L; Mallick, N. Biotechnol. Lett. 2005, 27, 59.
[23]   Shrivastav, A; Mishra, SK; Mishra, S. Int. J. Biol. Macromol. 2010, 46, 255.
[24]   Doi, Y; Kunioka, M; Nakamura, Y; Soga, K. Macromol. 1987, 20, 2988.
[25]   Kellerhals, MB; Kessler, B; Witholt, B. Biotechnol. Bioeng. 2000, 65, 306.
[26]   Valappil, SP; Peiris, D; Langley, GJ; Herniman, JM; Boccaccini, AR; Bucke, C; Roy, I. J. Biotechnol. 2007, 127, 475.
[27]   Page, WJ. FEMS Microbiol. Lett. 1992, 103, 149.
[28]   Yamane, T; Fukunaga, M; Lee, YW. Biotechnol. Bioeng. 1996, 50, 197.
[29]   Renner, G; Haage, G; Braunegg, G. Appl. Microbiol. Biotechnol. 1996, 46, 268.
[30]   Anderson, AJ; Dawes, EA; Haywood, GW; Bryom, D. Copolymer production. 1993. US Patent No. 5264546.
[31]   Sudesh, K; Abe, H; Doi, Y. Prog. Polymer Sci. 2000, 25, 1503.
[32]   Lynd, LR; Wyman, CE; Gerngross, TU. Biocommodity engineering. Biotechnol. Prog. 1999,15, 777.
[33]   Braunegg, G; Bona, R; Koller, M. Polymer-Plastics Technol Eng. 2004, 43, 1779.

[34] Steinbüchel, A; Wullbrandt, D; Fuchtenbusch, B. Appl. Microbiol. Biotechnol. 2000, 53,167-172.

[35] Choi, J; Lee, SY. Bioprocess. Eng. 1997, 17, 335-342.

[36] Taniguchi, I; Kagotani, K; Kimura, Y. Green. Chem. 2003, 5, 545.

[37] Hu, WF; Chua, H; Yu, PHF. Biotechnol. Lett. 1997, 19, 695.

[38] Fernandez, D; Rodriguez, D; Bassas, M; Vinas, M; Solanas, AM; Llorens, J; Marques, AM; Manresa, Biochem. Eng. J. 2005, 26, 159.

[39] Cho, KS; Ryu, HW; Park, CH; Goodrich, PR. Biotechnol. Lett. 1997, 19, 7.

[40] Son, H; Park, G; Lee, S. Biotechnol. Lett 1996, 18, 1229.

[41] Yu, PH; Chua, H; Huang, AL; Ho, KP. Appl Biochem Biotechnol, 1999, 77, 445.

[42] Wong, AL; Chua, H; Yu, PH. Appl. Biochem. Biotechnol. 2000, 84, 843.

[43] Dionisi, D; Carucci, G; Papini, MP; Riccardi, C; Majone, M; Carrasco, F. Water. Res. 2005, 39, 2076.

[44] Rafael, G. Electronic. J. Biotechnol. 2001, 4, 2.

[45] Rhu, DH;Lee, WH; Kim, JY; Choi, E Water Sci Technol. 2003, 48, 221.

[46] Lee, SY; Middelberg, APJ; Lee, YK. Biotechnol Lett, 1997, 19, 1033-1035.

[47] Wong, HH; Lee, SY. Appl. Microbiol. Biotechnol. 1998, 50, 30.

[48] Povolo, S; Casella, S. Macromol. Symp. 2003, 197, 1.

[49] Marangoni, C; Furigo, A; De Aragão, GMF. Process. Biochem. 2002, 38, 137.

[50] Liu, F; Li, W; Ridgway, D; Gu, T. Biotechnol. Lett. 1998, 20, 345.

[51] Yellore, V; Desai, A. Lett Appl Microbiol. 1998, 25, 391.

[52] Dhanasekar, R; Viruthagiri, T. Indian J Chem Technol. 2005, 12, 322.

[53] Salmiati, Z; Ujang, MR; Salim, MF; Din, M; Ahmad, MA. Water. Sci. Technol. 2007, 56, 179.

[54] Aldor, IS; Kim, SW; Jones, KL; Keasling, JD. Appl. Env. Microbiol. 2002, 68, 3848.

[55] Andreessen, B; Lange, AB; Robenek, H; Steinbuchel, A. Appl. Env. Microbiol. 2010, 76, 622.

[56] Ibrahim, MHA; Steinbuchel, A. J. Appl. Microbiol. 2010, 108, 214.

[57] Ashby, RD; Solaiman, DKY; Foglia, TA. J. Polymers Env. 2004; 12, 105.

[58] Shrivastav, A; Mishra, SK, Shethia, B; Pancha, I; Jain, D; Mishra, S. Int. J. Biol. Macromol. 2010, 47, 283.

[59] Ghosh, PK; Mishra, S; Gandhi, MR; Upadhyay, SC; Paul, P; Anand, PS; Popat, KM; Shrivastav, A; Mishra, SK; Ondhiya, N; Maru, RD; Dyal, G; Brahmabhatt, H; Boricha B; Chadhary, DR; Rebary, B; Zala, KS. Patent 1838/DEL/2009, Patent WO/2011/027353.

[60] Ciesielski, S; Pokoj, T; Klimiuk, E. J. Microbiol. Biotechnol. 2010, 20, 853.

[61] Zhu, C; Nomura, CT; Perrotta, JA; Stipanovic, AJ; Nakas, JP. Biotechnol. Prog. 2010, 26, 424.

[62] Ibrahim, MH; Steinbuchel, A. J. Appl. Microbiol. 2010, 108, 214.

[63] Nikel, PI; Pettinari, MJ; Galvagno, MA; Me´ndez, BS. Appl. Microbiol. Biotechnol. 2008, 77(6), 1337.

[64] Ramsay, JA; Hassan, MCA; Ramsay, BA. Canadian. J. Microbiol. 1995, 41, 262.

[65] Lee, SY. Bioprocess. Eng. 1998, 18, 397.

[66] Omar, S; Rayes, A; Eqaab, A; Viss, I; Steinbuchel, A. Biotechnol. Lett. 2001, 23, 1119.

[67] Mercan, N; Beyatli, Y. Zuckerindustrie. 2005, 130, 410.

[68] Yilmaz, M; Beyatli, Y. Zuckerindustrie. 2005, 130, 109.

[69] Solaiman, DKY; Ashby, RD; Hotchkiss, AT; Foglia, TA. Biotechnol. Lett, 2006, 28, 157.

[70] Yezza, A; Fournier, A; Halasz; Hawari, J. Appl. Microbiol. Cell. Physiol. 2006, 73, 211.

[71] Shunsaku, U; Seiji, M; Aya, T; Tsuneo, Y. Appl Env Microbiol. 1992,58,3574.

[72] Loo, CY; Lee, WH; Tsuge, T; Doi, Y; Sudesh, K. Biotechnol. Lett. 2005, 27, 1405.

[73] Alias, Z; Tan, IKP. Bioresource Technol. 2005, 96, 1229.

[74] Yezza, A; Halasz, A; Levadoux, W; Hawari, J. Appl. Microbiol. Biotechnol. 2007, 77, 269.

[75] Nehal, T; Ujjval, T; Patel, KC. Bioresource. Technol. 2005, 96, 1843.

[76] Jin, W; Han, QY. Appl. Microbiol.Physiol. 2007, 75, 871.

In: Polymer Synthesis
Editor: E. Kowsari

ISBN 978-1-61324-672-6
© 2012 Nova Science Publishers, Inc.

Chapter 12

Prospective to Produce Polyhydroxyalkanoic Acids from Cyanobacteria

Bhabatarini Panda, Rakesh Verma, Ranjana Bhati, Shilalipi Samantaray and Nirupama Mallick

Agricultural and Food Engineering Department, Indian Institute of Technology, Kharagpur, West Bengal, India

Abstract

Cyanobacteria are prokaryotic oxygenic photoautotrophs with different morphologies, ranging from unicellular to colonial and filamentous forms, and are found in almost every conceivable habitat on earth. These organisms with a short generation time need some simple inorganic nutrients such as phosphate, nitrate (not in case of N_2-fixers), magnesium, sodium, potassium and calcium as macro-, and Fe, Mn, Zn, Mo, Co, B and Cu as micronutrients for their growth and multiplication. Further, these organisms can successfully be cultivated in wastewaters due to their ability to use inorganic nitrogen and phosphorus for their growth and multiplication. Therefore, wastewater treatment involving cyanobacteria is quite attractive because of their ability to transform waste into useful biomass using sunlight as the energy source. Thus, these tiny photoautotrophs can be considered as alternative hosts for low-cost production of polyhydroxyalkanoates (PHAs), the most appropriate materials for alternative plastics. The photoautotrophic production of PHAs in cyanobacteria may encompass the high cost incurred due to feeding large amount of organic carbon and continuous oxygen supply during bacterial fermentation.

An overview is presented here on the progresses and possibilities of producing PHAs from cyanobacteria. The majority of efforts on photoautotrophic production of PHAs using CO_2 as the sole carbon source by cyanobacteria are available, but the contents, in general, are very low and amount less than 10% of dry cell weight (dcw) with sole exception to *Synechococcus* sp. MA19, where a much higher poly-β-hydroxybutyrate (PHB) content, 55% (dcw) was reported. Some cyanobacteria can accumulate PHB when grown mixotrophically with acetate, glucose or other exogenous carbon sources; accumulation more than 40% (dcw) was recorded for *Nostoc muscorum* and *Synecho-*

cystis PCC 6803. Recent study of our laboratory showed that a N_2-fixing cyanobacterium, *Aulosira fertilissima* can accumulate PHB up to 85% (dcw) under optimized condition. Further, *Nostoc muscorum* was also found to accumulate poly(3-hydroxybutyrate-*co*-3-hydroxyvalerate) or P(3HB-*co*-3HV) co-polymer ~60% (dcw) under specific condition. Uncouplers like carbonylcyanide-m-chloro phenylhydrazone (CCCP) and dicyclohexyl cabodiimide (DCCD) stimulated PHB accumulation, whereas supplementation of 3-(3,4-dichlorophenyl)-1,1-dimethylurea (DCMU) to the photoauto-trophically-grown cultures suppressed PHB accumulation.

Analysis of the material properties of these polymers by mechanical tests, surface analysis and differential scanning calorimetry (DSC) exhibited comparable material properties with the commercial polymers. Addition of exogenous carbon although found essential for stimulation of PHAs synthesis in cyanobacteria, the magnitude was significantly lower as compared to that of heterotrophic bacteria. Thus, low-cost production of PHAs polymers from cyanobacteria can be envisaged.

INTRODUCTION

Plastics have become one of the most widely used materials in the last 50 years. The history of plastics goes back to mid 19th century when the first thermoplastic resin, celluloid, was produced by the reaction of cellulose with nitric acid. Since then the types and qualities have greatly increased, producing superior materials such as epoxies, polycarbonates, teflon, silicones, polysulfones, etc., and replacing glass, wood and other construction materials, and even metals in many industrial, domestic and environmental applications. [1,2] Although plastics have had a remarkable impact on our day-to-day life, it has become increasingly obvious that there is a price to be paid for their use, as they are not biologically degradable.

Durability and resistance to degradation are desirable properties when plastics are in use, but these very desirable properties have now become their greatest problem. According to an estimate, more than 150 million tons of plastics are produced every year. Due to the worldwide use of plastic products, plastic wastes have become a major component of both industrial and municipal wastes. The massive scale per capita consumption of plastics at around 80 and 60 kg, respectively in USA and Europe has created a major threat to the solid waste management program. [3] In developed countries, most goods made of plastics are ended up after their useful life as discarded wastes, and occupy the environment as landfills. [4] Plastics being xenobiotic are recalcitrant to microbial degradation and persist in the environment for many years. [5] Excessive molecular size, large number of aromatic rings, and/ or halogen substitutions seems to be responsible for the resistance of these chemicals. [6,7]

Several thousand tons of plastics are discarded into the marine environment every year and approximately one million marine animals are killed every year either by choking on plastics they mistook as food or becoming entangled in non-degradable plastic debris. [8] Incineration has been one option in dealing with these evil materials, but it is expensive and dangerous because certain plastics such as polyvinyl chloride (PVC) are potential source of highly toxic dioxins when burned in municipal incinerators. Polystyrene foam products are often made with chlorofluorocarbons (CFCs) and hydrochlorofluorocarbons (HCFCs), both of which are ozone-destroying and greenhouse chemicals. [9] Recycling can be the second option, but it also has disadvantage, as it is difficult to sort out the wide varieties of plastics.

Moreover, recycling changes the material properties of plastic to a great extent, thus the application range of the recycled plastic is further limited. [8]

One of the safest and least expensive ways to deal with the disposal of these non-degradable plastics is to use municipal landfills. However, this has become a problem for municipalities worldwide because municipal landfills lose capacity because of accumulation of these environmentally unfriendly plastics. Thus, the increasing cost of solid waste disposal, decreasing capacity of municipal landfills and the potential hazards from plastic waste incineration are the major concerns for eco-friendly and sustainable development.

Over the past three decades there have been a growing public and scientific interest regarding the use and development of eco-friendly biodegradable polymers as an ecologically useful alternative to plastics, which must still retain the desired physical and chemical properties of conventional synthetic plastics; thus offering a solution for the existing grave problem of plastic wastes. The search for such biodegradable plastics has led to the introduction of a few types of polymers: Polyhydroxyalkanoates (PHAs), Polylactides (PLA), aliphatic polyesters, polysaccharides and/or blends of the above. Among these, the microbially-produced PHAs offer much potential for significant contributions as 'bioplastics'.

POLYHYDROXYALKANOATES (PHAS)

The property that distinguishes PHAs from the petroleum-based plastics is their biodegradability. PHAs are degraded upon exposure to soil, compost, or marine sediment. Biodegradation is dependent on a number of factors such as microbial activity in the environment and the exposed surface area, moisture, temperature, pH, etc. [10] Polymer composition, molecular weight, crystallinity and nature of the monomer units are also found to have significant effects on degradation. Co-polymers containing 4-hydroxybutyrate (4HB) monomer units are found to degrade more rapidly than poly-3-hydroxybutyrate (PHB) or 3-hydroxybutyrate-co-3-hydroxyvalerate [P(3HB-co-3HV)] co-polymer. [11,12] Micro-organisms secrete enzyme (PHA depolymerase) that break down the polymer into its monomer units, called hydroxyacids, which are utilized as carbon sources for growth of that microorganism. Biodegradation of PHAs under aerobic conditions results into carbon dioxide and water, whereas in anaerobic conditions the degradation products are carbon dioxide and methane. Studies have shown that 85% of PHAs are degraded in seven weeks. [5,13] PHAs have also been reported to degrade in aquatic environments (Lake Lugano, Switzerland) within 254 days even at temperatures not exceeding 6°C. [13] It is not only the biodegradability that makes the PHAs so fascinating but these are also derived from the renewable sources. Being the product of renewable carbon sources, their production and uses follow sustainable closed cycles. [14]

PHAs can be classified into three groups depending on the number of carbon atoms in the monomer units; short-chain-length (SCL), medium-chain-length (MCL) and long-chain-length (LCL) PHAs composed by hydroxyacids with 3-5, 6-14 or more than 14 carbon atoms, respectively. [15] Amongst SCL-PHAs, poly-β-hydroxybutyrate (PHB) is the most commonly found polymer in various bacteria. It is a stereo-regular polymer having asymmetric carbon atoms in R-stereochemical configuration. As a consequence, the polymer is isotactic, highly crystalline (crystallinity ranges between 60 to 80%) and relatively stiff; its glass-to-rubber transition temperature (T_g) and melting temperature (T_m) are comparable with the polypropylene as well as some mechanical properties like Young's modulus and tensile strength. On the other hand, MCL-PHAs are semicrystalline elastomers characterized by low T_m values, low tensile strength and high values of elongation-to-break. [16] This family of polyesters possesses two main drawbacks: that they usually soften at temperatures around 40 °C and they are characterized by very low crystallization rates. However, both the limitations can be overcome by incorporation of MCL monomer units into the PHB backbone, which alters the material properties of the polymer, making it suitable for commercial application. [17]

IMPORTANCE OF P(3HB-*co*-3HV) CO-POLYMER

Within the cell, PHB exists in a fluid, amorphous state. However, after extraction from the cell with organic solvents, PHB becomes highly crystalline [18] and in this state is a stiff but brittle material. Because of its brittleness, PHB is not very stress resistant. Also, the relatively high melting temperature of PHB (around 180 °C) is close to the temperature where this polymer decomposes thermally and thus limits the ability to process the homopolymer. Initial biotechnological developments were therefore, aimed at making PHAs that were easier to process. Detection of heteropolymeric PHAs consisting of 3-hydroxybutyric acid (HB), 3-hydroxyvaleric acid (HV), 3-hydroxyhexanoic acid (HX) and 3-hydroxyheptanoic acid (HH) in isolates of activated sludge was another milestone in PHA research, owing to the more desirable properties of heteropolymers. Numbers of patents were filed for production, extraction and blending processes during 1980s. The incorporation of 3HV into the PHB resulted in a poly(3-hydroxybutyrate-*co*-3-hydroxyvalerate) [P(3HB-*co*-3HV)] co-polymer that is less stiff and brittle than PHB, and can be used to prepare films with excellent water and gas barrier properties reminiscent of polypropylene, and also can be processed at a lower temperature while retaining most of the other excellent mechanical properties of PHB. [19] Thus, this co-polymer has become a commercial product. Co-polymer, P(3HB-*co*-3HV) produced from *Ralstonia eutropha*, now called as *Cupriavidus necator*, was sold under the trade name BIOPOL[®] by Monsanto [20], which has now been own by Metabolix (Massachusetts, USA). Other functional companies now engaged in production of bioplastics are BioMer (Germany) and Mitsubishi Gas Chemical (Japan) under the trade name of Biomer[®] and Biogreen[®], respectively.

PHAs PRODUCTION BY BACTERIAL FERMENTATION

Progresses in bacterial PHAs production have been well reviewed by many researchers. [2,14,21-37] However, the major drawback in commercialization of bacterial PHAs as a source of biodegradable plastics is its high cost of production. The cost of raw material itself accounts for ~50% of the total production cost. The cost of PHA using the natural producer *C. necator* is approximately US$16 per Kg, which is more than 15 times higher than polypropylene. [32]

Table 1. Reports on accumulation of PHB in various cyanobacterial species
(modified from Vincenzini and De Philippis[72])

Cyanobacterium	Culture condition	PHB content (% dcw)	Reference
Chlorogloea fritschii	mixotrophy (acetate)	10	Carr (1966)[42]
Gloeocapsa PCC 6501	photoautotrophy	P	Rippka *et al.* (1971)[43]
Chlorogloea fritschii Mitra	photoautotrophy	P	Jensen and Sicko (1973)[44]
Nostoc sp.(13 strains)	photoautotrophy	P	Jensen (1980)[45]
Microcystis aeruginosa	photoautotrophy	P	Jensen and Baxter (1981)[46]
Spirulina platensis	photoautotrophy	6	Campbell *et al.* (1982)[47]
Anacystis cyanea	photoautotrophy	P	Sicko-Goad (1982)[48]
Aphanocapsa PCC 6308	mixotrophy (acetate)	P	Allen (1984)[49]
Oscillatoria limosa	photoautotrophy	P	Stal *et al.* (1990)[50]
Spirulina jenneri NK1	photoautotrophy	P	Vincenzini *et al.* (1990a)[51]
Spirulina laxissima MG5	photoautotrophy	P	Vincenzini *et al.* (1990b)[52]
Spirulina maxima (3 strains)	photoautotrophy mixotrophy (acetate)	<1 3.2	Vincenzini *et al.* (1990a)[51] De Philippis *et al.* (1992)[53]
Spirulina platensis (4 strains)	photoautotrophy mixotrophy (acetate)	<1 2.9	Vincenzini *et al.* (1990a)[51] De Philippis *et al.* (1992)[53]
Gloeothece sp. PCC 6909	photoautotrophy	6.4	Stal (1992)[54]
Anabaena cylindrica 10 C	photoautotrophy nitrogen-starved	0.2	Lama *et al.* (1996)[55]
Synechococcus sp. MA19	photoautotrophy nitrogen-starved	27	Miyake *et al.* (1996)[56]
Synechococcus sp. MA19	photoautotrophy phosphate-starved	55	Nishioka *et al.* (2001)[39]
Synechocystis sp. PCC 6803	Mixotrophy (acetate) chemoheterotrophy (acetate)	7 10	Sudesh *et al.* (2002)[57]
Nostoc muscorum	photoautotropy chemoheterotrophy (acetate)	8.5 43	Sharma and Mallick (2005a)[58] Sharma and Mallick (2005b)[59]
Synechocystis sp. PCC 6803	mixotrophy (acetate + glucose) + P-limitation chemoheterotrophy (acetate)	29 22	Panda *et al.* (2006)[40] Panda and Mallick (2006)[60]
Nostoc muscorum	(acetate + glucose) + P-limitation	46	Sharma *et al.* (2007)[41]
Synechocystis sp. UNIWG	mixotrophy (acetate) + nitrogen- starved + CO$_2$	14	Toh *et al.* (2008)[61]
Synechocystis sp. PCC 6803	mixotrophy (acetate) + N and P-limitation	43	Panda (2008)[62]
Spirulina subsalsa	NaCl	5.9	Shrivastav *et al.* (2010)[63]
Aulosira fertilissima	photoautotrophy	6.5	Bhati *et al.* (2010)[64]
Calothrix sp.		6.4	
Scytonema sp.		7.4	
Aulosira fertilissima	Optimized condition	85	Samantaray and Mallick (2011)[65]

P: Present, not quantified.

Table 2. Reports on accumulation of P(3HB-co-3HV) co-polymer in cyanobacteria

Cyanobacterium	Culture condition	P(3HB-co-3HV) (% dcw)	Reference
Anabaena cylindrica 10 C	acetate (0.1%) + propionate (0.4%)	2	Lama *et al.* (1996)[55]
Nostoc muscorum	acetate (0.11%) + propionate (0.08%)	31	Mallick *et al.* (2007a)[66]
Synechocystis sp. PCC 6803	acetate (0.4%) + valerate (0.1%) + P and N limitations	45	Panda (2008)[62]
Nostoc muscorum	acetate (0.4%) + valerate(0.4%)+N-deficiency	60	Bhati and Mallick (2011)[67]

To reduce this limitation, alternative substrates (cheap carbon sources) such as molasses, whey, hemicellulose, palm oil, etc. are receiving much attention, but no major break-through in this area has been announced so far. [32,38]

CYANOBACTERIA AS PRODUCTION HOSTS

Another potential production system may be the cyanobacteria, which are oxygenic photoautotrophic prokaryotes. The advantages of using cyanobacteria in comparison to heterotrophic bacteria are enormous as these are oxygen evolving photoautotrophic organisms, so there is no need to supplement carbons for growth and oxygen in production units/ area. Some of them can fix atmospheric nitrogen, so no need to provide nitrogen source(s) for those species. These organisms with a short generation time need some simple inorganic nutrients such as phosphate, nitrate (not in case of N_2-fixers), magnesium, sodium, potassium and calcium as macro-, and Fe, Mn, Zn, Mo, Co, B and Cu as micronutrients for their growth and multiplication. Further, these organisms can successfully be cultivated in wastewaters due to their ability to use inorganic nitrogen and phosphorus for their growth, and wastewaters such as effluents of farm-yards, fish-farms, sewage treatment plants, etc. are rich sources of N and P, [58] and therefore, wastewater treatment involving microalgae is quite attractive because of its ability to transform waste into useful biomass using sunlight as the energy source. Thus, photoautotrophic production of PHAs in cyanobacteria may encompass the high cost incurred due to feeding of organic carbon and continuous oxygen supply during bacterial fermentation.

OCCURRENCE OF PHAS IN CYANOBACTERIA

The accumulation of PHAs in cyanobacteria has been studied by many researchers. Table 1 presents a detailed review on the presence of PHAs in various cyanobacterial species, but the contents, in general, are found to be very low and amount less than 10% of dry cell weight (dcw) under photoautotrophic growth conditions. In contrast to this, *Synechococcus* sp.

MA19, a unicellular cyanobacterium isolated from the volcanic rock of Japan, has been reported to accumulate PHB up to 55% (dcw) under phosphate-limited condition.[39] In *Synechocystis* sp. PCC 6803 an accumulation up to 29% (dcw) was recorded under mixotrophy with phosphate limitation. [40] *Nostoc muscorum* also exhibited a PHB accumulation up to 46% when subjected to chemoheterotrophy under phosphate limitation. [41] Recent study of our laboratory showed that a N_2-fixing cyanobacterium, *Aulosira fertilissima* can accumulate PHB up to 85% (dcw) under optimized condition. [65] Few cyanobacteria were also found to accumulate poly(3-hydroxybutyrate-*co*-3-hydroxyvalerate) or P(3HB-*co*-3HV) co-polymer under various specific conditions (Table 2).

ACCUMULATION AND MOBILIZATION OF PHB IN CYANOBACTERIA: A CASE STUDY WITH *SYNECHOCYSTIS* SP. PCC 6803

PHB metabolism has been studied in PHB accumulating bacteria, especially in *Alcaligenes eutrophus* (*C. necator*), where PHB is synthesized from acetyl-CoA via a three enzymatic reactions. [27]

$$2\,\text{Acetyl-CoA} \xrightarrow{\text{3-ketothiolase}} \text{Acetoacetyl-CoA} + \text{CoASH}$$

$$\text{Acetoacetyl-CoA} + \text{NADPH} \xrightarrow[\text{reductase}]{\text{acetoacetyl-CoA}} \text{3-Hydroxybutyryl-CoA} + \text{NADP}$$

$$\text{3-Hydroxybutyryl-CoA} \xrightarrow{\text{PHA synthase}} \text{3-Hydroxybutyrate} + \text{CoASH}$$

In *C. necator*, these three enzymes are constitutively synthesized. The first enzyme of the sequence, 3-ketothiolase controls the synthesis of PHB with free CoA acting as the key regulatory molecules. [68] In an *Acinetobacter* sp., which shows regulated PHB accumulation, PHB biosynthesis gene operon is controlled at the transcriptional level. [69] However, in a thermophillic cyanobacterium, *Synechococcus* sp. MA19, PHB synthesis is controlled by the enzyme PHB synthase and found to be dependent on the C/N ratio. [70] In a N_2-fixing cyanobacterium, *Nostoc muscorum*, acetyl phosphate is found to have a post-translational control over PHB synthase, and PHB synthesis is stimulated under accumulation of NADPH or a high ratio of reducing power to ATP in the cell. [71]

Since PHB plays a role as carbon and energy reserve material, [22,72] its synthesis is expected to be coupled with mobilization. In case of *Azotobacter beijerinckii* and *C. necator*, biosynthesis and mobilization of PHB usually occurs in a cyclic manner, where acetyl-CoA precursors are converted to PHB by the usual three sequential enzyme mediated reactions, and mobilization of PHB starts by the intracellular PHB depolymerase leading to acetyl-CoA generation by subsequent steps. [68,73]

$$CH_3COSCoA \xrightarrow[\substack{\text{3-enzyme} \\ \text{mediated reactions}}]{\text{Accumulation}} PHB \xrightarrow[\text{PHB depolymerase}]{\text{Mobilization}} CH_3COSCoA$$

The acetyl-CoA is further oxidized via tricarboxylic acid cycle to generate energy. On the other hand, occurrence of PHB in cyanobacteria, which lacks a complete TCA cycle, is very intriguing since the breakdown of this reserve polymer can play only a minor role in cell metabolism. The acetyl-CoA which arises from its depolymerization can be utilized neither for significant energy production nor for the synthesis of more than a few cell constituents. [74] Thus, the role of PHB in cyanobacterial metabolism is still unclear. Here, we will discuss our attempt to explore the role of NADPH to ATP ratio in PHB biosynthesis by disrupting the balance with the help of some specific metabolic inhibitors and to study PHB mobilization process in the test cyanobacterium, *Synechocystis* sp. PCC 6803 under various conditions with an aim to explore its role in cyanobacterial metabolism.

IMPACT OF UNCOUPLERS

In order to elucidate the impact of uncuplers on PHB biosynthesis, carbonylcyanide-m-chloro phenylhydrazone (CCCP) was added to the standard mineral medium. PHB accumulation in *Synechocystis* sp. PCC 6803 was found to be stimulated up to 22.6% (dcw) in 20 μM CCCP-treated cultures, from the basal PHB content of 4.5% (dcw, Figure 1).

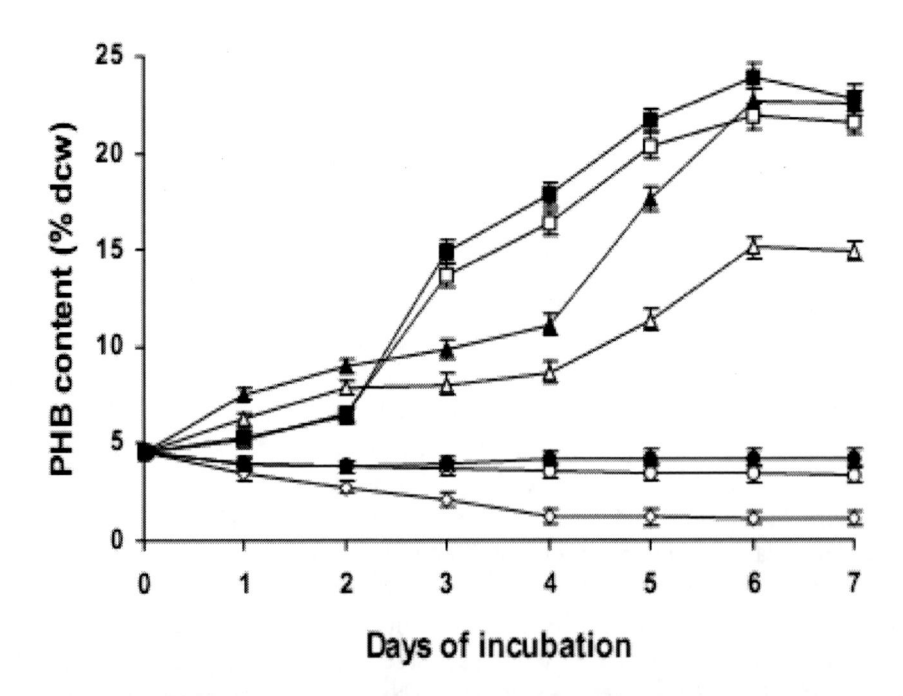

Figure 1. Effects of uncouplers on PHB accumulation potential of the photoautotrophically-grown *Synechocystis* sp. PCC 6803. Control (◊), CCCP 10 μM (∆), CCCP 20 μM (▲), DCCD 10 μM (□), DCCD 20 μM (■), DNP 1 mM (○) and DNP 2 mM (●). Source: Panda [62].

Dicyclohexyl cabodiimide (DCCD)-treated cultures also depicted a similar rise, which was however, not observed under 2,4-dinitrophenol (DNP) supplementation. Under dark condition, the CCCP-, DCCD- and DNP-treated cultures also showed insignificant rise in PHB content (Table 3). Similar trend was prevailed even in presence of glucose.

**Table 3. Effects of CCCP (20 μM), DCCD (20 μM) and DNP (2 mM)
on PHB accumulation potential of *Synechocystis* sp. PCC 6803
under dark on day 6 of incubation**

Treatment	PHB content (% dcw)
Dark	6.0 ± 0.18^a
Dark + CCCP	6.9 ± 0.22^a
Dark + CCCP + G	7.3 ± 0.27^a
Dark + DCCD	7.1 ± 0.24^a
Dark + DCCD + G	7.4 ± 0.26^a
Dark + DNP	5.7 ± 0.25^a
Dark + DNP + G	5.8 ± 0.23^a

G: glucose, 0.4% glucose was added.
All values are mean $\pm$ SE, n=3.
Values in the column superscripted by same letters are not significantly ($P > 0.05$) different from each other (Duncan's new multiple range test). Source: Panda [62].

As reported by Mallick et al. [71], uncouplers abolish phosphorylation without inhibiting electron transport. This may leads to a rise in the concentration of reducing power through the accelerated non-cyclic electron transfer of photosynthetic light reaction, as the electron transport, freed of the restraints imposed by the coupling mechanism, is often greatly accelerated. [75] ATP pool of *Synechocystis* sp. PCC 6803 under CCCP treatment was found to decrese significantly. [76] Under DCCD treatment, similar decline in ATP content in *Natronobacterium pharaonis* was also recorded. [77] Therefore, under such situations, an imbalance in the ratio between NAD(P)H and ATP was generated, which might be the cause for the enhanced PHB accumulation (Figure 1).

When *Synechocystis* sp. PCC 6803 was transferred to dark in presence of glucose (Table 3), growth of the cyanobacterium was supported by the breakdown of glucose via the oxidative pentose-phosphate cycle. As oxidative pentose-phosphate cycle is the main pathway for generation of NADPH, uncoupling the mechanism by CCCP, DCCD or DNP would result into rise in the pool of NADPH in the cell. Under such circumstances a stimulated accumulation of PHB is expected. However, these results did not exhibit any such significant rise. This could be explained in the light of the previous findings of Mallick *et al* [71], where they postulated that PHB accumulation in cyanobacteria might be linked to photophosphorylation rather than oxidative phosphorylation. Results of DNP treatment (both under light and dark conditions) give support to the above view as phenols are the classical uncouplers of oxidative phosphorylation and DNP even at 10^{-3} M has only mild effects on chloroplast function. [78]

DCMU SUPPLEMENTATION UNDER PHOSPHATE DEFICIENCY

Figure 2 presents the effect of 3-(3,4-dichlorophenyl)-1,1-dimethylurea (DCMU), a potent inhibitor of non-cyclic electron transport chain of photosynthesis (Z scheme), on PHB accumulation potential of *Synechocystis* sp. PCC 6803, grown photoautotrophically and under phosphate-deficient condition. In contrast to P-deficient control, where the accumulation reached up to 11% (dcw) on day 10 of phosphate deficiency, the DCMU-supplemented cultures depicted a declining trend. Further, no rise in PHB content was evident when glucose was supplemented along with DCMU (Table 4).

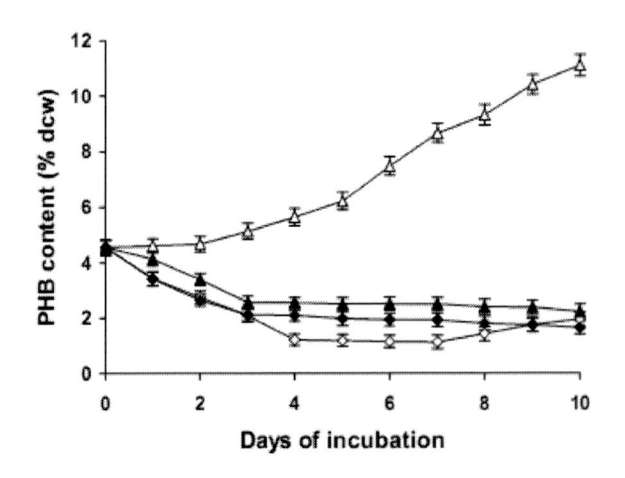

Figure 2. Effect of DCMU (10 mM) on PHB accumulation potential of the photoautotrophically-grown *Synechocystis* sp. PCC 6803. Control (◊), P-deficient (Δ), control + DCMU (♦) and P-deficient + DCMU (▲). Source: Panda [62].

Table 4. Effect of DCMU (10 mM) on PHB accumulation potential of *Synechocystis* sp. PCC 6803 in presence of glucose on day 10 of incubation

Treatment	PHB content (% dcw)
Control	2.0 ± 0.16^{a}
Control + DCMU	1.6 ± 0.15^{a}
Control + DCMU + G (0.1%)	2.1 ± 0.27^{a}
Control + DCMU + G (0.2%)	2.2 ± 0.23^{a}
Control + DCMU + G (0.4%)	2.4 ± 0.26^{a}
P-deficient (P_0)	11.1 ± 0.21^{b}
P_0 + DCMU	2.2 ± 0.26^{a}
P_0 + DCMU + G (0.1%)	2.5 ± 0.22^{a}
P_0 + DCMU + G (0.2%)	2.7 ± 0.24^{a}
P_0 + DCMU + G (0.4%)	2.8 ± 0.27^{a}

G: glucose, All values are mean ± SE, n=3.

Values in the column superscripted by same letters are not significantly ($P > 0.05$) different from each other (Duncan's new multiple range test). Source: Panda [62].

As postulated by De Philippis *et al.*[52,53], PHB synthesis would be stimulated under phosphate starvation, when reducing power may be in excess, because ATP synthesis is known to decrease markedly with the onset of phosphate limitation, while reduction of NADP through non-cyclic photosynthetic electron flow is not inhibited. [79,80] Therefore, NADPH synthesis was suppressed in phosphate-starved cells by DCMU treatment (Figure 2) . DCMU inhibits photosystem II through binding to the Q_B binding site *vis-à-vis* the non-cyclic electron flow and NADPH production. [81] In *Chlorella vulgaris* and *Thalassiosira rotula*, NADPH concentration was found to decrease significantly under DCMU treatment. [82] In *Synechocystis* sp. PCC 6803, PHB accumulation showed a rising trend under P-deficiency, and the accumulation reached up to 11.1% on day 10 of phosphate deficiency. On the other hand, a reverse trend was evident in the DCMU-supplemented medium. This is in agreement with De Philippis *et al.* [53,54] that in photoautotrophically-grown cultures, NADPH synthesis through non-cyclic electron transfer plays a major role in PHB accumulation. Results from Table 4, where addition of glucose into the DCMU-treated P-deficient as well as control cultures has not been found stimulatory for PHB accumulation give further support to the view of Mallick *et al.*[71] that PHB synthesis in cyanobacteria might be linked to photophosphorylation rather than oxidative phosphorylation.

EFFECT OF 2,3-BUTANEDIONE

Figure 3 presents the effect of 2,3-butanedione on PHB accumulation potential of the test cyanobacterium grown under phosphate limitation. PHB accumulation reached up to 11.1% (dcw) in phosphate-deficient cells (Figure 3A). A PHB pool of 25.8% (dcw) was recorded in acetate-supplemented P-deficient cultures (Figure 3B). In both the cases, supplementation of 10 mM 2,3-butanedione resulted into complete suppression of PHB accumulation. Partial restoration was however, recorded on acetyl phosphate supplementation (Table 5).

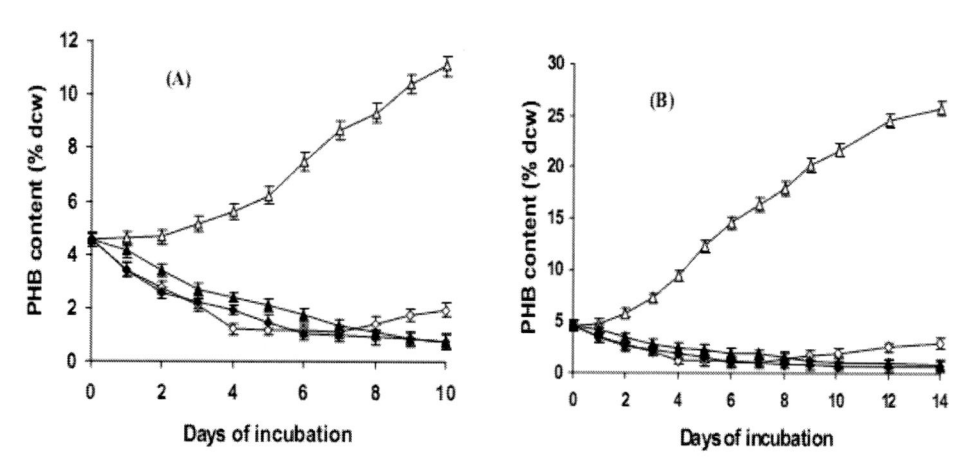

Figure 3. Effect of 2,3-butanedione (10 mM) on PHB accumulation potential of (A) P-deficient cells of *Synechocystis* sp. PCC 6803. Control (◊), P-deficient (Δ), control + 2,3-butanedione (♦) and P-deficient + 2,3-butanedione (▲). (B) P-deficient + acetate-grown cells of *Synechocystis* sp. PCC 6803. Control (◊), P-deficient + acetate (Δ), control + 2,3-butanedione (♦) and P-deficient + acetate + 2,3-butanedione (▲). 0.4% acetate was used. Source: Panda [62].

Table 5. Interactive effects of 2,3-butanedione and acetyl phosphate on PHB accumulation potential of *Synechocystis* sp. PCC 6803 on day 10 of incubation

Treatment	PHB content (% dcw)
P-deficiency (P_0)	11.1 ± 0.19^a
P_0 + 2,3-butanedione	0.8 ± 0.06^b
P_0 + 2,3-butanedione + AP (10 mM)	6.1 ± 0.13^c
P_0 + 2,3-butanedione + AP (20 mM)	9.5 ± 0.17^a
P_0 + A	25.8 ± 0.26^d
P_0 + A + 2,3-butanedione	0.8 ± 0.08^b
P_0 + A + 2,3-butanedione + AP (10 mM)	10.2 ± 0.14^a
P_0 + A + 2,3-butanedione + AP (20 mM)	17.7 ± 0.21^e

A: acetate (0.4%), AP: acetyl phosphate.

All values are mean ± SE, n=3.

Values in the column superscripted by different letters are significantly ($P < 0.05$) different from each other (Duncan's new multiple range test). Source: Panda [62].

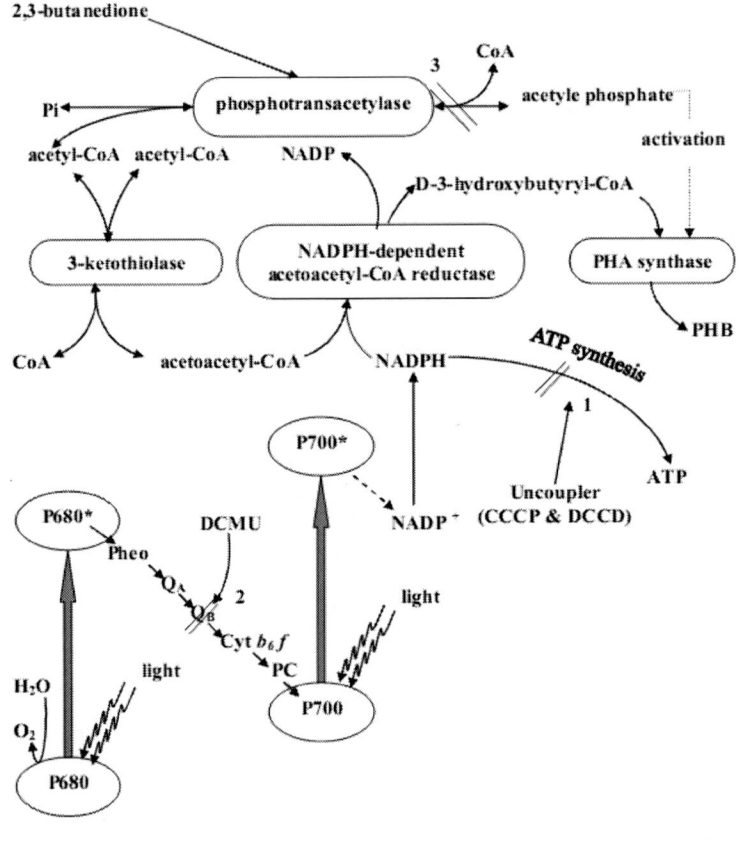

Figure 4. Hypothesized schematic regulation of PHB accumulation by metabolic inhibitors in *Synechocystis* sp. PCC 6803 (modified from Asada *et al.*[83]). 1. Uncouplers (CCCP and DCCD) increased NADPH accumulation thus increased PHB synthesis. 2. DCMU inhibited NADPH synthesis thus decreased PHB accumulation. 3. 2,3-butanedione inhibited acetyl phosphate synthesis thus decreased PHB accumulation. Source: Panda [62].

MOBILIZATION OF PHB IN *SYNECHOCYSTIS* SP. PCC 6803

In the test cyanobacterium, under photoautotrophic condition, PHB accumulation reached the maximum value, i.e. 4.5% (dcw) at the stationary phase of growth. As this value was not high enough to get a clear observation of mobilization pattern, PHB accumulation up to 9.5 and 22.4% (dcw) was achieved by subjecting the cells, respectively to N-deficiency and chemoheterotrophy.

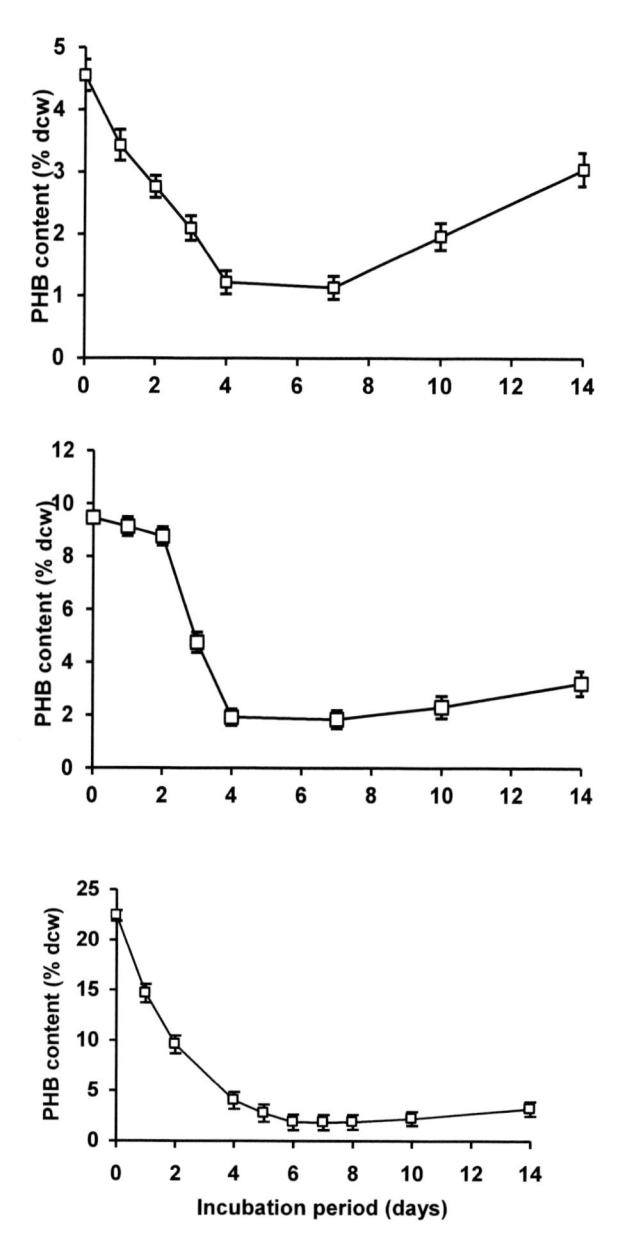

Figure 5. Time-course of PHB mobilization in *Synechocystis* sp. PCC 6803 incubated under the usual light-dark (14: 10 h) cycles. Cultures (A) pre-grown in normal BG-11 medium, (B) pre-grown in N-deficient BG-11 medium and (C) pre-grown under chemoheterotrophic condition. Source: Panda [62].

P-DEFICIENCY AND PHB MOBILIZATION

Mobilization of PHB in P-deficient BG-11 medium under usual light-dark condition is shown in Table 6. The N-deficient and chemoheterotrophic cultures exhibited a slower PHB mobilization in P-deficient medium. Contrary to this, in photoautotrophic cultures a rising trend was evident.

Table 6. Effects of illumination, dark incubation and P-deficiency on PHB mobilization in *Synechocystis* sp. PCC 6803 pre-grown under photoautotrophic, N-deficient and chemoheterotrophic conditions

Treatment	PHB content (% dcw)		
	Photoautotrophy	N-deficiency	Chemoheterotrophy
Biosynthesis	4.5 ± 0.18^b	9.5 ± 0.16^a	22.4 ± 0.95^a
Light-dark cycles (Control)	1.2 ± 0.08^d	1.9 ± 0.13^d	1.9 ± 0.12^c
Continuous illumination	1.5 ± 0.11^d	2.3 ± 0.12^d	2.3 ± 0.12^c
Dark incubation	2.6 ± 0.15^c	6.5 ± 0.14^b	8.6 ± 0.15^b
P-deficiency	$*5.6 \pm 0.15^a$	5.0 ± 0.13^c	8.8 ± 0.22^b

All values are mean ± SE, n=3.
Values in the column superscripted by different letters are significantly ($P < 0.05$) different from each other (Duncan's new multiple range test).
Separate analysis was done for each column. * denotes stimulation.
Cells pre-grown under photoautotrophic and N-deficient conditions were analyzed for PHB content on day 4, whereas chemoheterotrophically-grown cells were harvested on day 6. Source: Panda [62].

PHB mobilization was found much faster under illuminated/ light-dark cycles than under dark condition (Table 6). This indicates that *Synechocystis* cell utilizes PHB simultaneously with the photosynthetic activity. The retarded mobilization under dark condition could be due to the reduced availability of energy, as the enzyme PHB depolymerase, consumes ATP for PHB mobilization. [23] Interestingly, the rate of PHB mobilization under light-dark cycles was similar to that under continuous illumination, thus depicting that PHB mobilization in *Synechocystis* requires only minimal amount of energy which is sufficiently generated by the natural light-dark cycles. [85] PHB mobilization was however, profoundly slower in P-deficient medium for cultures pre-grown under N-deficient and chemoheterotrophy (Table 6). This may be resulting from the unavailability of energy, as ATP production decreases with the onset of phosphate limitation. [79,80] Contrary to this, the photoautotrophically-grown cells exhibited accumulation of PHB under P-deficiency. This is supported by the earlier report of Sharma and Mallick [58] for *Nostoc muscorum*, where NADPH accumulation under P-deficiency was found to stimulate PHB accumulation.

MOBILIZATION OF PHB IN PRESENCE
OF GLUCOSE AND ACETATE

Mobilization of PHB was remarkably reduced under glucose- and acetate-supplementation (Figure 6). This indicates the role of PHB as a carbon reserve. In *C. necator*, Handrick *et al.*[86] observed a much faster PHB mobilization in the absence of exogenous carbons, which was however, drastically slower down when carbon was supplemented. Interestingly, at 0.2% acetate supplementation the PHB content exceeded the initial PHB values of N-deficient and chemoheterotrophic cultures (Figure 6), thus exhibiting the biosynthesis of PHB at higher acetate concentration. At lower carbon doses, a concentration-dependent mobilization of PHB was however, clearly evident. This indicates a simultaneous synthesis mobilization of PHB in *Synechocystis* sp. PCC 6803 reported for *C. necator.* [87]

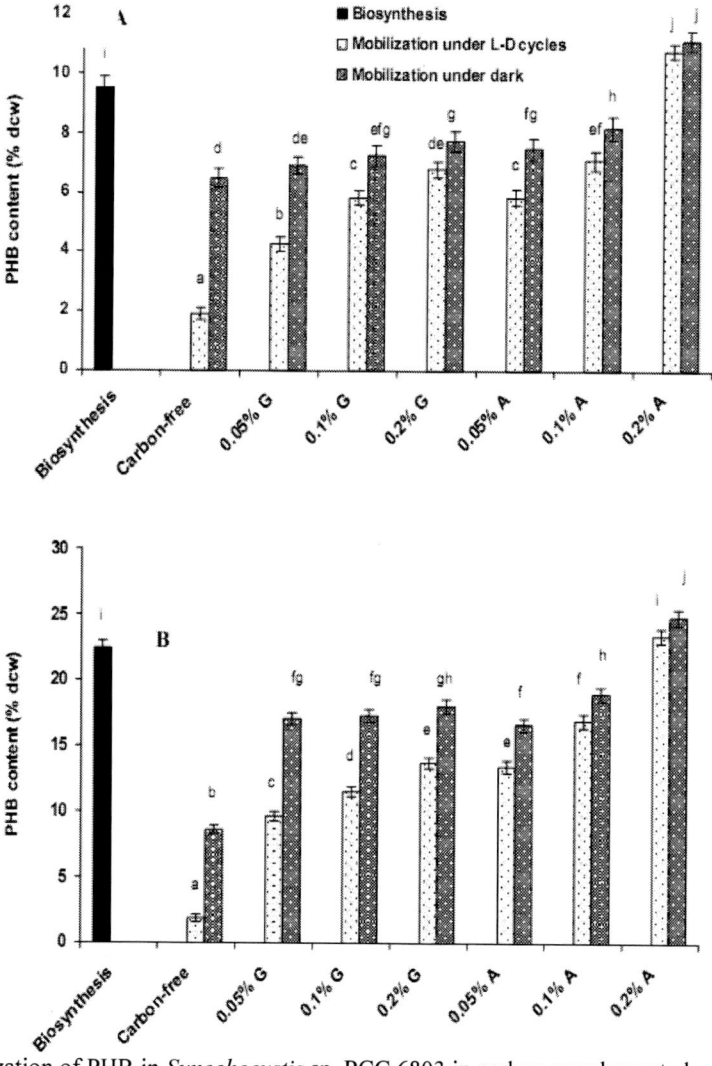

Figure 6. Mobilization of PHB in *Synechocystis* sp. PCC 6803 in carbon-supplemented medium under dark and the usual light-dark cycles. (A) pre-grown in N-deficient BG-11 medium, and (B) pre-grown under chemoheterotrophic condition. A: acetate and G: glucose. Source: Panda [62].

CHARACTERIZATION OF PHAS FILMS
SURFACE STUDY

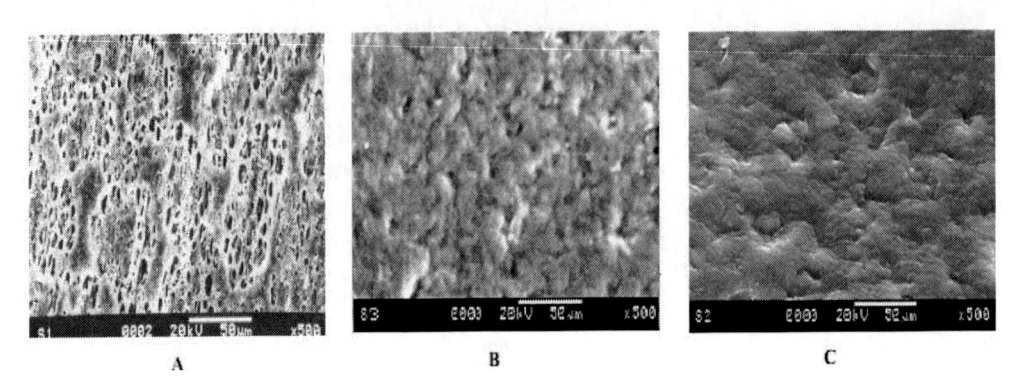

Figure 7. Scanning electron micrographs of the upper surface of A) PHB from *Synechocystis* sp. PCC 6803, B) P(3HB-*co*-13% 3HV) and C) P(3HB-*co*-27% 3HV) membranes obtained from *Nostoc muscorum* (magnification at 500x).

The upper surface of the film, i.e. the film facing to air was analyzed by scanning electron microscope (Figure 7). The homopolymer PHB film possessed large number of voids, which decreased significantly with increasing 3HV mol%. The co-polymer film with 27% mol% HV content showed comparatively more regular and compact arrangements than the film with 13 mol% HV.

MECHANICAL AND THERMAL PROPERTIES

The mechanical properties of PHB homopolymer and P(3HB-*co*-3HV) co-polymer from *Aulosira fertilissima, Nostoc muscorum* and *Synechocystis* sp. PCC 6803 are studied using the stress-strain measurement (Table 7). The Young's modulus of the co-polymer decreased significantly as compared to the PHB homopolymer. The tensile strength of the co-polymer also decreased markedly when compared with PHB homopolymer. At the same time, the elasticity was found to be much higher than that of the homopolymer and a profound rise was observed for *Synechocystis* sp. PCC 6803.

The thermal behaviour of the homo- and co-polymer samples of cyanobacterial species was investigated with the help of DSC thermograms. The melting temperature (T_m) of P(3HB-*co*-3HV) co-polymer decreased from 176 to 152, 148.8 and 155°C, respectively for *Synechocystis* sp. PCC 6803, *Nostoc muscorum* and *Aulosira fertilissima*. Similarly, the glass transition temperature (T_g) values were also decreased from 0.8 to -5.7, -4.3 and -5.5 °C, respectively. The crystallinity of the co-polymer samples thus demonstrated a significantly lower value than that of the homopolymer of PHB.

The homopolymer of PHB is a stiff and relatively brittle thermoplastic. Its meting point is just slightly lower than its degradation temperature (180 ^{0}C); this makes its processing a difficult task. A significant reduction in melting temperature with incorporation 3HV units widens its processing window. [15]

Table 7. Comparative account on the properties of PHB and P(3HB-co-3HV) co-polymers from cyanobacterial sources with the commercially available polymers

Property	Commercial polymer (Aldrich, USA)		Synechocystis sp. PCC 6803		Nostoc muscorum Agardh		Aulosira fertilissima	
	PHB	P(3HB-co-3HV) 75:25	PHB	P(3HB-co-3HV) 75:25	PHB	P(3HB-co-3HV) 78:22	PHB	P(3HB-co-3HV)* 78:25
T_g (°C)	0-5	-6	0.8	-5.7	0.8	-4.3	0.6	-5.5
T_m (°C)	171-182	137	176	152	176	148.8	174	155
X_c (%)	60-80	40	62.4	45	62.4	nr	60.7	46.1
Young's modulus (GPa)	3.5	0.7	4.2	0.7	3.9	0.8	3.4	0.6
Mechanical strength (MPa)	40	30	32.4	18.0	32.4	21.5	37.6	18.1
Elongation to break (%)	5	nr	4.7	324.7	4.7	85.1	4.9	87.2
Reference	Vincenzini and De Philippis (1999)[72], Galego et al. (2002)[88], Ojumu et al. (2004)[33], Sankhala et al. (2010)[89]		Panda (2008)[62]		Bhati et al. (2010)[64], Bhati and Mallick (2011)[67]		Samantaray and Mallick (2011)[65]	

nr: not reported.

* Unpublished data

Highly crystalline polymers are usually brittle and find a narrow range of applications. Therefore, a lower degree of crystallinity may be related to the high molar mass of the co-polymer than PHB, since high molar mass polymers usually crystallize more slowly, leading to smaller crystals and to a lower degree of crystallinity. [90] Thus, the properties of the polymers from cyanobacterial sources are found to be comparable with the commercial PHA polymers, which advocate its potential application in various fields.

CONCLUSION

Accumulation of PHAs reached up to 40-55% (dcw) in a few cyanobacterial species. [39,41,62] In *Aulosira fertilissima* PHB content even reached up to 85% (dcw) under optimized condition. [65] Addition of exogenous carbons although found necessary for stimulation of PHAs in these cyanobacterial species, the amount was profoundly lower as compared to that of heterotrophic bacteria. [91,92] Cyanobacteria, thus, do have potential for low-cost PHAs production. However, the major bottle-neck in cyanobacterial PHAs production is the lack of an economically feasible mass cultivation system. Ongoing mass

254   Nirupama Mallick, Bhabatarini Panda, Silalipi Samantaray et al.
cultures of cyanobacteria have productivity only of 10-15 g dry wt. m^{-2} day^{-1} in sunlight receiving area. [93] Therefore, further research should be directed for the development of high performance photobioreactors to achieve greater productivity, which may lead to the introduction of PHAs films from photosynthetic microorganisms into various fields.

REFERENCE

[1]   Poirier, Y.; Nawrath, C.; Somerville, C. Biotechnology 1995, 13, 142-150.

[2]   Lee, S. Y. Biotech. Bioeng. 1996a, 49, 1-4.

[3]   Kalia, V. C.; Raizada, N.; Sonakya, V. J. Sci. Ind. Res. 2000, 59, 433-445.

[4]   Stein, R. S. Proc. Natl. Acad. Sci. 1992, 89, 835-838.

[5]   Brandl, H.; Gross, R. A.; Lenz, R. W.; Fuller, R. C. (1990) In: Flechter, A. (ed.) Advances in biochemical engineering/ biotechnology. Springer Verlag, New York, USA, 41, pp. 77-93.

[6]   Alexander, M. Science 1981, 211, 132-138.

[7]   Atlas, R. M.; Bartha, R. (1993) Microbial ecology: fundamentals and applications. 3rd edn., The Benjamin/Cummings Pub. Co., Menlo Park, CA 96, p. 564.

[8]   Fiechter, A. (1990) Plastics from bacteria and for bacteria: poly (β-hydroxyalkanoates) as natural, biocompatible, and biodegradable polyesters. Springer-Verlag, New York, pp. 77 - 93.

[9]   Chiras, D. D. (1994) Environmental science. The Benjamin/Cumming Pub. Co. Inc., Redwood, California, USA, p. 611.

[10]   Boopathy, R. Biores. Technol. 2000, 74, 63-67.

[11]   Kunioka, M.; Kawagushi, Y.; Doi, Y. Appl. Microbiol. Biotechnol. 1989, 30, 569-573.

[12]   Doi, Y.; Segawa, A.; Kunioka, M. Int. J. Biol. Macromol. 1990, 12, 101-111.

[13]   Johnstone, B. Far East. Econ. Rev. 1990, 147, 62-63.

[14]   Braunegg, G.; Gilles, L.; Klaus, F. J. Biotechnol. 1998, 65, 127-161.

[15]   Steinbüchel, A. (1996) In: Doi, Y. and Fukuda, K. (eds.), Biodegradable plastics and polymers. Elsevier Science, New York, USA, pp. 362-364.

[16]   De Koning, G. Can. J. Microbiol. 1995, 41, 303-309.

[17]   Matsusaki, H.; Abe, H.; Doi, Y. Biomacromolecules 2000, 1, 17 - 22.

[18]   Doi, Y. Macromol. Symp. 1995, 98, 585-599.

[19]   Marchessault, R. H. Trends Polym. Sci. 1996, 4, 163-168.

[20]   Luzier, W. D. Proc. Nat. Acad. Sci. 1992, 89, 839-842.

[21]   Lafferty, R. M.; Korsatko, B. and Korsatko, W. (1988) In: Rehm, H. J. and Reed, G. (eds.), Biotechnology: special microbial processes. Wiley-VCH Pub., Weinheim, Germany, pp. 135-176.

[22]   Anderson, A. J.; Dawes, E. A. Microbiol. Rev. 1990, 54, 450-472.

[23]   Doi, Y. (1990) Microbial polyesters. VCH Publishers, New York, USA, p. 166.

[24]   Fuller, R. C.; Lenz, R. W. Nat. History 1990, 5, 82-84.

[25]   Lee, S. Y. Tibtech. 1996b, 14, 431-438.

[26]   Sasikala, C. (1996) In: Neidleman, S. L. and Laskin, A. I. (eds.), Advances in applied microbiology. Academic Press, California, USA, 42, 97-218.

[27] Steinbüchel, A.; Schlegel, H. G. Mol. Microbiol. 1991, 5, 535-542.

[28] Lee, S. Y.; Choi, J.; Wong, H. W. Int. J. Biol. Macromol. 1999, 25, 31-36.

[29] Sudesh, K.; Abe, H.; Doi, Y. Prog. Polym. Sci. 2000, 25, 1503-1555.

[30] Ishizaki, A.; Tanaka, K.; Taga, N. Appl. Microbiol. Biotechnol. 2001, 57, 6-12.

[31] (31) Serafim, L. S.; Lemos, P. C.; Ramos, A. M.; Crespo, G. P.; Ries, M. A. (2001) In: Chillini, E., Mendes Gil, M. H., Braunegg, G., Buchert, J., Gatenholm, P. and van der Zee, M. (eds.), Biorelated polymer: sustainable polymer science and technology. Kluwer Academic Pub., Coimbra, Protugal, pp. 147-177.

[32] (32) Reddy, C. S. K.; Ghai, R.; Rashmi; Kalia, V. C. Bioresour. Technol. 2003, 87, 137-146.

[33] Ojumu, T. V.; Yu, J.; Solomon, B. O. Afr. J. Biotechnol. 2004, 3, 18-24.

[34] Khanna, S.; Srivastava, A. K. Process Biochem. 2005, 40, 607-619.

[35] Lenz, R. W.; Marchessault, R. H. Biomacromolecules 2005, 6, 1-8.

[36] Valappil, S. P.; Boccaccini, A. R.; Bucke, C.; Roy, I. Antonie Van Leeuwenhoek. 2007, 91, 1-17.

[37] Suriyamongkol, P.; Weselake, R.; Narine, S.; Moloney, M.; Shah, S. Biotechnol. Adv. 2007, 25, 148-175.

[38] Alias, Z.; Tan, K. P. I. Bioresour. Technol. 2005, 96, 1229-1234.

[39] Nishioka, M.; Nakai, K.; Miyake, M.; Asada, Y.; Taya, M. Biotechnol. Lett. 2001, 23, 1095-1099.

[40] Panda, B.; Jain, P.; Sharma, L.; Mallick, N. Biores. Technol. 2006, 97, 1296-1301.

[41] Sharma, L.; Singh, A. K.; Panda, B.; Mallick, N. Bioresour. Technol. 2007, 98, 987-993.

[42] Carr, N. G. Biochim. Biophys. Acta. 1966, 120, 308-310.

[43] Rippka, R.; Neilson, A.; Kunisawa, R.; Cohen-Bazire, G. Arch. Mikrobiol. 1971, 76, 341-348.

[44] Jensen, T. E.; Sicko, L. M. Cytologia 1973, 38, 381-391.

[45] Jensen, T. E. Microbios. Lett. 1980, 11, 117-125.

[46] Jensen, T. E.; Baxter, M. Cytobios 1981, 32, 129-137.

[47] Campbell, J.; Stevens, S. E. Jr.; Bankwill, D. L. J. Bacteriol. 1982, 149, 361-366.

[48] Sicko-Goad, L. Protoplasm 1982, 111, 75-86.

[49] Allen, M. M. Annu. Rev. Microbiol. 1984, 38, 1-25.

[50] Stal, L. J.; Heyer, H.; Jacobs, G. (1990) In: Dawes, E. A. (ed.), Novel biodegradable microbial polymers. NATO ASI series, Kluwer Academic Pub., Dordrecht, The Netherlands. E 186, pp. 435-438.

[51] Vincenzini, M.; Sili, C.; De Philippis, R.; Ena, A.; Materassi, R. J. Bacteriol. 1990, 172, 2791-2792.

[52] De Philippis, R.; Ena, A.; Guastini, M.; Sili, C.; Vincenzini, M. FEMS Microbiol. Rev. 1992a, 103, 187-194.

[53] De Philippis, R.; Sili, C.; Vincenzini, M. J. Gen. Microbiol. 1992b, 138, 1623-1628.

[54] Stal, L. J. FEMS Microbiol. Rev. 1992, 103, 169-180.

[55] Lama, L.; Nicolaus, B.; Calandrelli, V.; Maria, M. C.; Romano, I.; Gambacorta, A. Phytochemistry 1996, 42, 655-659.

[56] Miyake, M.; Erata, M.; Asada, Y. J. Ferment. Bioeng. 1996, 82, 512-514.

[57] Sudesh, K.; Taguchi, K.; Doi, Y. Int. J. Biol. Macromol. 2002, 30, 97-104.

[58]   Sharma, L.; Mallick, N. Bioresour. Technol. 2005a, 96, 1304-1310.

[59]   Sharma, L.; Mallick, N. Biotechnol. Lett. 2005b, 27, 59-62.

[60]   Panda, B. and Mallick, N. Lett. Appl. Microbiol. 2006, 44, 194-198.

[61]   Toh, P. S. Y; Jau M. H.; Yew S.P.; Abed R. M. M.; Sudesh, K. J. Biosci. 2008, 19, 21-38.

[62]   Panda, B. Accumulation of polyhydroxyalkanoates in a unicellular cyanobacterium Synechocystis sp. PCC 6803. Ph. D. Thesis, Indian Institute of Technology, Kharagpur, India, 2008, p.144.

[63]   Shrivastav, A.; Mishra, S. K.; Mishra, S. Int. J. Biolog. Macromol. 2010, 46, 255-260.

[64]   Bhati, R.; Samantaray, S.; Sharma, L.; Mallick, N. Biotechnol. J. 2010, 5, 1181-1185.

[65]   Samantaray, S.; Mallick, N., J. Appl. Phycol. 2011 (In Press).

[66]   Mallick, N.; Gupta, S.; Panda, B.; Sen, R. Biochem. Eng. J. 2007a, 37, 125-130.

[67]   Bhati, R.; Mallick, N., J. Chem. Tech. Biotech. 2011 (accepted).

[68]   Oeding, V.; Schlegel, H. G. Biochem. J. 1973, 134, 239-248.

[69]   Schembri, M. R.; Bayly, R. C.; Davies, J. K. J. Bacteriol. 1995, 177, 4501-4507.

[70]   Miyake, M.; Kataoka, K.; Shirai, M.; Asada, Y. J. Bacteriol. 1997, 179, 5009-5013.

[71]   Mallick, N.; Sharma, L.; Singh, A. K. J. Plant Physiol. 2007b, 164, 312-317.

[72]   Vincenzini, M.; De Philippis, R. (1999) In: Cohen, Z. (ed.), Chemicals from microalgae. Taylor and Francis Inc., USA, pp. 292-312.

[73]   Senior, P. J.; Dawes, E. A. Biochem. J. 1973, 134, 225-238.

[74]   Smith, A. J. (1982) In: Carr, N. G. and Whitton, B. A. (eds.), The biology of cyanobacteria. Blackwell Scientific Pub., Oxford, UK, pp. 47-85.

[75]   Izawa, S.; Good, N. E. (1972) In: Pietro, A. S. (ed.), Methods in enzymology- XXIVB. Academic Press, New York, USA, pp. 355-377.

[76]   Ryu, J. Y.; Song, J. Y.; Lee, J. M.; Jeong, S. W.; Chow, W. S.; Choi, S. B.; Pogson, B. J.; Park, Y-II. J. Biol. Chem. 2004, 279, 25320-25325.

[77]   Avetisyan, A. V.; Kaulen, A. D.; Skulachev, V. P.; Feniouk, B. A. Biochemistry 1998, 63, 625-628.

[78]   Neumann, J.; Jagendorf, A. T. Arch. Biochem. Biophys. 1964, 107, 109-119.

[79]   Bottomley, P. J.; Stewart, W. D. P. Arch. Microbiol. 1976, 108, 249-258.

[80]   Konopka, A.; Schnur, M. J. Phycol. 1981, 17, 118-122.

[81]   Rippka, R. Arch. Microbiol. 1972, 87, 303-322.

[82]   Steigenberger, S.; Terjung, F.; Grossart, H.-P.; Reuter, R. EARSeL Proc. 2004, 3, 18-25.

[83]   Asada, Y.; Miyake, M.; Miyake, J.; Kurane, R.; Tokiwa, Y. Int. J. Biol. Macromol. 1999, 25, 37-42.

[84]   Iyer, P. P.; Ferry, J. G. J. Bacteriol. 2001, 183, 4244-4250.

[85]   Jau, M. H.; Yew, S. P.; Toh, P. S. Y.; Chong, A. S. C.; Chu, W. L.; Phang, S. M.; Najimudin, N.; Sudesh, K. Int. J. Biol. Macromol. 2005, 36, 144-151.

[86]   Handrick, R.; Reinhardt, S.; Jendrossek, D. J. Bacteriol. 2000, 182, 5916-5918.

[87]   Taidi, B.; Mansfield, D. A.; Anderson, A. J. FEMS Microbiol. Lett. 1995, 129, 201-206.

[88]   Galego, N.; Miguens, F. C.; Sanchez, R. Polymer 2002, 43, 3109-3114.

[89]   Sankhla, I. S., Bhati, R.; Singh, A. K.; Mallick, N. Biores. Technol. 2010, 101, 1947-1953.

[90] Olieviera, F.C.; Dias, M. L.;Castilho, L. R.; Freire, D. M. G. Biores. Technol. 2007, 98, 633-638.

[91] Chen, G. Q.; Zhang, S. J.; Park, S. J.; Lee, S. Y. Appl. Microbio. Biotechnol. 2001, 57, 50-55.

[92] Shang, L.; Jiang, M.; Chang, H. N. Biotechno. Lett. 2003, 25, 1415-1419.

[93] Belay, A. (2004) In: Vonshak, A. (ed.), Spirulina platensis (Arthrospira): Physiology, Cell-Biology and Biotechnology. Taylor and Francis, London, pp. 131-158.

INDEX

D

O

P

S

T